Don & Clay McGuire
October 1996

Arnold Koerte

Two Railway Bridges of an Era
Zwei Eisenbahnbrücken einer Epoche

Firth of Forth
and Firth of Tay

Technical Progress, Disaster and New Beginning
in Victorian Engineering

Technischer Fortschritt, Desaster und Neubeginn
in der Viktorianischen Ingenieurbaukunst

Birkhäuser Verlag
Basel · Boston · Berlin

Contents

- 6 Preface

Chapter I
The "Tay Bridge Disaster" – Its History and Consequences
- 11 In Retrospect
- 12 The "Tay Bridge Disaster"
- 16 Significance of the Tay Bridge – Before the Disaster
- 19 Preceding Battles
- 22 "Rainbow Bridge"
- 23 Times of Change
- 24 Eyth's "Enno Bridge"
- 26 First Proposals
- 28 Commission
- 29 "Final Plan"
- 30 Alterations
- 33 Cornerstone and Caissons
- 35 Piers
- 35 First Setbacks
- 37 Further Alterations
- 42 Girders
- 44 High Girder Going Down
- 45 Excerpts
- 46 Trial Run and Opening
- 48 Judgement: "Early-American"
- 51 Building Costs
- 52 Precursors
- 53 Load Assumptions and Guesswork
- 56 December 28, 1879
- 60 The Final Hours
- 66 Eyewitnesses
- 82 The Day After
- 86 Diving Manoeuvres
- 88 Public Inquiry
- 95 Beaumont's Egg
- 97 Henry Noble's Role
- 102 Bouch's Testimony
- 104 Wind Pressure
- 107 Wind Pressure Upon Rolling Loads?
- 107 The Verdict

Chapter II
The Interim Between the Bridges
- 111 Sequel
- 116 Reconstruction
- 116 Previous Bridge Failures
- 118 The Angers Disaster
- 120 Practicioners versus Theoreticians – National Differences
- 121 The Lessons of the "Tay Bridge Disaster"

Chapter III
The Railway Bridge Across the Firth of Forth
- 123 Background
- 126 Age of Wild Projects
- 128 The Role of James Anderson
- 130 "Railway Mania"
- 132 Bouch's Suspension Bridge Across the Forth
- 134 A New Beginning
- 140 The Principle of the Cantilever Bridge
- 144 The Design by Fowler and Baker
- 146 Bridge Esthetics
- 150 Tides, Wind and Weather
- 152 Wind Pressure Experiments
- 154 Final Structure
- 159 Sequence of Construction
- 161 Electrical Matters
- 162 Novel Caissons
- 167 Raising the Viaducts
- 168 The Steel
- 171 Bed Plates and "Skewbacks"
- 171 The Tubular Girders
- 175 Erection of the Towers
- 176 The Cantilevers
- 182 Press Commentaries
- 183 Corrective Liftings – "A Trifle Beyond"
- 187 Sun Radiation
- 187 The Central Girders – Two Bridges within the Bridge
- 191 Forces of Nature and "Keystone"
- 194 Loading Tests
- 195 Official Opening
- 196 The Workmen
- 203 The Everlasting Painting Job
- 203 The Creators of the Bridge
- 203 Sir John Fowler (1817–1898)
- 205 Sir Benjamin Baker (1840–1907)
- 207 Sir William Arrol (1839–1913)
- 210 A Few Figures

- 211 Conclusions
- 220 Bibliography
- 221 Table of Illustrations

Inhaltsverzeichnis

8 Vorwort

Kapitel I
«Tay Bridge Disaster» – Seine Vorgeschichte und Folgen
11 Rückblick
12 «Tay Bridge Disaster»
16 Bedeutung der Tay-Brücke – Vor dem Stichtag
19 Vorkämpfe
22 «Regenbogenbrücke»
23 Zeit des Umbruches
24 Eyth's «Ennobrücke»
26 Erste Entwürfe
28 Auftrag
29 «Endgültiger» Plan
30 Änderungen
33 Grundstein und Caissons
35 Pfeiler
35 Erste Rückschläge
37 Erneute Änderungen
42 Fachwerkträger
44 Mittelträger versinkt
45 Auszüge
46 Probelauf und Einweihung
48 Urteil «Frühamerikanisch»
51 Baukosten
52 Vorboten
55 Lastannahmen und «Guesswork»
55 Der 28. Dezember 1879
60 Die letzten Stunden
66 Augenzeugen
83 Der Tag danach
87 Tauchmanöver
89 Öffentliche Untersuchung
95 «Beaumont's Egg»
97 Henry Noble's Rolle
101 Bouch's Zeugnis
104 Windbelastung
107 Winddruck auf rollende Lasten?
108 Das Verdikt

Kapitel II
Zwischen den Brücken
111 Nachspiel
116 Wiederaufbau
116 Frühere Brückeneinstürze
118 Das Angers-Desaster
120 Empiriker gegen Theoretiker – Nationale Unterschiede
121 Die Lehren aus dem «Tay Bridge Disaster»

Kapitel III
Die Eisenbahnbrücke über den Firth of Forth
123 Vorgeschichte
126 Zeit wilder Projekte
128 Die Rolle von James Anderson
130 «Railway Mania»
132 Bouch's Hängebrücke über den Forth
138 Neuanfang
140 Prinzip der Ausleger-Brücke
143 Der Entwurf von Fowler und Baker
146 Brücken-Ästhetik
150 Gezeiten, Wind und Wetter
152 Winddruck-Experimente
154 Endgültige Konstruktion
159 Bauablauf
161 Elektrisches
162 Neuartige Caissons
167 Viadukt-Anhebungen
168 Der Stahl
171 Lagerplatten und «Skewbacks»
173 Die Rohrträger
175 Montage der Pfeilertürme
178 Die Ausleger
182 Presse-Kommentare
183 Korrektive Hebungen – «a trifle beyond»
187 Sonneneinstrahlung
187 Die Mittelträger – Zwei Brücken in der Brücke
191 Naturkräfte und «Schlußstein»
195 Belastungsprobe
196 Feierliche Eröffnung
196 Die Arbeiter
203 Der ewige Anstrich
203 Die Schöpfer der Brücke
203 Sir John Fowler (1817–1898)
206 Sir Benjamin Baker (1840–1907)
208 Sir William Arrol (1839–1913)
210 Ein paar Zahlen

211 Ein Schlußwort
220 Bibliographie
221 Verzeichnis der Abbildungen

Preface

A double anniversary has occasioned this book. It commemorates the so-called "Tay Bridge Disaster" of December 28, 1879, that rocked Victorian Society some 112 years ago, and also the opening of the railway bridge across the Firth of Forth on March 4, 1890, which signalled the beginning of a new era in bridge construction, 101 years ago. Both events, at the transition of the "cast-iron age" to the "steel age" are linked to each other in more ways than one: geographically, by the proximity of their sites on Scotland's east coast; in terms of time, in the sense that each was part of that decade of transition from 1880 to 1890; and in terms of their effect upon history, for each did much to change the public attitude toward major feats of engineering.
The pivotal event of the Tay Bridge Disaster is our point of departure from which we shall first move back in time, reconstructing the historic origin of the bridge. Built in the years 1871 to 1877 by the renowned bridge builder Sir Thomas Bouch, this structure was already considered a major engineering achievement and was the longest span of its time. With its failure, more than just a bridge had collapsed. It meant the end of an almost messianic self-confidence among engineering circles, and also the demise of free experimentation in bridge construction by the early pioneers. The new beginning after the shock helped to bring about that "edifice of a century" – the famous Firth-of-Forth Railway Bridge, 1882 to 1889. The triumvirate formed by its builders – the engineers Fowler and Baker as well as the contractor Arrol – excelled by an ingenious combination of novel methods in design, construction and organisation – spearheading future bridge design on a scientific basis in Great Britain.
However, it is not the bridges' historic significance alone which acounts for their near-magnetic hold on admirers (and which called this continental author back to Scotland time and time again). Rather, both bridges radiate a special fascination which only those early examples of European technology have been able to evoke. Here, a most immediate sensory perception coincides with the spectacle of a magnificent natural setting. It is the purpose of this book to bring to life and to illustrate the force of this experience. For this reason, the historic scenes and events will be

juxtaposed with images of how these settings appear today; wherever possible, the same view points as in the original photographs have been used. Naturally, there have been changes in the course of 100 years, but considering the upheavals elsewhere on the continent, the Scottish landscape still strikes us as being incredibly stable and homogeneous. It almost seems as if the quiet harmony between nature's course and human everyday life simply refuses to be disturbed by new technology or environmental interference – provided we leave today's Sellafield out of the picture for a moment. At the Firth of Forth, we are struck by the lonely silence of the scene. The belching smoke stacks of Dundee's former jute industry have long since disappeared; the landscape appears less austere and wild than in the early pictures, being more densely populated now – and most of all, it carries more vegetation than before. Today's Tay Bridge traces the course of the old one, sweeping widely across the Firth. Considering all those tales about the old Tay Bridge we may have heard before, it transmits a strangely restrained aura today. Indeed, nothing could be more peaceful than this scene. That rather matter-of-fact crossing of an estuary between green hills recalls more a classical viaduct than a bridge.

But at the Firth of Forth, the powerful lines of the structure correspond ideally to the lines of the landscape. Up to this day, that heavy-boned iron web marches like some primeval creature across the Firth – having become itself part of the natural setting. More recently, the modern road suspension bridge nearby has joined the old rail bridge about 1 km upstream, occupying the same narrows that seem predestined for a crossing. At this spot, nature and man-made works together culminate in a synthesis, which neither would have achieved in isolation. Even if the iron structures of early industrialization appeared to someone like William Morris as "apparatures becoming uglier and uglier all the time", today we may rightly speak of a *"Gesamtkunstwerk"* (total work of art) at the Forth.

Yet this would never have come about without the experiences gained at the Tay. Therefore, the specific circumstances, observations and findings derived from this disaster shall be linked to the masterpiece of the Forth Bridge. This story is also, in no small part, about people, the living and working conditions of the bridge builders, eye-witness reports and the fate of the engineers involved – as far as they were concerned with the bridges. The complete biographies of such diverse characters as Thomas Bouch, John Fowler, Benjamin Baker and William Arrol would each fill volumes of their own. Using the Tay Bridge Disaster as a historic turning point, the events leading up to it will be examined first (Chapter I), to be followed by a study of the interlude of reappraisal and the new start (Chapter II), and concluding with the final construction of the Forth Bridge (Chapter III).

Vorwort

Der unmittelbare Anlaß für dieses Buch ist ein zweifaches Jubiläum: Das sogenannte «Tay Bridge Disaster» vom 28. Dezember 1879 hat vor nunmehr 112 Jahren die viktorianische Gesellschaft erschüttert. Und die Eröffnung der Eisenbahnbrücke über den Firth of Forth am 4. März 1890, vor 101 Jahren, markiert den Anfang einer neuen Zeit im Brückenbau. Beide Ereignisse, am Übergang von der «Gußeisenzeit» zur «Stahlzeit», sind in mehrfacher Hinsicht miteinander verknüpft: Örtlich durch die geographische Nähe an der Ostküste Schottlands; zeitlich durch jenes Jahrzehnt des Umbruches von 1880 bis 1890; und wirkungsgeschichtlich durch eine Wandlung in der Rezeption technischer Großbauwerke.
Das Schlüsselereignis des Brückeneinsturzes am Tay ist die Wendemarke, von der aus zunächst rückwärts die Entstehungsgeschichte der Tay-Brücke aufgerollt wird: 1871–1877 von dem gefeierten Brückenbauer Sir Thomas Bouch errichtet, galt sie bereits als technische Großtat und längste Brücke der Zeit. Mit ihrem Einsturz brach für die Epoche mehr als nur eine Brücke zusammen: Er bedeutete das Ende eines messianischen Selbstvertrauens der Ingenieure, aber auch das Erlöschen des freien Experimentierens im Brückenbau der frühen Pioniere. Der Neuanfang nach dem Schock führte direkt zum Jahrhundertwerk der berühmten Firth-of-Forth-Eisenbahnbrücke, 1882–1889. Das Dreigespann ihrer Erbauer – die Ingenieure Fowler und Baker sowie der Unternehmer Arrol – glänzte durch eine ingeniöse Verbindung neuer Berechnungs-, Bau- und Organisationsmethoden, welche den wissenschaftlich fundierten Brückenbau in Großbritannien begründen halfen.
Wenn allerdings jemand ‹vom Kontinent› sooft nur wegen zweier Brücken nach Schottland kommt, dann nicht nur wegen deren historischer Bedeutung. Vielmehr geht von beiden Bauwerken eine besondere Faszination aus, wie sie nur den frühen Beispielen der europäischen Technikgeschichte zueigen ist. Hier trifft sich eine ganz unmittelbare sinnhafte Erfahrung mit dem Schauspiel einer grandiosen Natur: Die Kraft dieser Wahrnehmung erlebbar und anschaulich zu machen, ist ein Anliegen dieses Buches. Weshalb den historischen Szenen und Begebenheiten die authentischen Schauplätze in ihrer heutigen Gestalt gegenübergestellt werden. Soweit möglich geschieht das

von denselben Standorten wie in den alten Fotografien. Natürlich hat es Veränderungen gegeben im Laufe von 100 Jahren – aber im Vergleich zu den Umwälzungen anderswo muß uns die schottische Landschaft immer noch erstaunlich beständig und homogen erscheinen. Fast sieht es so aus, als ob der ruhige Gleichklang zwischen Naturablauf und menschlichem Alltag sich durch keinerlei neue Technik oder Umwelt-Eingriffe stören läßt – wenn wir vom heutigen «*Sellafield*» einmal absehen. Am Firth of Tay beeindruckt uns die verlassene Stille des Schauplatzes: Die rauchenden Schlote des ehemaligen Textilzentrums Dundee sind verschwunden, die Landschaft nicht mehr so karg und wild wie in den frühen Aufnahmen, dafür dichter besiedelt und vor allem mit mehr Vegetation versehen als ehedem. Die heutige Tay-Brücke folgt dem Lauf der alten in weitem Bogen über den Meeresarm und ist – angesichts dessen, was man über die Tay-Brücke gehört haben mag – von seltsam verhaltener Wirkung. Nichts könnte friedlicher sein als diese Szene; und die eher undramatische Überquerung eines Wasserweges zwischen grünen Hügeln erinnert heute an einen klassischen Viadukt, aber kaum an eine Brücke.

Wogegen am Firth of Forth die machtvollen Konturen des Bauwerkes im Profil der Landschaft ihre ideale Entsprechung finden: Noch immer schreitet dieses schwer-knochige Eisengerippe wie ein vorzeitliches Ungetüm über die Bucht – und ist heute selbst zu einem Bestandteil dieser Natur geworden. Dazu gesellt sich in neuerer Zeit die moderne Straßen-Hängebrücke etwa 1 km flußaufwärts und an derselben Engstelle gelegen, welche wie geschaffen für einen Übergang erscheint: Hier steigern sich Natur und Menschenwerk gegenseitig zu einer Wirkung, die keines von beiden für sich allein erreicht hätte. Auch wenn die Eisenbauten aus der Frühzeit der Industrialisierung in den Worten des Zeitgenossen William Morris wie «Apparaturen» erscheinen mußten, die «immer häßlicher und häßlicher» werden, so darf man heute am Forth getrost von einem *Gesamtkunstwerk* sprechen.

Dieses wäre aber ohne die Erfahrungen am Tay nicht zustandegekommen – und im Folgenden sollen die näheren Umstände, Beobachtungen und Erkenntnisse aus jener Katastrophe verknüpft werden mit dem Meisterwerk der Forth-Brücke. Nicht zuletzt geht es dabei um Personen, nämlich die Lebens- und Arbeitsbedingungen der Brückenbauer, die Berichte der Augenzeugen, und die Schicksale der maßgeblichen Ingenieure, soweit sie die Brücken betreffen. Denn die Biographien so unterschiedlicher Charaktere wie Thomas Bouch, John Fowler, Benjamin Baker und William Arrol würden eigene Bände füllen. – Vom «*Tay Bridge Disaster*» als historischem Wendepunkt wird zunächst dessen Vorgeschichte beleuchtet (Kapitel I), gefolgt von einem Zwischenspiel über Rückbesinnung und Neuanfang (Kapitel II), bis zum endgültigen Bau der Forth-Brücke (Kapitel III).

Chapter I
The "Tay Bridge Disaster" – Its History and Consequences

In Retrospect

Looking back today at the major engineering feats of the closing 19th century, some 100 years hence, we tend to see light rather than shade. We seem to sense more a future-oriented optimism among the contemporaries, than doubts about the blessings of technology. However, already in the last century, anxiety and fear accompanied technical development time and again, as disasters are not an invention of today. Indeed, spectacular accidents went hand in hand with every new example of progress, with every technological achievement, as soon as they touched – albeit "innocently" – upon the limits of what was then considered possible.

During the early phase of European industrialization, an untamed euphoria and paralyzing horror often went hand in hand – illustrating as it were the professional image of the enterprising engineer. Still, public opinion hardly ever questioned technology as such, but rather the practical ability of the engineer, the validity of the plans or the supervision during execution. By and large it was held that nothing could really go wrong, provided "it was properly done". In this vein, even those accidents that did occur were rationalized and thus seemed "under control", both in terms of their extent, i. e. the number of victims, and the fact that they were limited to a single locality.

Existential fear came into being only with the advent of the global disasters of the 20th century. This fear is beyond any attempt at rational explanation; it sees technical progress itself, having long since broken the shackles of human control, as the arch-evil of the time. With that, the handy separation disappears between a rational response to actual accidents and an emotional anxiety that is no longer receptive to rational argumentation.

Bridge builders of the 19th century could still accept so-called "black days" as a necessary component of progress. Indeed, many findings about the new material, iron, were in fact obtained only after bridge failures: cast-iron, prone to breakage, was first followed by wrought iron, and finally by steel. Only then could the large spans, newly possible, generate new insights into permissible loads and, in particular, into the behaviour during oscillation. The new structures themselves led the way, indirectly, to decisive progress in calculating, statics and solid scientific theory – and away from the empirical guesswork of the pioneer days.

Nonetheless, the last third of the 19th century witnessed engineering disasters of such magnitude which even it (with its "black days philosophy") was no longer capable of explaining away as some kind of inherent consequence or even necessity. On the contrary, it was that night of the accident at the Firth of Tay in Scotland, near Dundee, which acted like a beacon of doom. Its circumstances, coming to light only with time, demonstrated how this event could well have been avoided, as far as humanly possible; far from being an act of fate, it was an accident caused – yes even provoked – by man himself. Still, this did not make it any less appalling for those who experienced the tragedy than it does for us today.

The "Tay Bridge Disaster"

The two rail bridges across the Forth and Tay are linked not just by their proximity to each other on Scotland's east coast, nor by their construction dates – a mere 11 years separated construction commencement of each project. Rather, the failure of the Tay Bridge helped to create the scene for erecting the Forth Bridge. That is to say, both engineering feats of this time are closely linked in terms of their technological development. For this reason, any description of their origin must first invoke the memory of a disaster which was to be a historic turning point for many families in Scotland, as well as for Victorian Great Britain. Even abroad, this disaster had shocked the professional world.

The event in question is the one which, during a winter gale of exceptional ferocity, was to become known as the Tay Bridge Disaster. This stormy night is being commemorated this year for the 112th time.

It was on December 28th and the last Sunday of the year 1879, shortly past seven o'clock at night, when nearly 1 km of this bridge's length crashed into the sea – taking with it a fully-occupied train which had just entered the "High Girders", as the central portion of the bridge was called. Just as eerie as the scene itself,

auch jene handliche Trennung zwischen einem rationalen Reagieren auf tatsächliche Unglücksfälle und einer emotionalen Lebensangst, die Argumenten nicht mehr zugänglich ist.

Nun konnte man noch im Brückenbau des 19. Jahrhunderts sogenannte «schwarze Tage», vor oder nach der Vollendung, als entwicklungsgeschichtliche Notwendigkeiten sehen: Denn viele Erkenntnisse über das neue Baumaterial, Eisen, wurden in der Tat erst durch Einstürze gewonnen: Auf das bruchanfällige Gußeisen folgte zuerst Schmiedeeisen und schließlich Stahl. Die erst damit möglichen großen Spannweiten brachten neue Erkenntnisse über zulässige Lastannahmen und insbesondere über das Schwingungsverhalten. Und die neuen Konstruktionen selbst führten indirekt zu einer entscheidenden Weiterentwicklung der Baustatik, von den empirischen Erfahrungswerten der Frühzeit bis hin zu einer theoretisch fundierten Wissenschaft.

Und dennoch, auch das letzte Drittel des 19. Jahrhunderts erlebte bereits technische Katastrophen eines Ausmaßes, die man schon damals nicht mehr mit einer Art von immanenter Folgerichtigkeit oder gar Notwendigkeit erklären konnte. Im Gegenteil, es war jene Unglücksnacht am *Firth of Tay* in Schottland, vor Dundee, welche wie ein Fanal wirkte. Ihre erst nach und nach zutage tretenden Begleitumstände zeigten, daß jenes Ereignis nach menschlichem Ermessen sehr wohl hätte vermieden werden können – also weniger ein schicksalhaftes Unglück, als ein vom Menschen selbst verursachter, ja herausgeforderter Vorfall war. Das allerdings macht ihn nicht weniger unheimlich für die Zeitgenossen, wie für uns heute.

«Tay Bridge Disaster»

Die beiden Eisenbahnbrücken über den Firth of Forth und den Forth of Tay sind nicht nur durch ihre geografische Nachbarschaft an der Ostküste Schottlands miteinander verbunden. Ebensowenig ist der geringe zeitliche Abstand ihres jeweiligen Baubeginns von nur 11 Jahren die einzige Gemeinsamkeit: Vielmehr hat der seinerzeitige Einsturz der Tay-Brücke erst die Voraussetzungen für den Bau der Forth-Brücke geschaffen. Das heißt, diese beiden Großbauwerke der Zeit sind in ihrer technischen Entwicklung miteinander verknüpft. Deshalb muß auch jede Beschreibung der Entstehungsgeschichte zunächst den Nachhall eines Brückeneinsturzes zu Gehör bringen, der für viele Familien Schottlands, aber auch des viktorianischen Englands, zu einer historischen Wendemarke werden sollte. Denn diese Katastrophe hatte die Fachwelt – nicht nur in Großbritannien – wie ein Schock getroffen.

was the fact that although there were several witnesses, almost none of them were later on able to report precisely on the actual course of the disaster. When they appeared before the court, their testimony corresponded only to the extent that all had witnessed some signal lamps, fiery flashes, flying sparks and then "a mass of fire falling toward the water like a comet". The time at which these meagre observations were made coincided with an unusually fierce gale – estimated later at 90 m.p.h. – which had wrought destruction (albeit less severe) also elsewhere along the Tay around 7:20 p.m. Since the fall of the bridge had also ripped down the telegraph wires, the horrible suspicion was confirmed only hours later: there were 75 casualties and no survivors *(Fig. 1)*[1].

For the majority of the Scottish population, this event was nothing less than an act of divine judgement: "If there is one voice louder than others in this terrible event, it is that of God, determined to guard his Sabbath with jealous care!" – so a minister of the Free Church of Scotland.

In all of Great Britain, the shock of this disaster was very grave indeed – although in southern England one may have been less inclined to see it as the just punishment for the sin of travelling on Sundays. Even today such an incident may still horrify human imagination, especially when the details are little known. The shock cannot be accounted for merely by the particulars of the disaster itself. If this were the case, the event would have been long overshadowed by much larger catastrophes since then.

Perhaps the Tay Bridge Disaster is so clearly embedded in memory because it had an immense impact on the complacency prevailing at the time, on the smug self-confidence of the Victorians in their industrial superiority, on their arrogant pride for having built the longest and largest bridge of the world. Disasters were frequent enough in the era of commercial and imperial growth – and military disasters in colonial battlefields could always be compensated by subsequent victories: just 11 months earlier, 600 British soldiers had been killed in the Zulu uprising at Isandhlwana, and their sacrifice had evoked grief as much as pride. But now on the Tay, the greatest blow was to national pride itself, a higher price than the 75 dead. Such a

---

[1] Perhaps we have to take it as late tremors of this disaster when in such a well-documented event the accounts relating to date and number of casualties still differ up to this day. Although most sources agree on the date of Sunday, December 28, 1879, some reports quote the 27th, others the 29th of December. The number of deaths is generally held to be 75, but 76 and 77 have been mentioned as well.

Es ist die Rede von jenem Ereignis, das in einem winterlichen Orkan von unerhörter Heftigkeit als *Tay Bridge Disaster* bekanntgeworden ist. Und diese Sturmnacht jährt sich heute zum hundertzwölften Male:

Es war der 28. Dezember und letzte Sonntag des Jahres 1879, kurz nach 7 Uhr abends, als fast 1 Kilometer Brückenlänge ins Meer stürzte – und einen vollbesetzten Eisenbahnzug mit sich riß, der soeben in die hochliegenden Mittelträger eingefahren war. Gespenstisch wie die Szene war auch der Umstand, daß es zwar eine Handvoll Zeugen gab, sie aber später wenig Genaues über den wirklichen Hergang des Einsturzes sagen konnten. Als sie vor Gericht erschienen, stimmten die meisten Aussagen nur insoweit überein, daß sie Signallampen, grelle Feuerblitze, starken Funkenflug und dann «eine Masse Feuers, das kometenhaft gen Wasser fiel» erblickt hatten. Diese mageren Beobachtungen trafen zeitlich mit einem besonders heftigen Windstoß – später mit 145 km/h rekonstruiert – zusammen, der gegen 19.20 Uhr auch andernorts am Tay beträchtliche Verwüstungen angerichtet hatte, obgleich keine so folgenschweren. Da mit dem Einsturz auch die Telegrafendrähte gerissen waren, wurde der grausige Verdacht erst Stunden später zur Gewißheit: Es gab 75 Tote und keine Überlebenden *(Abb. 1)*[1].

Für einen Großteil der schottischen Bevölkerung war das Ereignis nichts weniger als ein Gottesurteil: «Wenn es hier eine Stimme gibt, die stärker als alles andere in diesem schrecklichen Ereignis ertönt», sagte ein Pfarrer der Freien Kirche Schottlands, «so ist es die Stimme Gottes, entschlossen, seinen Sabbath mit eifersüchtiger Sorge zu bewahren!»

In ganz Großbritannien saß der Schock dieser Katastrophe in der Tat sehr tief – wenn sie auch im südlichen England weniger als die gerechte Strafe für das verwerfliche Reisen an Sonntagen gewertet wurde. Aber ein solches Ereignis läßt noch heute die menschliche Vorstellungskraft erschaudern, gerade wenn die Details weniger bekannt sind. Auch kann die Erschütterung nicht allein durch den Unglücksfall als solchen erklärt werden. Wenn dem so wäre, dann wäre dieses Ereignis längst von Katastrophen größeren Ausmaßes überschattet worden.

---

[1] Vielleicht müssen wir es dem Nachbeben jener Katastrophe zuschreiben, daß selbst bei einem so gut dokumentierten Ereignis die Angaben zum Datum wie auch zur Zahl der Todesopfer bis heute auseinandergehen. Obgleich die meisten Quellen im Datum des Sonntags vom 28. 12. 1879 übereinstimmen, ist in einigen Berichten vom 27., in anderen vom 29. 12. die Rede. Die Zahl der Todesopfer wird allgemein mit 75 angegeben, stellenweise aber auch mit 76 bzw. 77.

1
The scene of the Tay Bridge Disaster on the day after December 28, 1879. Divers searching for victims of the accident; in *The Illustrated London News* of January 10, 1880 [25, p. 231].

1
«*Tay Bridge Disaster*» – der Unglücksort am Tage nach dem Brückeneinsturz vom 28. Dezember 1879. Taucher suchen nach den Opfern der Katastrophe; dargestellt in «*The Illustrated London News*» vom 10. 1. 1880 [25, S. 231].

posture exemplifies that form of conceit which today gloats in its achievements and tomorrow discards them as mere trifles, compared to what is to follow.
The pride of the nation never recovered from the fall of the Tay Bridge, nor from the dismal revelations of the court proceedings to follow. In retrospect, this shock to the sheer exuberance of the new industrial age was in many ways what the sinking of the *Titanic* would be for the world at large, about one generation thereafter. Of course, there had been serious train accidents before in Britain which, what with their increasing number of casualties, were all named "disasters" every time.
In the "Dee Bridge Disaster" of 1847 *(Fig. 2)*, a cast-iron girder of the bridge built by Robert Stephenson had broken under the weight of a locomotive, with the loss of 5 lives [23, p. 27][2]. For short-span rail bridges, simple cast-iron girders were being used at that time, reinforced by wrought-iron bars at the line of tension stress in the lower flansh. The structural principle was sound in itself and similar to our reinforced or prestressed concrete beams of today. Unfortunately, how-

2 Robert Stephenson was a famous bridge builder and the son of George Stephenson, pioneer of the steam engine.

Das *Tay Bridge Disaster* steht vielleicht deshalb so deutlich in der Erinnerung, weil es eine gewaltige Wirkung hatte auf die Selbstgefälligkeit der Zeit, auf das allzu glatte Vertrauen der Viktorianer in ihre industrielle Überlegenheit, auf ihren arroganten Stolz, die längste und größte Brücke der Welt gebaut zu haben. Denn Katastrophen gab es ja häufig genug im Zeitalter des kommerziellen und imperialen Wachstums – und die militärischen Desaster auf kolonialen Kriegsschauplätzen konnten immer wieder durch nachfolgende Siege wettgemacht werden: Erst 11 Monate vorher waren 600 britische Soldaten im Zulu-Aufstand bei Isandhlwana gefallen und ihr Opfer hatte sowohl Trauer als auch Stolz hervorgerufen. Jetzt am Tay aber war das größte Opfer der Nationalstolz selbst – weniger die 75 Menschen. Es war jene Überheblichkeit, die sich heute ihrer Errungenschaften rühmt und sie morgen schon als bloße Nichtigkeiten abtut, im Vergleich zu dem, was noch folgen soll; aber der Eigendünkel der Nation hat sich vom Einsturz der Tay-Brücke nie wieder erholt, ebensowenig wie von den entnervenden Enthüllungen des Gerichtsverfahrens, welches folgen sollte. – Im Rückblick war dies ein ähnlicher Schock für die unbändige Aufbruchstimmung des neuen Industriezeitalters, wie es eine Generation

ever, the method of fastening the bars to the beams was faulty. Already this disaster had sent tremors through the engineering world – at a time when railway and bridge construction still lay in one set of hands, with the same engineers often executing both types of work as a matter of course. A consequence of this accident was that, already there, the use of cast-iron for bridge construction fell into disfavour.

But now, 32 years later, a train accident coincided with a bridge failure of the then longest rail bridge of the world (over 2 miles long) on the Tay. Bridge and train, the two symbols of an era, lay wrecked in shambles. With this disaster, much more than just a bridge had collapsed for Victorian society: its unlimited faith in technological progress, so grandly demonstrated by Paxton's Crystal Palace and Brunel's Great Western Railway, was instantaneously shattered [25, p. 232].

später der Untergang der *Titanic* für die Weltöffentlichkeit sein würde. Natürlich hatte es schon früher in England schlimme Eisenbahnunglücke gegeben – die mit sich steigernden Zahlen von Todesopfern jedesmal als «*Disaster*» bezeichnet wurden:

Im *Dee Bridge Disaster* von 1847 *(Abb. 2)* war ein gußeiserner Träger der von Robert Stephenson erbauten Brücke unter dem Gewicht einer Lokomotive zusammengebrochen und hatte 5 Menschenleben gefordert [23, S. 27][2]. Für Eisenbahnbrücken kürzerer Spannweite pflegte man damals einfache gußeiserne Träger zu verwenden, die mit schmiedeeisernen Stangen zur Aufnahme der Zugkräfte im unteren Flansch abgebunden waren. Das Prinzip war an sich richtig und ähnelt dem heutigen bei armierten oder vorgespannten Betonträgern. Aber die Methode der Befesti-

[2] Robert Stephenson war ein berühmter Brückenbauer und Sohn des Bahnbrechers der Dampflokomotive, George Stephenson.

2
The Dee Bridge Disaster of 1847 on the Chester-Holyhead line, after one of the cast-iron girders had collapsed. Taken from the *Illustrated London News,* showing the wreckage of the train which entombed five victims [23, plate 6].

2
«*Dee Bridge Disaster*» von 1847 auf der Chester-Holyhead-Strecke, nach dem Bruch eines gußeisernen Trägers. Dargestellt in «*The Illustrated London News*»: Die Trümmer des Zuges mit 5 Toten liegen im Wasser [23, Tafel 6].

But even on the bridge itself, both time and place had conjoined in several simultaneous and fateful ways. Following the familiar pattern of disastrous events known as a "sequence of unlucky circumstances", in today's American adage known as "Murphy's Law" (If anything can go wrong, it will go wrong – and it will do that in the worst possible moment), here two highly unfavourable sets of circumstances crossed at a singular point of place and time. There was the instant of a speeding train coinciding with the instant of a particularly fierce gale – a rare encounter of utmost vertical and horizontal stress – and this occurred precisely at that critical point of the structure where the low girders met the "High Girders" of the bridge's central portion. The account of this disaster by the *Vossische Zeitung* of December 30, 1879, had inspired the German writer Theodor Fontane (1819–1898) to write his famous ballad "Die Brück' am Tay". This incident was to remain the worst train accident till the end of the century.

Significance of the Tay Bridge – Before the Disaster

Before we turn to the actual cause of the accident (due less to the underestimated wind pressure than to some hair-raising faults in material and workmanship) we ought to recall the significance of this bridge at the time, and the nature of its creator. Today we may find it hard to believe that this seemingly unimpressive structure, erected between 1871 and 1877 by Sir Thomas Bouch, was once considered to be a world miracle. With its regular sequence of piers and girders marching rather like a viaduct far across the waters, we can hardly make out the raised central portion – the actual bridge – in the distance *(Fig. 3)*.

Nevertheless, the former Tay Bridge (as compared to the heavier appearance of today's new Tay Bridge at the same location) was not altogether without visual appeal. Its graceful frame borne by 85 slender piers of iron tubing "stalks spider-legged across the Tay" [25, p. 230]. With a total length of over 2 miles (3 km) it traversed the immense Firth – first heading out straight northward, later leading over to the east into a large bend near Dundee. This spectacle underlined perhaps the most unusual, even extravagant nature of this endeavour: the bridge meant to cross step by step, quite soberly and without respect, an expanse of open sea, two miles across at this point, with shores too far apart to be taken in by a single glance. It meant the total surrender to the forces of tide, wind and weather in a new, unheard-of way, altogether different from crossing a ravine in a single leap. Comparing this to Telford's

gung der Stangen am Träger war leider verkehrt. Bereits dieses Unglück hatte die Fachwelt aufhorchen lassen – zu einer Zeit, als der Eisenbahn- und Brückenbau noch ganz selbstverständlich in einer Hand lagen, oft von denselben Ingenieuren ausgeführt. Eine Folge des Unglückes war, daß eigentlich schon damals die Verwendung von Gußeisen im Brückenbau viel an Glaubwürdigkeit verloren hatte.

Nun aber, 32 Jahre später, waren am Firth of Tay auf spektakuläre Weise ein Zugunglück und der Einsturz der damals längsten Eisenbahnbrücke der Welt (von über 3 km Länge) zuammengekommen: Brücke und Eisenbahn, die beiden Symbole einer Epoche, lagen in Trümmern. Mit diesem Desaster war für die viktorianische Gesellschaft mehr als eine Brücke zusammengebrochen: Ihr unbegrenzter Glaube an den technischen Fortschritt, der sich so grandios in Paxton's Londoner Kristallpalast und in Brunel's *Great Western Railway* bestätigt sah, war plötzlich dahin [25, S. 232].

Aber auch auf der Brücke selbst waren Zeit und Ort gleich mehrfach und in fataler Weise aufeinander getroffen: Nach dem bekannten Muster verhängnisvoller Abläufe, das hierzulande als eine «Verkettung unglücklicher Umstände» gilt, und das man im heutigen Amerika schlicht mit «Murphy's Law» erklären würde («If anything can go wrong, it will go wrong – and it will do that in the worst possible moment») – waren hier in der Tat gleich zwei denkbar ungünstige Zeitpunkte mit einem speziellen «Ortspunkt» zusammengestoßen: Einmal der Zeitpunkt eines mit überhöhter Geschwindigkeit einfahrenden Zuges mit dem Augenblick einer besonders heftigen Orkanböe – beides schon ein sehr seltenes Zusammentreffen maximaler vertikaler und horizontaler Belastung – und das gerade an der Stelle, wo die Konstruktion in die weitgespannten Hochträger des Mittelteiles übergeht. Der Bericht über diese Katastrophe in der «Vossischen Zeitung» vom 30. 12. 1879 hat Theodor Fontane (1819–1898) zu seiner berühmten Ballade «Die Brück' am Tay» angeregt. Es war das schwerste Zugunglück bis zum Ende des Jahrhunderts.

Bedeutung der Tay-Brücke – Vor dem Stichtag

Bevor wir uns der eigentlichen Unglücksursache zuwenden – es waren weniger der unterschätzte Winddruck, als einige haarsträubende Material- und Konstruktionsfehler – gilt es, sich der Bedeutung dieser Brücke in jener Epoche und der Person ihres Erbauers zu erinnern. Es fällt uns heute schwer zu glauben, daß diese eher undramatisch wirkende Brücke, 1871 – 1877 von Sir Thomas Bouch erbaut, einmal als eine Art von Weltwunder gegolten haben soll. Denn mit ihrer

3
The railway bridge across the Firth of Tay in its original appearance, before the collapse of 1879. View from the north shore near Dundee towards the south. Photographer unknown [30, p. 492].

3
Die Eisenbahnbrücke über den Firth of Tay im Originalzustand, vor dem Einsturz von 1879. Blick vom Nordufer bei Dundee nach Süden, Fotograf unbekannt [30, S. 492].

famous Menai Suspension Bridge of 1826 *(Fig. 4)*, the difference could not be more evident: The issue here was no longer to overcome an obstacle by a single leap, "dry-footedly" so to speak, as classical bridges were supposed to do – but to literally wade across the open sea; it meant reaching the central portion only after going a long distance on "wet feet". The center section consisted of 13 parallel lattice girders of 245 ft (74.68 m) span each, supported on 12 cast-iron tubular piers. By raising the central girders above the rail bed, the trains travelled through them like in a tunnel, leaving below a clear height of 88 ft (26.82 m) for sea-going vessels, while all the other and shorter spans (two thirds of the bridge's total length) rested below the tracks, allowing trains to pass over on top.

The overall appearance of the bridge was thus "practical", but not especially harmonious. Commenting on the bridge in 1880, a German architect in public service, Mr. Havestadt, did not mince his words:

> "Aside from length and economy of this structure, there is nothing particularly remarkable to be found here, generally speaking. Both the spans and the heights of piers stay within rather modest limits. In brief, the completed structure

gleichmäßigen Reihung von Stützen und Fachwerkträgern marschiert sie eher in der Art eines Viaduktes soweit über das Wasser hinaus, daß man erst in großer Ferne den hochgesetzten Mittelteil – also die eigentliche Brücke – zu erkennen vermag *(Abb. 3)*.

Dennoch ist die damalige Tay Brücke (im Gegensatz zum wuchtigeren Erscheinungsbild der heutigen Brücke an gleicher Stelle) nicht ohne Reiz für das Auge: Mit ihrer grazilen Eisenkonstruktion auf 85 schlanken Eisenrohrpfeilern «stackst sie spinnenbeinig über den Tay» [25, S. 230]. Auf eine Gesamtlänge von über 3 km überquert sie – zunächst in gerader Linie nordwärts und erst vor Dundee in eine weite Ostkurve übergehend – die gewaltige Meeresbucht. Dieser Anblick unterstreicht vielleicht das völlig Unübliche, ja für die Zeitgenossen Ungeheuerliche des Vorhabens: Nämlich Schritt für Schritt, ganz nüchtern, fast respektlos ein Stück offenes Meer zu überschreiten, das an dieser Stelle immerhin mehr als drei Kilometer breit ist und die beiden Ufer gar nicht mehr als Einheit wahrnehmen läßt! Es bedeutet, sich dem Meer mit seinen Gezeiten, Wind und Wetter in einer neuartigen, waghalsigen und ganz anderen Weise auszuliefern, als etwa nur eine Schlucht in kühnem Schwung zu überwinden: Wenn man an Telfords be-

appears, in light of the majestic firth, like a conglomerate of various systems of piers and bridges, derived from smallish means and conceived without any regard for a magnificent and powerful impression." [19, p. 212]

Due to its great length, the rather monotonous course of the bridge hardly helps to arrest the eye at any one point, nor does the bridge seem to lead anywhere in particular. Widely reaching out across the open sea, it rather augurs some "uncertain passage". But quite possibly, it may have been precisely the peculiar appearance of this "non-bridge" without any visible bridgeheads which so fascinated the contemporaries:

> "The entire structure is of a highly pleasing and nimble appearance. It is so long, so airy, so thin that... the sight of a train rolling across the bridge evokes some instinctive anxiety."

Thus the *Vossische Zeitung* described the then longest railway bridge of Europe [30, p. 493]. In any case, there simply was no structure around to compare with that one: it was the pride of Dundee and was regarded in its day as a magnificent piece of engineering [8, p. 27].

rühmte Menai-Hängebrücke von 1826 denkt *(Abb. 4)*, wird der Unterschied besonders deutlich: Es geht nicht mehr darum, ein Hindernis durch einen einzigen Sprung «trockenen Fußes» zu beseitigen, wie es einer klassischen Brücke geziemt, sondern es buchstäblich zu durchwaten! Was bedeutet, erst nach langer Wegstrecke, mit vielen dem Wasser ausgesetzten Fußpunkten, den eigentlichen Mittelteil zu erreichen: Dieser besteht aus 13 parallelen Fachwerkträgern von je 245 Fuß (74,68 m) Länge, von 12 gußeisernen Rohrstützen gehalten. Dabei sind diese Mittelträger so hochgesetzt, daß der Zug tunnelartig durch sie hindurchfährt und darunter erst die notwendige Durchfahrtshöhe von 88 Fuß (26,82 m) für die Schiffahrt erreicht wird. Wogegen alle übrigen, kürzeren Träger (sie machen zwei Drittel der Brückenlänge aus) unter den Schienen der Bahn liegen, die auf der Träger-Oberseite fährt.

Das Gesamterscheinungsbild der Brücke ist damit ein zwar «praktisches», aber nicht sonderlich harmonisches. Wenig gnädig äußert sich dazu ein deutscher Zeitgenosse, der Regierungsbaumeister Havestadt im Jahre 1880:

> «Außer Länge und Ökonomie des Bauwerks ist an demselben im allgemeinen nichts sonderlich Bemerkenswertes aufzufinden. Die Spannweiten und Pfeilerhöhen bleiben in ziemlich bescheidenen Grenzen. In kurzen Worten ausgesprochen erscheint das fertige Bauwerk gegen-

4
The chain suspension bridge across the Straits of Menai: Thomas Telford's masterpiece with a clear span of 539 ft (164 m) and built already 50 years before, from 1819 to 1826. In a drawing of 1826 by the German contemporary Karl Friedrich Schinkel [5, p. 23].

4
Die Ketten-Hängebrücke über die Meeresstraße von Menai: Bereits 50 Jahre früher von Thomas Telford in den Jahren 1819–1826 erbaut – und mit 539 Fuß (164 m) Spannweite sein Meisterstück. In einer Zeichnung des Zeitgenossen Karl Friedrich Schinkel von 1826 [5, S. 23].

Preceding Battles

For this proud work, there had been a long and fierce struggle in which even Thomas Bouch himself was engaged *(Fig. 5)*. Born in 1822 as the son of a sea-captain in Thursby, Cumberland, he rose to be chief engineer and manager of the Edinburgh & Northern Railway, at only 27 years of age. Railway construction had progressed with some speed in the lowlands of Southern England. Yet in Scotland, progress was hampered time and again by natural obstacles, such as steeper gradients and large estuaries cutting deeply into the country. Particularly Scotland's eastern coastline, between Edinburgh and Dundee 46 miles (74 km) to the north, is interrupted by the wide estuaries of the Firth of Forth and the Firth of Tay – the Tay being Scotland's longest river with 120 miles (193 km) *(Fig. 6)*.

Looking across the two Firths for the first time, Bouch is said to have already stated emphatically that these obstacles could and should be bridged – quite oblivious to the laughter of railway shareholders. After having built and carried across successfully his "floating railways" with the help of loading ramps and steam ferries at the Forth, he set up business for himself as a consulting engineer. He began diligently to erect the many smaller bridges, viaducts and rail lines which gave him – so one assumed at least – the knowledge and experience necessary to throw an iron bridge across the Tay, later on. It must be remembered, though, that Bouch had designed nearly 300 miles (480 km) of railway in Scotland and northern England, building more bridges than anybody else in his time.

über der mächtigen Meeresbucht als ein, ohne Rücksicht auf eine großartige und kraftvolle Wirkung entworfenes, von kleinen Verhältnissen entlehntes Konglomerat verschiedener Pfeiler- und Brückensysteme.» [19, S. 212]

Wegen der großen Länge bietet in der Tat der eher monotone Brückenverlauf dem Auge kaum feste Haltepunkte und erscheint auch nicht sonderlich zielgerichtet: Mit der weit ausholenden Wegführung über offenes Wasser wird eher eine «Fahrt ins Ungewisse» angedeutet. Möglicherweise war es aber gerade das absonderliche Erscheinungsbild einer «Nichtbrücke» ohne erkennbare Uferpunkte, das die Zeitgenossen faszinierte:

«Der ganze Bau hat ein höchst gefälliges und leichtes Aussehen. Er ist so lang, so luftig, so dünn, daß ... der Anblick eines über die Brücke dahin rollenden Eisenbahnzuges unwillkürlich nervöse Unruhe verursacht.»

So beschrieb die «Vossische Zeitung» die seinerzeit längste Eisenbahnbrücke Europas [30, S. 493]. Wie auch immer, es gab bislang kein vergleichbares Bauwerk. Es war der Stolz von Dundee und galt gemeinhin als ein Beispiel hoher Ingenieurbaukunst [8, S. 27].

Vorkämpfe

Um dieses stolze Werk war aber lange Zeit hart gerungen worden – nicht zuletzt von Thomas Bouch selbst *(Abb. 5)*. 1822 als Sohn eines Kapitäns in Cumberland geboren, war er bereits mit 27 Jahren Oberingenieur und Manager der *Edinburgh and Northern Railway* ge-

5
Sir Thomas Bouch, 1822–1880.
Born in Cumberland as the son of a sea-captain, Thomas Bouch was already an engineer with the Stockton & Darlington Railway with 17 years of age. At the age of 27, he became chief engineer and manager of the Edinburgh & Northern Railway. During his long career he designed new railway viaducts in the highlands, such as the Beelah, Deepdale and Tees Viaducts. These were followed by the popular railway ferries at the Forth and Tay, as well as a tramway system for Edinburgh. All these undertakings contributed to his securing the commission for the Tay Bridge, and later for the Forth Bridge as well.
Despite his rather shadowy qualifications, he excelled by virtue of his considerable breadth of imagination. He became a member of the British Association of Civil Engineers and was finally knighted by Queen Victoria for having built the Tay Bridge. He died in Moffat, barely 10 months after the Tay Bridge Disaster. – J. Murray reports that Bouch, according to new findings about his life and work, was what would nowadays be described as a "chancer" [20].

5
Sir Thomas Bouch, 1822–1880.
Als Sohn eines Kapitäns in Cumberland geboren, war Thomas Bouch bereits mit 17 Jahren als Ingenieur bei der Stockton & Darlington Railway tätig. Mit 27 Jahren wurde er Oberingenieur und Manager der Edinburgh & Northern Railway. Im Laufe seiner langen Karriere entwarf er neuartige Eisenbahnviadukte im Bergland, so den Beelah-, Deepdale- und Tees-Viadukt. Dazu kamen die vielbeachteten Eisenbahn-Fähren am Forth und Tay sowie ein Straßenbahnnetz für Edinburgh: All diese Arbeiten halfen, ihm den Auftrag für die Tay-Brücke und dann auch noch für die Forth-Brücke zu verschaffen.
Trotz seiner «rather shadowy qualifications» zeichnete er sich durch beträchtlichen Ideenreichtum aus, wurde Mitglied des britischen Zivilingenieur-Verbandes und schließlich von Queen Victoria für den Bau der Tay-Brücke geadelt. Er starb in Moffat, 10 Monate nach dem Einsturz der Tay-Brücke. – J. Murray schreibt, daß Bouch nach neueren Erkenntnissen über seinen Lebensweg das war, was man heute einen «chancer» nennen würde, also jemand, der das Risiko sucht oder es zumindest nicht scheut [20].

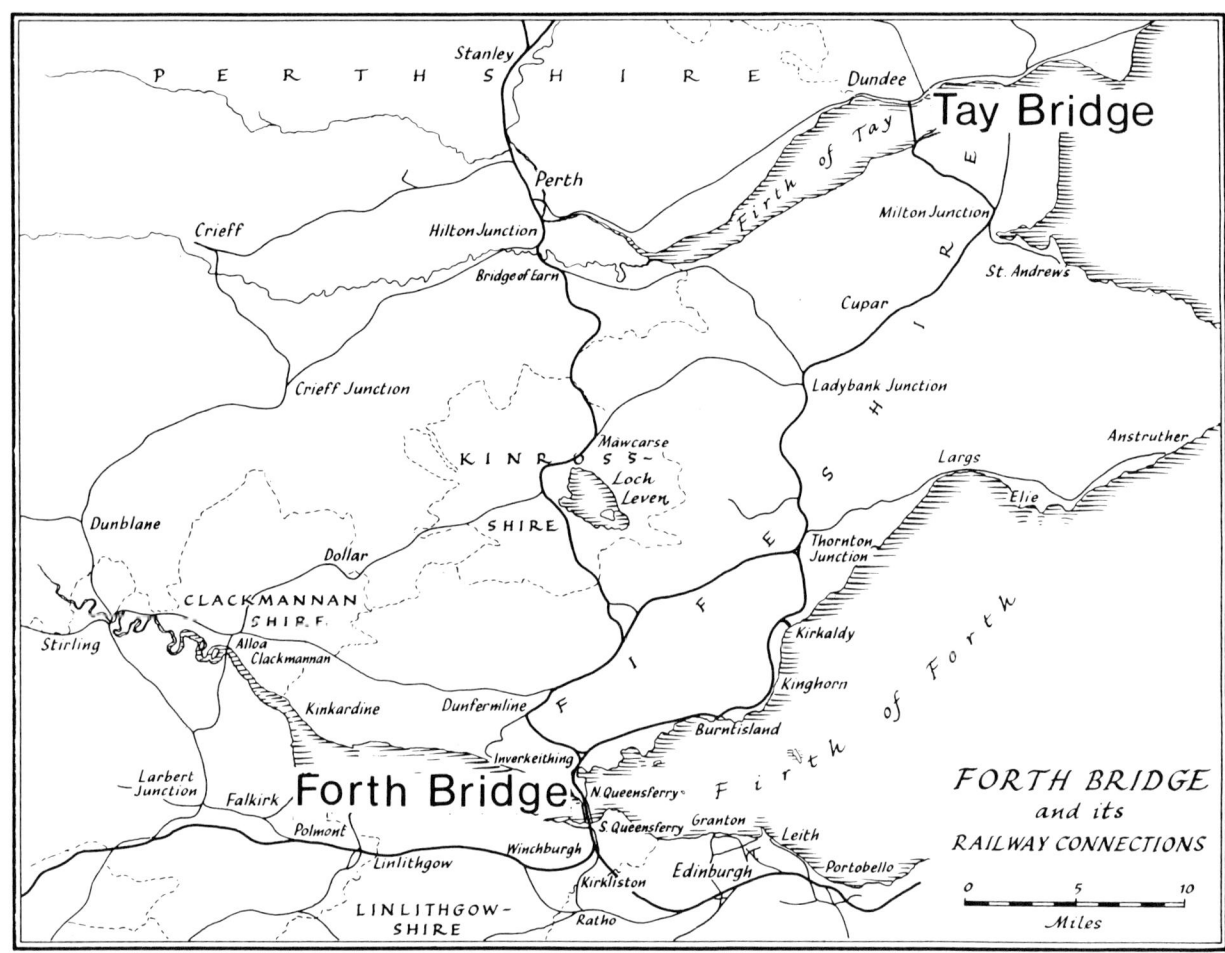

6
Location of the railway bridges at the Firth of Forth and the Firth of Tay [23, p. 20].

6
Übersichtskarte zur Lage der Eisenbahnbrücken am Firth of Forth und am Firth of Tay [23, S. 20].

worden. Der Eisenbahnbau war in den Niederungen Süd-Englands recht flott vorangekommen. In Schottland aber hatte er immer wieder mit erheblichen natürlichen Hindernissen zu kämpfen, wie größeren Steigungen und tief ins Land einschneidenden Meeresbuchten. Namentlich die Ostküste Schottlands wird zwischen der Hauptstadt Edinburgh und dem 74 km weiter nördlich gelegenen Dundee von den großen fjordartigen Meeresarmen des Firth of Forth und des Firth of Tay unterbrochen. Dabei ist der Tay mit 193 km Schottlands längster Fluß *(Abb. 6)*.

Beim Anblick der Meeresarme des Forth und Tay soll Bouch bereits mit Emphase erklärt haben, daß man diese Hindernisse überbrücken könne und müsse – was die Eisenbahn-Aktionäre nur mit Gelächter quittierten. Nachdem er aber bereits «schwimmende Eisenbahnen» erfolgreich am Forth mit Hilfe von Verladebrücken und Dampffähren gebaut und übergesetzt hatte, machte er sich als beratender Ingenieur

Whatever the personal ambitions of Thomas Bouch may have been, the railway companies were already pressing their own plans. The North British Railway Company (called NBR henceforth), so successful in the south, was unwilling to cede the money-making Scottish routes to their arch-rival, the Caledonian Railway Company (CR for short) alone. Suddenly this bridge venture, daring as it may have seemed, offered the last chance to the NBR to avoid extinction by the CR. Although the NBR recognized very late, almost too late, the necessity for bridging the Tay, there had been good reasons long before to construct such a bridge. A railway journey from Edinburgh to Dundee in the 1850's was an experience which, made once, nobody ever wanted to repeat if there was any alternative. Despite a distance of only 46 miles (74 km) between the two cities, 3 hours and 12 minutes were required for this trip under normal circumstances – and much longer in stormy weather on the Forth or the Tay. The fastest train in those days left Edinburgh's Waverly Station by 6:25 a.m.; in Granton at the Forth the passengers changed to a paddlewheel steamferry for the journey across to Burntisland; from there, a second train carried them through Fife to Tayport, where yet another ferry brought them across the Tay to Broughty Ferry, east of Dundee. Finally, a third train, albeit belonging to the CR, transported them to Dundee. Far more convenient than this ordeal was CR's longer route via Perth *(Fig. 6)*.

Thus it was evident even without Bouch's vision that a direct line from Edinburgh to Dundee would open up Scotland's northeast to the transport of goods and passengers, and offer new prosperity to the NBR. Yet still in 1854, with the take-over of the Edinburgh and Northern Railway by the NBR, the new directors ventured the remark to Bouch that it was "the most insane idea that could ever be propounded" to build such a bridge of two miles in length. Even those who were enamoured by the proposal, sensing victory at last in their bitter struggle with the CR, were astonished at the costs, estimated at 200,000 Pounds Sterling, because in that year the NBR had been forced to declare a loss of about 3,000 Pounds, while the CR had made a profit of more than 20,000 Pounds.

selbständig: Er begann, mit Eifer die vielen kleineren Brücken, Viadukte und Bahnlinien zu bauen, welche ihm, – so dachte man wenigstens – die notwendige Kenntnis und Erfahrung gaben, später eine eiserne Brücke über den Tay zu schlagen. Allerdings muß man daran erinnern, daß Bouch annähernd 300 Meilen (480 km) an Eisenbahnen in Schottland und Nord-England gebaut hatte, und damit mehr Brücken als irgendjemand sonst zu seiner Zeit. Was auch immer die persönlichen Ziele von Thomas Bouch gewesen sein mögen, die Eisenbahngesellschaften verfolgten bereits ihre eigenen Pläne.

Die im Süden so erfolgreiche *North British Railway Company* – im Weiteren *NBR* genannt, war nicht gewillt, ihrem Erzrivalen, der *Caledonian Railway Company*, kurz *CR,* die lukrativen schottischen Routen allein zu überlassen. Nun bot sich mit einem solchen Brückenprojekt, so waghalsig es auch erscheinen mochte, für die *NBR* die letzte Chance, ihrer Ausschaltung durch die *CR* zu entgehen. Auch wenn die *NBR* sehr spät, fast zu spät, die Notwendigkeit einer Überquerung des Tay erkannte, hatte es schon lange vorher gute Gründe für eine solche Brücke gegeben. Denn eine Bahnfahrt von Edinburgh nach Dundee war in den 1850er Jahren eine Erfahrung, die einmal gemacht, niemand ein zweitesmal auf sich nahm, wenn es eine Alternative gab. Obgleich die Entfernung beider Städte nur 74 km beträgt, benötigte man für diese Reise *3 Stunden und 12 Minuten* unter normalen Bedingungen – und wesentlich länger bei Sturm am Forth oder Tay. Der schnellste Zug jener Tage verließ Waverly Station in Edinburgh um 6.25 Uhr morgens; in Granton am Firth of Forth stiegen die Passagiere in eine Raddampfer-Fähre um und setzten nach Burntisland über; dort brachte sie ein zweiter Zug durch die Grafschaft Fife bis nach Tayport, wo eine weitere Fähre die Fahrgäste über den Firth of Tay nach Broughty Ferry transportierte; schließlich brachte sie ein dritter und letzter Zug – der allerdings zur *CR* gehörte – endlich nach Dundee. Ungleich vorteilhafter als diese Strapaze war die längere Fahrt mit der *CR* über Perth *(Abb. 6)*.

So war es auch ohne die Vision von Bouch einleuchtend, daß eine direkte Bahnlinie von Edinburgh nach Dundee den Nordosten Schottlands für den Güter- und Personenverkehr erschließen und damit auch der *NBR* bessere Zeiten bescheren würde. Aber noch 1854, als die *Edinburgh and Northern Railway* von der *NBR* übernommen wurde, erklärte man Bouch seitens der neuen Direktoren, eine solche Brücke von 2 Meilen Länge zu bauen, sei «die verrückteste Idee, die man je hätte vorbringen können». Selbst diejenigen, welche von dem Vorschlag angetan waren und einen Sieg im

"Rainbow Bridge"

A considerable segment of public opinion held that such a bridge would most certainly collapse, either through faulty engineering or through an act of God. The few who did consider the undertaking technically feasible, were still convinced that the investors would never live to see a profit. And some rather poetic or spiteful critics even called it "that rainbow bridge". The Dundee Harbour Board had become prosperous by high levies collected from shipping and was alarmed now at the thought of goods being transported by rail across a bridge that was to be located dangerously close to the harbour, according to first plans. Also, the neighbouring town of Perth, an important rival of trade, was not amused and declared that the bridge would block off its sea traffic. This objection was met later on by the promise to raise the central spans, allowing for free passage of seagoing vessels underneath. – Above all that row of protest, the thin but piercing voice of the Dundee Rights of Way Association was heard: "If there must be a bridge, it should carry a public footpath!"

But in the main, it was the fierce struggle between the great rivals, the North British and the Caledonian Railway in the east and west of Scotland which postponed the start of construction by many years. This battle had gone so far as to see the trainmen beat each other up in public, while the rolling stock had declined to pitiful conditions and the fares to suicidal levels. Shareholders of the NBR watched in helpless horror the depreciation of their investment. On one occasion the shareholders received, in lieu of a dividend, a tastefully printed map of their railway network – as if to show them where their money was being squandered.

The supporters of the CR contended the NBR was deliberately preventing people from travelling to the countryside on their day of leisure, since the NBR ran no trains on Sundays – with the evil result that large amounts of spiritous liquors were consumed on the Sabbath! The reason for this latest escalation was that the NBR had begun to take over the entire traffic of the Northern Highlands after it had aquired the Edinburgh, Perth & Dundee Railway. With the construction of the Tay Bridge, the NBR would finally gain triumphant supremacy over all of eastern Scotland, which the CR in turn tried to counter with their smash-and-grab tactics. The wide-spread fragmentation into so many private railway lines, all competing without coordination, was the result of that short but crazy period of "Railway Mania" of 1845/50. It was to be the height of railway speculation, but also the result of Scottish-

bitteren Kampf mit der *CR* witterten, waren entsetzt über die geschätzten Kosten von 200 000 Pfund Sterling. Denn in jenem Jahr mußte die *NBR* einen Verlust von etwa 3 000 Pfund erklären, während die *CR* über 20 000 Pfund Gewinn gemacht hatte.

«Regenbogenbrücke»

Ein beträchtlicher Teil der öffentlichen Meinung blieb dabei, daß eine solche Brücke mit Sicherheit einstürzen würde, sei es durch Fehlbarkeit der Ingenieure, sei es durch göttliche Fügung. Die wenigen, welche das Unternehmen für bautechnisch möglich hielten, waren selbst dann noch überzeugt, daß es den Investoren nie einen Profit abwerfen würde. Und eher poetische oder spöttische Kritiker nannten sie sogar «die Regenbogenbrücke». Das *Dundee Harbour Board* war durch hohe Abgaben der Schiffahrt reich geworden und jetzt alarmiert bei dem Gedanken, daß Güter auf dem Schienenweg transportiert werden würden, über eine Brücke, die nach ersten Plänen gefährlich nahe am Hafen liegen sollte. Und die Nachbarstadt Perth, ein wichtiger Handelsrivale, erklärte unwillig, die Brücke würde ihre Seewege blockieren – ein Einwand, dem man später durch das Versprechen, die Mittelträger für freie Schiffahrt der Segler höher zu legen, begegnete. Über all dem Protestgeschrei war schließlich noch die dünne, aber schrille Stimme der *Dundee Rights of Way Association* zu vernehmen: «Wenn es eine Brücke geben muß, dann soll sie einen öffentlichen Fußweg erhalten!»

In der Hauptsache aber waren es die erbitterten Grabenkämpfe zwischen den beiden großen Rivalen *NBR* und *CR,* im Osten und Westen Schottlands, welche den Baubeginn noch um viele Jahre verzögerten. Denn dieser Krieg war soweit gekommen, daß sich die jeweiligen Eisenbahner öffentlich prügelten, der Wagenpark auf einen erbärmlichen Zustand und die Fahrpreise auf selbstmörderische Beträge absanken. Bürger, die Aktien der *NBR* besaßen, verfolgten in hilflosem Schrecken den Wertverlust ihrer Anlagen. Die Aktionäre erhielten einmal, wie es scheint, anstelle einer Dividende eine geschmackvoll gedruckte Karte ihres Eisenbahnnetzes – wie um ihnen zu zeigen, wo ihr Geld verschleudert wurde.

Die Vorkämpfer der *CR* behaupteten, die *NBR* würde die Leute davon abhalten, an ihrem freien Tage aufs Land zu fahren, denn es gab sonntags keinen Zugverkehr auf der *NBR* – so daß ein großer Whisky-Konsum am Sabbath die üble Folge sei! Der Grund für diese letzte Eskalation war, daß die *NBR* mit der Übernahme der *Edinburgh, Perth and Dundee Railway* begonnen hatte, den gesamten Verkehr der nordöstlichen High-

7
"The Great Triumvirate" of British railway engineers: Isambard Kingdom Brunel, Robert Stephenson and Joseph Locke [23, plates 2–4].

English antagonism. Only with the NBR's takeover of the Edinburgh & Glasgow Railway, was the tide turned, allowing for the founding of a centralized railway network at last.

Times of Change

We now approach the actual effects and influence which this bridge exerted upon its time. Although it still stands today under the looming shadow of the Tay Bridge Disaster in our collective psyche, merely figuring as the unlucky forerunner of the famed Forth Bridge, the Tay Bridge did have a significance all its own for this era, long before the accident took place. Within the great age of the industrial revolution in Britain, those years between 1860 to 1880 – and, in turn, the period of this bridge's origin – were times of change. With the advent of the initial phase of British railway construction known as the "heroic age" and its formidable railway engineers, seemingly omnipotent in their trade, the foundations for new developments had been set. Specifically, the great "triumvirate" *(Fig. 7)* of the early pioneers, Isambard Kingdom Brunel, Joseph Locke and Robert Stephenson [23, plates 2–4], had prepared the ground from which the famous bridges of Thomas Telford would spearhead the world renown of British engineering. Telford was later to become the first President of the Institution of Civil Engineers in London.

lands an sich zu ziehen. Und mit dem Bau der Tay-Brücke würde sie endlich den Triumph der Gesamtherrschaft über das östliche Schottland einfahren. Was wiederum die *CR* mit ihren «*smash-and-grab tactics*» – dem schnellen Aufkauf vieler kleiner Linien – zu verhindern wußte: Die weite Zersplitterung in zahlreiche konkurrierende Privatlinien ohne Koordination war das Ergebnis jener kurzen, aber verrückten Zeit der «*Railway Mania*» von 1845/50, der Blüte der Eisenbahnspekulation, aber auch nationaler, schottisch-englischer Antagonismen. Erst mit der Übernahme der *Edinburgh and Glasgow Railway* durch die *NBR* wandte sich das Blatt und konnte endlich der Grundstein für ein zentrales Eisenbahnnetz gelegt werden.

Zeit des Umbruches

Damit nähern wir uns der eigentlichen Wirkungsgeschichte dieser Brücke. Wenngleich sie heute unter dem Schatten des «*Tay Bridge Disaster's*» im Bewußtsein steht und lediglich als der unglückselige Vorläufer für die berühmte Firth-of-Forth-Brücke gilt, so hatte sie – lange vor dem Ereignis – sehr wohl ihre eigene Bedeutung für die Zeit. Innerhalb der großen Epoche der industriellen Revolution in England waren jene Jahre zwischen 1860 und 1880 – und damit die engere Entstehungszeit der Brücke – selbst wiederum eine Zeit des Umbruches. Mit der Frühphase des britischen Eisenbahnbaues, dem «heroischen Zeitalter» und seinen unerschrockenen «*Railway Engineers*» – die uns heute wahrhaftig als Alleskönner erscheinen müssen – waren die Grundlagen der neuen Entwicklung bereits gelegt. Namentlich das «große Triumvirat» *(Abb. 7)* der frühen Pioniere Isambard Kingdom Brunel, Joseph Locke und Robert Stephenson [23, Abb. 2 – 4] hatte den Boden bereitet, auf dem die berühmten Brücken von Thomas Telford die Weltgeltung britischer Ingenieurbaukunst begründen halfen. Telford war später der erste Präsident der *Institution of Civil Engineers* in London geworden.

Bis dahin hatte es sich im Brückenbau meist um die Überwindung relativ schmaler, wenn auch gefährlicher Hindernisse wie Flußläufe, Schluchten, Kanäle etc. gehandelt, also überschaubarer Wasserwege. Daß man nun daran ging, ganze Meeresarme zu überbrücken, war eine neue Sache. Dabei mußte das zunächst mit den damals gebräuchlichen Baumaterialien geschehen, nämlich Gußeisen und allenfalls Schmiedeeisen für besonders beanspruchte Bauteile. Doch bald zeigte sich, daß die allgemeine Entwicklung in der Eisenherstellung nicht Schritt gehalten hatte mit den rasanten Fortschritten im Brückenbau, mit seinen immer gewagteren Spannweiten und dynami-

7
«Das große Triumvirat» der britischen *Railway Engineers:* Isambard Kingdom Brunel, Robert Stephenson und Joseph Locke [23, Tafeln 2–4].

Up to this point in time, bridge construction had mostly dealt with crossing relatively narrow if dangerous obstacles, such as rivers, ravines, channels etc. – obstacles that could easily be comprehended. But now setting out to bridge entire estuaries was quite a different matter. And this had to be achieved at first with the conventional materials then in use, such as cast iron and, occasionally, wrought iron (at best) for members under heavy stresses. But soon it became evident how the general development of iron production had been unable to keep pace with the rapid progress in bridge building, reaching out for ever-more audacious spans and dynamic loads. That is to say, one had met until then the new requirements with the materials of the old times, those of the cast-iron age. Initial successes with cast-iron short-span bridges had lulled the profession into a false sense of security.

Over the years, an exclusive circle of renowned and wordly autodidacts had emerged, which was now being confronted by a new breed of scientifically trained theoreticians and designers. These new engineers set out with an increasing lack of respect, to challenge the established but unsophisticated rules of "practical engineering" – derived rather more from empirical experience than from scientific knowledge or reasoning. – It is this time of change, coinciding with the turbulent history of the origin of the Tay Bridge, which is to be illuminated here.

In going over the literature available, we sometimes even come across some rather off-beat sources which skillfully combine the spirit of the new time with the lucid account of a prominent event, as suggested below.

Eyth's "Enno Bridge"

In this context, it is useful to recall Max Eyth (1836–1906), the German engineer and poet of technology from Kirchheim/Teck in Schwaben who, along with the better-known Theodor Fontane, may be seen as an important witness of his time *(Fig. 8)*. In his handsome and at the time widely-read book *Hinter Pflug und Schraubstock* ("Behind plow and bench-vice") [4][3] he describes in captivating detail the rise and fall of his so-called "Enno Bridge". The ciphered meaning of this name largely escaped the attention of engineering circles, thanks to an authorial trick: by altering all

3 Here we are concerned with the last chapter of this book, entitled "Berufstragik" (Tragedy of profession), with page numbers referring to the 1918 edition. This work has since been re-edited with a slightly altered subtitle and includes an introductory essay by Theodor Heuss about Max Eyth.

schen Belastungen. Das heißt, man war bislang den Anforderungen neuer Art mit den Materialien alter Zeit, der Gußeisenzeit, begegnet. Die anfänglichen Erfolge mit gußeisernen Brücken kleinerer Spannweiten hatten die Fachwelt in eine falsche Sicherheit gewiegt.

Denn mit den Jahren hatte sich eine Kaste hochangesehener und welterfahrener Autodidakten herausgebildet, denen jetzt der neue Typus des wissenschaftlich geschulten Theoretikers und Statikers gegenübertrat. Die neuen Ingenieure begannen zunehmend respektloser, die bewährten, aber hemdsärmeligen Regeln des «*practical engineering*» in Frage zu stellen; denn sie beruhten eher auf praktischer Erfahrung als auf wissenschaftlicher Grundlage. Es ist diese Zeit des Umbruches, welche mit der wechselvollen Entstehungsgeschichte der Tay-Brücke zusammentrifft und hier näher beleuchtet werden soll. Dabei sind es manchmal Glücksfälle, die abseits der bekannten Fachliteratur zu Quellen führen, welche den Geist jener neuen Zeit zu verbinden wissen mit der prägnanten Wiedergabe eines besonderen Ereignisses:

Eyth's «Ennobrücke»

Es ist die Rede von *Max Eyth* (1836 – 1906), jenem schwäbischen Ingenieur und Technikpoeten aus Kirchheim/Teck, der neben dem bekannteren Theodor Fontane als ein wichtiger Zeitzeuge zu gelten hat *(Abb. 8)*. In seinem ansehnlichen, seinerzeit viel gelesenen Buch «*Hinter Pflug und Schraubstock*» [4][3] beschreibt er auf fesselnde Weise den Bau und Untergang der sogenannten «Ennobrücke». Deren schlüsselhafte Bedeutung ist allerdings durch einen Kunstgriff Eyth's der Fachwelt kaum, wenn überhaupt, bekannt geworden: Mit der romanhaften Verfremdung aller Eigennamen breitet dieser Augenzeuge den Mantel der Verschwiegenheit über die Rolle der Beteiligten *(Abb. 9)*[4].

3 Es geht hier um das letzte Kapitel dieses Buches, betitelt «Berufstragik», wobei sich die Seitenzahlen auf die Ausgabe von 1918 beziehen. Inzwischen ist das Werk mit etwas verändertem Untertitel und mit einem einleitenden Essay von Theodor Heuss über Max Eyth neu herausgegeben worden.

4 Letztlich gibt Eyth selbst, allerdings nur in einer Fußnote (S. 386) den entsprechenden Hinweis: «Es möge hier ausdrücklich betont sein, daß sämtliche Namen, auch die Ortsnamen, die sich in irgendeiner Weise auf die ‹Ennobrücke› beziehen, aus naheliegenden Gründen erfunden sind. Dagegen ist die technische Geschichte des Unternehmens auch im einzelnen den Tatsachen entsprechend erzählt.»

8
Max Eyth (b. 1836 in Kirchheim/Teck, d. 1906 in Ulm, Germany), was engaged the world over, as an engineer and co-inventor of the steam-plows, by the British engineer John Fowler. Later on, Eyth became the founder of the German Agricultural Society and came to be well-known as a gifted "poet of technology" with numerous works about the technological development of his time [4].

8
Max Eyth, geboren 1836 in Kirchheim/Teck, gestorben 1906 in Ulm, war weltweit als Ingenieur und Miterfinder der Dampfpflüge des englischen Ingenieurs John Fowler tätig. Später wurde er zum Begründer der «Deutschen Landwirtschaftlichen Gesellschaft» und trat als ein begabter «Dichteringenieur» mit mehreren Werken zur technischen Entwicklung seiner Zeit hervor [4].

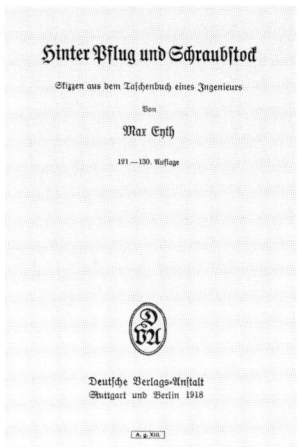

9
Front page of the 1918 edition. In the chapter *"Berufstragik"* (Tragedy of profession) the real story about the rise and fall of the Tay Bridge is related, all names having been altered. Certain key passages by this eye-witness will be transcribed hereafter. The selections recall, aside from the sorrows and fates of engineers involved, something about the ambience of this era, as well as Eyth's hymnically poetic style in expressing the belief in technical progress.

9
Titelseite der Ausgabe von 1918. Im Kapitel «Berufstragik» wird, nach Verfremdung aller Eigennamen, die tatsächliche Geschichte vom Bau und Einsturz der Tay-Brücke geschildert.
Einzelne Schlüsselpassagen dieses Augenzeugen sind im folgenden wiedergegeben: Die Auswahl mag neben den Sorgen und Schicksalen der beteiligten Ingenieure auch etwas vom Zeitgefühl der Epoche vermitteln sowie vom hymnisch/poetischen Sprachstil des technischen Fortschrittsglaubens

names in romancier's style, this eye-witness casts a shroud of anonymity over the roles of all actors involved *(Fig. 9)*[4].
Once, however, the reader has made the appropriate connections in Eyth's accurate and concise narrative of the bridge's erection, there can be no doubt that the "Ennobrücke" is in actual fact the Tay Bridge – nor that a certain "Sir William Bruce" stands in for no one else but Sir Thomas Bouch himself; just as "the consulting work of this famous civil engineer" did not center on the fictitious "Northflintshire Railway", but on the North British Railway, and so on.
Eyth describes for instance "the King of our profession... who in getting ready to leave puts on his straw-yellow gloves while his six draughtsmen... residing in a large bright hall are designing six bridges for five continents.... A handsome and imposing man in a white waistcoat with a mighty gold-blond beard that he tends to stroke with affection whenever his mind wanders off. One can hardly surpress a sense of instinctive reverence in his presence.... This famous Mr. Bruce keeps a business office in London and has built bridges in all parts of the world for the last twenty years." ("Bruce" = Bouch) [4, p. 387].
There simply is no source equal to this one on the Tay Bridge, recounting in the final chapter "Berufstragik" so succinctly the course of events, even the hopes and fates of his fellow engineers, for Max Eyth had himself been a witness to the major phases of construction during his professional work in Britain. His key figure is a certain "Harold Stoss", an Austrian engineer and Bouch's son-in-law, said to be the engineer and designer responsible for the bridge in Bouch's office. Eyth had kept up a close correspondence with him over many years. Today we may only guess whether this fictitious figure may in fact have been the resident engineer at the site, Albert Groethe [8, p. 27]. There could be some justification for this since according to Eyth, the British workers did have some trouble in pronouncing the engineer's foreign name, preferring to call him "Sir Harold" instead; this would not be too far off from the real "Sir Albert", as far as the sound goes. On the other hand, the same Albert Groethe, according to Hammond [8, p. 31] and others, later acted as a witness at the public inquiry, whereas Eyth had him perish with the train on that fatal night.

4 In the end, Eyth himself offers the decisive hint, albeit hidden in a footnote (p. 386): "It may be emphasized here explicitly that all names, even place-names relating in any way to the 'Enno Bridge', have been invented for obvious reasons. The technical account of the undertaking, on the other hand, follows the facts, down to details."

Hat man aber erst in Eyth's wahrheitsgetreuer und sehr präziser Baubeschreibung die wirklichen Zusammenhänge erkannt, bleibt kein Zweifel, daß mit seiner «Ennobrücke» die Tay-Brücke gemeint sein muß – und daß z. B. ein gewisser «Sir William Bruce» niemand anderer als Sir Thomas Bouch ist! Wie denn auch «die beratende Tätigkeit dieses berühmten Zivilingenieurs» nicht der fiktiven *«Northflintshire»*-Eisenbahn, sondern der *North-British*-Eisenbahn gegolten hat, usw.
Eyth schildert z. B. «*... den König unseres Berufes,... wie er (eben im Gehen begriffen) seine strohgelben Handschuhe anzog, und seine sechs Zeichner... in einem großen hellen Saal hausen und sechs Brücken für fünf Weltteile entwarfen... Ein schöner stattlicher Mann in weißer Weste, mit einem gewaltigen goldblonden Bart, den er liebevoll streichelt, so oft ihm die Gedanken stillstehen. Man kann sich in seiner Gegenwart vernunftloser Hochachtung kaum erwehren... Dieser berühmte Herr Bruce hat sein Geschäftsbureau in London und hat seit 20 Jahren in aller Welt Brücken gebaut.»* usw. («Bruce» = Bouch) [4, S. 387].
Es gibt keine vergleichbare Quelle zur Tay-Brücke, die im Schlußkapitel «Berufstragik» mit so genauer Sachkenntnis den Ablauf der Vorgänge, aber auch die Hoffnungen und Schicksale der Berufskollegen zu vermitteln weiß: Denn Max Eyth war selbst Augenzeuge der entscheidenden Bauphasen während seiner beruflichen Tätigkeit in England gewesen. Die eigentliche Schlüsselfigur bei Eyth ist ein gewisser «Harold Stoß», österreichischer Ingenieur und späterer Schwiegersohn Bouch's, der als maßgeblicher Konstrukteur und Statiker der Brücke in Bouch's Büro bezeichnet wird. Mit ihm war Eyth über viele Jahre in freundschaftlichem Briefwechsel verbunden. Es muß heute offen bleiben, ob mit dieser literarischen Figur vielleicht der leitende Ingenieur an der Baustelle, Albert Groethe [8, S. 27], gemeint sein kann (?). Dafür spräche, daß nach Eyth die britischen Bauarbeiter mit dem fremdländischen Namen des Ingenieurs Schwierigkeiten hatten und ihn stattdessen immer nur *«Sir Harold»* nannten: Von da wäre es nicht weit zum wirklichen *«Sir Albert»*, allein schon vom Klang her. Andererseits war nach Hammond [8, S. 31] und anderen derselbe Albert Groethe später noch bei der öffentlichen Untersuchung als Zeuge zugegen, während Eyth ihn in jener Nacht mit dem Unglückszug untergehen läßt.

First Proposals

In the end, it was thanks to the remarkable perseverance of Thomas Bouch himself – who was later to be knighted – that the Tay Bridge could be built after all. But it still took 27 years, almost one generation of designs, plans, alterations, stopping and restarting of preliminary work, just to set the corner-stone for this task. Yet Bouch never relented: "Among the 100 projects in all parts of the world, time and again the 'Enno Bridge' (Tay Bridge) appeared like a ghost of the future. This sets us engineers apart from other humans: Our ideas do not originate from the world of the past, but from the world to come. Yet they torment us no less" [4, p. 415].

Bouch's first proposal of 1854 was estimated at 150,000 Pounds Sterling[5]. But as years went by and the railway companies could not be won over to this undertaking, a new company was founded for constructing the bridge which was to be sold later to the railway companies. Businessmen from Dundee together with share-holders of the NBR now approached the public with the Tay Bridge and Dundee Union Railway Undertaking. The flowering jute industry headed by Dundee's "Jutecrats" firmly backed the endeavor for the well being of their city *(Fig. 10)*.

In October of 1864 – Bouch was now over forty years of age – there was a memorable public meeting at City Hall. There Bouch described his bridge as "a very ordinary undertaking", adding that he himself had already constructed several far more stupendous and greater bridges – without mentioning where. In addition, he staked his professional reputation to keeping the total costs within 180,000 Pounds Sterling. The people present accepted this declaration with considerable enthusiasm, as well as the following resolution:

> "Be it resolved that it would be for the public advantage, and tend greatly to the traffic of the north of Scotland and especially the town and trade of Dundee, were the present inconvenient and expensive route to the south improved by the construction of a bridge across the River Tay, and were suitable provision made for a general passenger station at Dundee; and that a Committee be appointed to consider and promote the scheme" [8, p. 26].

---

5 These cost figures are of course not terribly meaningful without knowing the living expenses in Scotland at that time. Still, they are included here so the reader may compare the increasing cost estimates.

10
View of Dundee from Law Hill around 1903: The city had grown prosperous by the jute industry and was being called "Jute City" or even "Jutopolis", capital of a then booming textile manufacture. Around 1865, about 120 smoke stacks of jute mills could still be counted from this spot. Behind the city, the reconstructed Tay Bridge by Barlow [2, p. 66].

10
Blick auf Dundee vom Law Hill um 1903: Die Stadt war durch den Jute-Handel reich geworden, man sprach von «Jute City» oder gar «Jutopolis», Hauptstadt einer damals blühenden Leinenindustrie. Um 1865 konnten von diesem Standort noch die 120 Schornsteine der Flachsspinnereien gezählt werden. Hinter der Stadt die von Barlow wiederaufgebaute Tay-Brücke [2, S. 66].

But thanks to the unrelenting opposition by the CR, also the 1860's slipped by in the process of clearing parliamentary hurdles. Not just the design but also the site had to be altered more than once. In 1865 the NBR teamed up with the Tay Bridge Building Company, taking over control in the end. Twice, Parliament denied approval and the undertaking seemed doomed, until in 1866 John Stirling of Kippendavie became the new chairman of NBR. He was one of the most influential men in railway history, a person of gentility and honesty in a trade dominated at times by rogues and charlatans. At the age of 55 he took charge of the ailing company and encouraged its frightened directors and shareholders to fight for survival.

In combining enthusiasm, cunning and persuasion, he removed all obstacles. Even the harbour trustees of Dundee slowly succumbed to the charm of "Old Kippendavie". He was able to captivate the self-confidence of the nation, while loosening its purse-strings. His vision "to weave iron and masonry through two miles of air and water" inspired the imagination of people who were well aware that everything that was new in this century of scientific progress, had to be done by Great Britain. Pride and profit were powerful allies and finally succeeded in obtaining Parliament's approval in March of 1870.

Alas, one final and macabre voice of opposition arose from Dundee in the form of letters written by an elderly constituent, Patrick Matthew. He lived at the Carse of Gowrie and dismissed the Parliament's approval: What could they know of bridges and engi-

Passagierbahnhof in Dundee getroffen würden; und wenn ein Kommittee ernannt würde zur weiteren Prüfung und Förderung dieses Vorhabens.» [8, S. 26]

Aber die unversöhnliche Opposition der CR erreichte, daß auch noch die sechziger Jahre damit verstrichen, die parlamentarischen Schwierigkeiten aus dem Weg zu räumen. Dabei wurde nicht nur der Bauplan, sondern auch der Bauplatz mehr als einmal gewechselt. 1865 verband sich die NBR mit der Tay-Brückenbaugesellschaft und übernahm diese schließlich. Zweimal noch verweigerte das Parlament seine Zustimmung und das Vorhaben schien bereits verloren, bis John Stirling of Kippendavie 1866 Vorsitzender der NBR wurde. Er war einer der einflußreichsten Männer der Eisenbahngeschichte, eine Persönlichkeit von Noblesse und Ehrlichkeit in einem Geschäft, das zeitweise von Schelmen und Scharlatanen beherrscht zu sein schien. Mit 55 Jahren übernahm er die kränkelnde Gesellschaft und ermutigte die verschreckten Direktoren und Aktienbesitzer, ums Überleben zu kämpfen.

Mit einer Mischung aus Enthusiasmus, List und Überredung beseitigte er alle Hindernisse; sogar die Hafengesellschafter von Dundee begannen langsam, auf das geänderte Klima zu reagieren, und erlagen schließlich dem Charme von «Old Kippendavie». Er verstand es, das Selbstvertrauen der Nation zu beflügeln und gleichzeitig ihre Geldbörse zu lockern: Seine Vision, «Eisen und Mauerwerk mit zwei Meilen von Luft und Wasser zu verweben» inspirierte die Ein-

neering works, as even the masonry of their new building at Westminster was already crumbling? The old man died the same year – but not without leaving behind a last and terrible warning:

> "In the case of accident, with a heavy passenger train on the bridge, the whole of the passengers will be killed. The eels will come to gloat over in delight the horrible wreck and banquet!" [21, S. 65]

Commission

In 1871, Charles de Bergue & Co, a renowned contractor from London, Cardiff and Manchester, was awarded the contract to span a single-track rail bridge across the Firth of Tay. Old Mr. de Bergue had tendered the lowest bid with 217,000 Pounds Sterling. From the beginning, "he was in love with that bridge" [4, p. 425]. Besides, he had already done some preliminary soundings for the piers, five years before and on the order of Bouch. It was felt to be extremely reassuring, that the same man who had carried out these very important soundings, now himself had to go on building on top of them.

De Bergue's resident engineer was Albert Groethe, a burly, stoical man with a kind heart; from the beginning on he supported a form of union for his workmen, the Tay Bridge Co-operative Association, as well as the erection of a canteen, reading-room, kitchen and dormitories. At the time, this constituted some unusually generous "worker benefits" to workers. Most of the 70 men and youngsters starting construction came from Fife and Forfar, but there were also itinerant labourers from Ireland and England. They had their quarters at the southern shore near Wormit – a small but noisy settlement of workshops, stores and huts. Aside from all his work, Mr. Groethe was moved to entertain his presumably patient readers with meticulous descriptions of the building process in the pious weekly "Good Words" [7][6].

---

6 Although his name originally reads Groethe, he spells himself Grothe as a British author.

bildungskraft von Menschen, die wohl wußten, daß alles, was in diesem Jahrhundert wissenschaftlicher Errungenschaften neu war, von Großbritannien getan werden mußte. Stolz und Profit waren mächtige Reizmittel und erreichten schließlich, daß das Parlament in London im März 1870 seine Zustimmung gab.

Eine letzte und makabre Stimme der Opposition aus Dundee erhob sich in den Leserzuschriften eines alten Mannes, Patrick Matthew. Er wohnte auf dem *Carse of Gowrie* und hielt nichts von der Billigung des Parlaments: Was könne es schon wissen von Brücken und Ingenieurbauwerken, wenn doch das Mauerwerk seines eigenen Neubaues (Westminster) bereits zerbröckelte? Der alte Mann starb im gleichen Jahre – mit einer letzten furchtbaren Warnung:

> «Im Falle eines Unglückes, mit einem schweren Personenzug auf der Brücke, würden alle Fahrgäste zu Tode kommen. Die Aale werden kommen, sich lustvoll zu weiden an dem schrecklichen Wrack, und festlich tafeln!» [21, S. 65]

Auftrag

Im Jahre 1871 erhielt Charles de Bergue & Co, eine alteingeführte Baufirma aus London, Cardiff und Manchester, den Auftrag, eine eingleisige Brücke über den Firth of Tay zu bauen. Der alte Mr. de Bergue hatte mit 217 000 Pfund Sterling das billigste Angebot abgegeben – «und war von Anfang an verliebt in diese Brücke» [4, S. 425]. Außerdem hatte er bereits 5 Jahre früher, 1866, im Auftrage Bouch's die Sondierungen für die Pfeilerfundamente vorgenommen: Man fand es außerordentlich beruhigend, daß nun derselbe Mann, der diese wichtigen Untersuchungen in Händen gehabt hatte, auf denselben auch weiterbauen mußte.

Der leitende Ingenieur de Bergue's auf der Baustelle war Albert Groethe, ein wuchtiger Mann mit stoischer Miene und gütigem Herzen; er setzte sich von Anfang an für eine Art Gewerkschaft seiner Arbeiter – *Tay Bridge Co-operative Association* – ein, sowie für die Einrichtung von Kantine, Leseraum, Küche und Wohnheimen. Das bedeutete damals eine recht unübliche Fürsorge für Werktätige. Die meisten der 70 Arbeiter und Jungen, welche mit dem Bau begannen, kamen aus Fife und Forfar, aber es gab auch Wanderarbeiter aus Irland und England. Sie hatten ihr Quartier am Südufer bei Wormit – eine kleine, aber lärmende Ansiedlung von Werkstätten, Geschäften und Hütten. Neben aller Arbeit liebte es Mr. Groethe, in dem frommen Blättchen *«Good Words»* seinen vermutlich

"Final Plan"

There seemed to be a curse on this project from the outset. First, the originally selected site had to be abandoned in favor of one two miles further inland. The Firth there is wider but the ground promised to consist of rock in reasonable depth, at least according to initial soundings by the contractors. And there was no reason to doubt that the matter would proceed satisfactorily. Mr. de Bergue came from an old Hugenot family and was known as a most reliable and experienced man; "besides, the foundations were not the engineer's business – his work only started above the bottom of the river" [4, p. 417]. This separation of responsibilities was to have dire consequences.

The "final" plan of autumn 1868 envisioned parallel girders with crossed diagonals of wrought iron for almost the whole extent of the bridge, most of the girders running continuously across four spans each [19, p. 212]. All in all, 89 spans were planned, consisting mostly of a series of shorter spans of 120 ft (36.58 m) and only the central portion consisting of longer spans of 200 ft (60.96 m) each. Thus the bridge is the accumulation – at least in this planning phase – of the following, rather monotonous, sequence of figures: Starting at the southern end with 6 piers, the tracks curve into a northerly direction; this is followed by 22 piers with spans of 120 ft (36.58 m) each in a straight northerly direction and slowly rising; the central portion follows with (the at that time planned) 15 piers carrying the high girders of 200 ft (60.96 m) span each, in order to gain a clear shipping passage of 88 ft (26.82 m) above high tide. This is followed by a further 16 piers with 120 ft (36.58 m) spans, similar to the southern portion, to be followed by a wide curve to the east (almost a quarter circle) with 25 spans of 66 ft (20.17 m) each. Finally there is a parabolic arch girder of 160 ft (48.77 m) allowing passage for smaller vessels near the shore, to be concluded by 6 piers of 67 ft (20.42 m) spans. Altogether, this amounts to 89 spans with a total length of bridge of 10,321 ft (3.146 m) – or almost two miles (1 mile = 1,609.34 m)!

«Endgültiger» Plan

Das Projekt schien von Anfang an mit einem Fluch beladen zu sein. Zunächst hatte man den ursprünglich gewählten Bauplatz um zwei Meilen weiter landeinwärts verlegen müssen. Die Bucht ist dort zwar breiter, aber der Boden versprach, zumindest nach den Probebohrungen des Unternehmers, aus Fels in erreichbarer Tiefe zu bestehen. Und es gab keinen Grund zu zweifeln, daß die Sache in zuverlässiger Weise vor sich gehen würde. Mr. de Bergue stammte aus einer alten Hugenottenfamilie und war als ein ebenso zuverlässiger wie umsichtiger Mann bekannt: «Außerdem ging sie (die Gründung) den Ingenieur nichts an – seine Arbeit begann erst über der Sohle des Strombettes» [4, S. 417]. Diese Trennung der Verantwortungen sollte Folgen haben.

Der «endgültige» Bauplan vom Herbst 1868 sieht für nahezu den gesamten Brückenverlauf Parallelträger mit gekreuzten Diagonalen aus Schmiedeeisen vor, wobei sich diese Fachwerke meist kontinuierlich über vier Felder erstrecken [19, S. 212]. Es sind insgesamt 89 Öffnungen geplant, wobei der Großteil aus einer Serie kürzerer Spannweiten von 120 Fuß (36,58 m) und nur der Mittelteil aus längeren Weiten von 200 Fuß (60,96 m) bestehen soll. Damit addiert sich die Brücke, zumindest in der Planungsphase, zu folgendem, etwas monoton wirkendem Zahlenspiel: Sie beginnt am Südende mit 6 Pfeilern, über die sich der Schienenweg nach Norden biegt; es folgen 22 Pfeiler in Abständen von 120 Fuß (36,58 m) in gerader nördlicher Richtung und langsam ansteigend; dann kommt der mittlere Teil mit (damals noch) 15 Pfeilern und den hochliegenden Parallelträgern von je 200 Fuß (60,96 m) Spannweite, um so die erforderliche Durchfahrtshöhe über Flutniveau von 88 Fuß (26,82 m) zu gewinnen. Hierauf folgen weitere 16 Pfeiler mit wieder 120 Fuß (36,58 m) langen Trägern ähnlich dem Südteil; es schließt sich eine große Ostkurve (fast ein Viertelkreis) an mit 25 Feldern von je 66 Fuß (20,17 m); schließlich kommt noch ein Sprengwerk (Parabelträger) von 160 Fuß (48,77 m) als Durchlaß für kleinere Schiffe in Ufernähe; und den Abschluß bilden wieder 6 Pfeiler im Abstand von 67 Fuß (20,42 m): alles zusammen ergibt das 89 Öffnungen mit einer Brückenlänge von 10 321 Fuß (3 146 m) – oder fast zwei englischen Meilen (1 Meile = 1 609,34 m)!

geduldigen Lesern akkurate Beschreibungen des Brückenbauwerks zum Besten zu geben [7][6].

---

[6] Obwohl sich sein Name eigentlich Groethe schreibt, nennt er sich Grothe als englischer Autor.

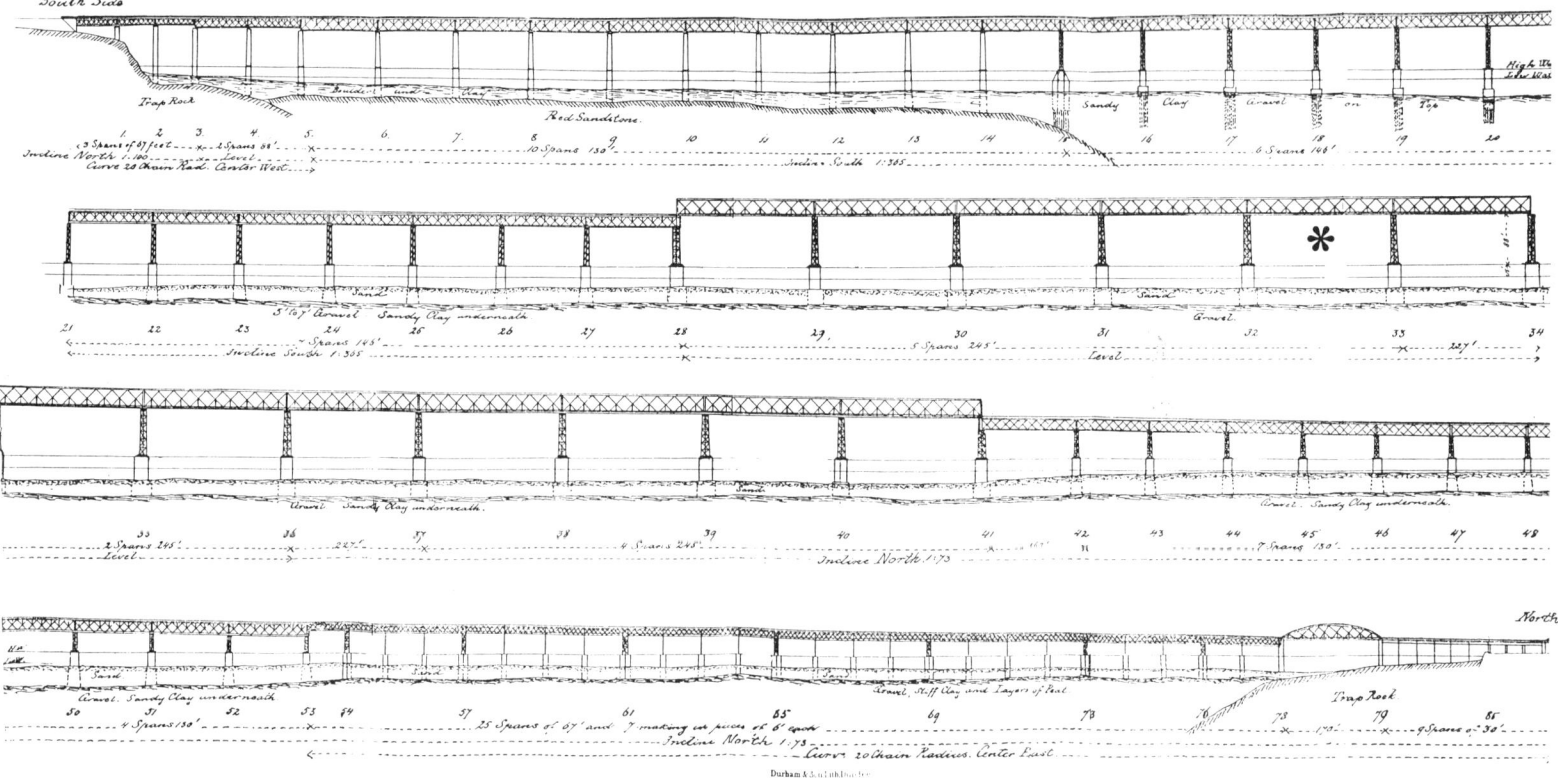

Alterations

In a departure from this "final" plan which was approved, the central girders with their piers will later be altered – at a time when the bridge is already well under construction *(Fig. 11)*. But also the many shorter spans turn out to be so uneven as to be comprehensible only in a special chart at best. In the end, there are 10 different spans ranging from 30 ft (9.14 m) to 245 ft (74.68 m) distributed over a total of 85 spans which together account for the overall length of about two miles. According to the various spans, also the heights of the girders vary from 16½ ft (5.03 m) in the short spans to 27 ft (8.23 m) for the central spans. Generally, the trains travel on top of the girders. Only in the parabolic arch and the high girders do the rails run along the lower flanshes, being enclosed within the main girders on either side *(Fig. 12)*.

Another alteration had been decided already at the outset of construction in order to save costs, and it was to have much greater consequences. Originally, piers of stone had been planned throughout, which

Änderungen

Abweichend von diesem «endgültigen» Plan, dem Genehmigungsplan also, werden später die Mittelträger samt Pfeilerstellung geändert – zu einer Zeit, als die Brücke längst im Bau ist *(Abb. 11)*. Aber auch die vielen kürzeren Spannweiten fallen so unterschiedlich aus, daß man sich allenfalls in einer Tabelle eine ungefähre Übersicht verschaffen kann: Es gibt zuletzt 10 verschiedene Spannweiten von mindestens 30 Fuß (9,14 m) bis maximal 245 Fuß (74,68 m), verteilt über insgesamt 85 Öffnungen. Damit wird schließlich die obige Gesamtlänge von über 3 km in etwa wieder erreicht. Den verschiedenen Spannweiten entsprechend variieren auch die Höhen der Fachwerkträger von 16,5 Fuß (5,03 m) für die kürzeren Felder bis zu 27 Fuß (8,23 m) für die Mittelträger. Dabei liegt die Fahrbahn in der Regel auf der Oberseite der Träger, beim Parabelträger und bei den hochgesetzten Mittelträgern aber in Höhe des Untergurtes – also eingeschlossen zwischen den beidseitigen Hauptträgern *(Abb. 12)*.

11
Elevation of the Tay Bridge – revised version showing the High Girders extended to 245 ft (74.68 m) spans and only 12 in number. ✻ marks the fifth of the central girders where the train was found after the disaster [6].

11
Seitenansicht der Tay-Brücke – revidierte Fassung mit den auf 245 Fuß (74,68 m) verlängerten Mittelträgern und nur noch 12 Mittelträgern. ✻ markiert den 5. Mittelträger, worin der Eisenbahnzug nach dem Einsturz gefunden worden ist [6].

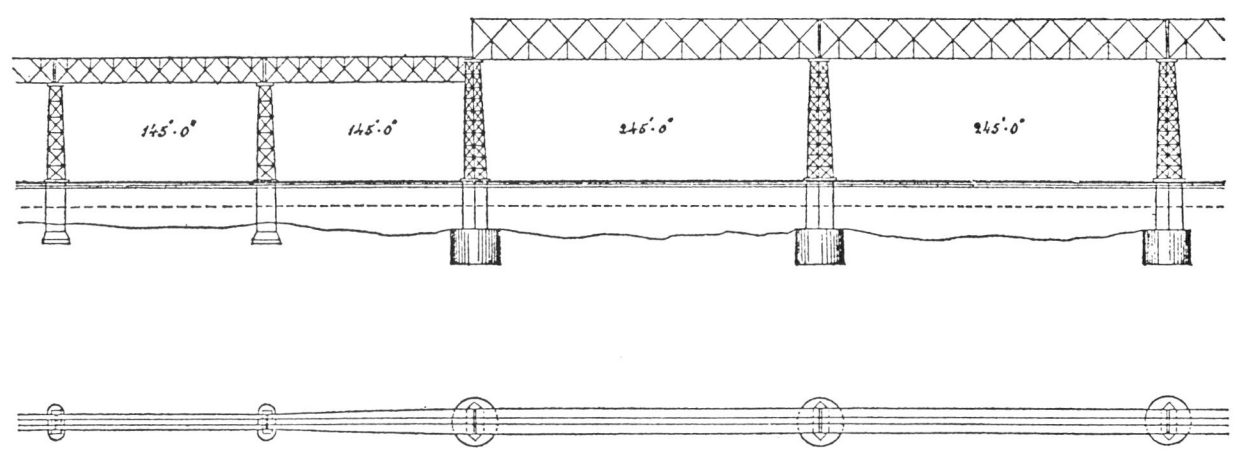

12
The point of transition from the shorter girders to the High Girders of the Tay Bridge. Drawing by Edgar Gilkes presented in a lecture at the Cleveland Institution of Civil Engineers on November 6, 1876 [26, p. 55].

12
Der Punkt des Übergangs von den kürzeren Fachwerkträgern in die hochliegenden Mittelträger der Tay-Brücke. Zeichnung von Edgar Gilkes für einen Vortrag beim Ingenieurverband von Cleveland am 6. 11. 1876 [26, S. 55].

had already at an early stage pushed the cost estimate for the bridge to 250,000 Pounds. This appeared exorbitant to the trustees of NBR. Only when the Parliamentary bill needed for so large a bridge had failed once again, did Bouch embark on new sketches for iron piers instead of stone piers as was customary then, leaving the calculations to his engineer.
Bouch possessed "the technological fantasy of a steam engine with precision steering and the working energy of a young elephant, but calculus is not his preference, whereas the large projects thrust upon him can not totally be resolved by the feeling between thumb and forefinger alone. In any case, he was delighted when I had calculated in two days the same figures which he had figured out intuitively in just two minutes. Frequently enough I was astonished how much bridges to him were mere matters of intuition, especially girder bridges" [4, p. 413].
Thus the idea was born to extend the stone piers only up to the high water level and to build them in iron from there onwards. Eight cast-iron tubular columns connected by wrought-iron bracings were to replace the stone piers from this point up. These 80 ft (24.38 m) tall spidery legs got to be ever more slender, simple and safe – resulting in the end, by progressive calculation, in a saving of 70,000 Pounds of total construction cost. This was a respectable amount indeed!
As we know, Bouch had promised not to exceed a contract sum of 180,000 Pounds. Nevertheless, those iron

Folgenschwerer sollte aber dann eine andere Änderung sein, die unter dem Kostendruck bereits vor Baubeginn erfolgt war: Ursprünglich waren durchweg Steinpfeiler geplant, was die Kostenschätzung der Brücke schon frühzeitig auf 250000 Pfund Sterling gebracht hatte und dem Verwaltungsrat der *NBR* viel zu hoch erschienen war. Erst als die Parlamentsakte, die man für eine so große Brücke brauchte, wieder einmal durchgefallen war, machte Bouch neue Skizzen für eiserne Pfeiler anstatt der bis dahin üblichen gemauerten Pfeilertürme, und überließ die Berechnungen seinem Ingenieur.
Denn Bouch hatte zwar «*die technische Phantasie einer Dampfmaschine mit Präzisionssteuerung und die Arbeitskraft eines jungen Elephanten, aber Rechnen ist nicht seine Liebhaberei, und schließlich lassen sich die großen Projekte, mit denen er überhäuft wird, nicht ganz mit dem Gefühl zwischen Daumen und Zeigefinger abmachen. Jedenfalls freute er sich, wenn ich in zwei Tagen dasselbe herausrechnete, was er in zwei Minuten herausgefühlt hatte. Oft genug war ich starr vor Erstaunen... wie sehr Brücken bei ihm Gefühlssache sind, namentlich Gitterbrücken*» [4, S. 413].
So entstand der Gedanke, die Steinpfeiler nur bis zur Höhe des Flutniveaus zu führen und von da an aufwärts in Eisen zu bauen: Acht gußeiserne, säulenartige Röhren, mit schmiedeeisernen Kreuzen verbunden, sollten von hier an die Steinpfeiler ersetzen. Diese 80 Fuß (24,38 m) hohen Spindelbeine gestalte-

13
Robert Stephenson's Britannia Bridge across the Straits of Menai spanning 460 ft (140.20 m), was built between 1846–1850, and is situated one mile further south of Telford's chain bridge (see *Fig. 4*). The powerful pylons and stylistic elements reminiscent of classical antiquity were regarded as prototypes for modern bridge construction by the engineering generation before Bouch [19, p. 61].

13
Robert Stephenson's «Britannia-Brücke» über die Meeresstraße von Menai, mit 460 Fuß (140,20 m) Spannweiten (1846–1850) – 1 Meile weiter südlich von Telfords Hängebrücke gelegen (vergl. *Abb. 4*). Die gewaltigen Pylonen und antikisierenden Stilelemente der Britannia-Brücke galten der Ingenieur-Generation vor Bouch als Vorbild im modernen Brückenbau [19, S. 61].

piers meant a radical departure from all the rules of bridge-building then in force, given that the majestic piers of Robert Stephenson's "Britannia Bridge" across the Menai (1846–1850) were still regarded as exemplary by most experts *(Fig. 13)*.

ten sich mit fortlaufender Berechnung immer zierlicher, einfacher, sicherer und mußten schließlich eine Reduzierung der Brückenkosten um 70 000 Pfund Sterling bedeuten – eine stattliche Summe! Denn Bouch hatte sich ja bei der Auftragserteilung für 180 000 Pfund Sterling verbürgt. Nichtsdestoweniger bedeuteten diese Eisenpfeiler eine radikale Abkehr von allen damals geltenden Regeln der Brückenbaukunst: Immer noch galten die gewaltigen Pfeilertürme von Robert Stephenson's «Britannia-Brücke» über den Menai (1846–1850) vielen Fachleuten als Vorbild *(Abb. 13)*.

Cornerstone and Caissons

On July 22, 1871, the cornerstone was set at the southern shore near Wormit. Eyth recounts this peculiar ceremony "in the Scottish wilderness" in vivid contrast to similar festivities customary in his native European homeland:

> "Here we stood, fifteen opened-up umbrellas, without song nor sound around a water-filled hole in the ground, as if to bury a rectangular suicide victim. Only the closest confidants could have guessed that this was indeed a joyful festivity, celebrating the sinking of the first founding stone of the great Enno Bridge (Tay Bridge)."

The first stone [4, p. 423]

The foundations for the piers were set in the usual manner with caissons of 9 ft (2.74 m) in diameter and 7 ft (2.13 m) in height, skillfully using the tide in transporting the caissons on barges, as well as the large girders later on.

> "Most of it is done for us by the tide, raising or lowering the water level at this spot by 6 to 10 ft, twice daily. There is something deeply stirring in watching for the first time how a mysterious force of nature emanating from the distant moon grasps our giant blocks silently but with a horrible all-crushing determination, while we just need to watch for not missing the right moment. It's frightening to see those huge masses rise with groans as a matter of course, to embark on their journey across the water and to sink gurgling into the depth to fulfill their new duties down below."

Mysterious force of nature [4, pp. 428, 429]

Caisson work was the most taxing for the men in those days. They had to dig down in those iron cages 20 ft (6.09 m) and more into the riverbed, in foul air, poor lighting, great heat, continuous humidity and high atmospheric pressure in order to keep out the water pressing in from all sides. They were in constant danger of accidents — there was no such thing as risk compensation in those days — and all that for an hourly wage of eight pence! But Bouch had promised them steady work for three years, which is what mattered. So men and youngsters, sometimes mere children encased in narrow steel chambers dug with picks, spades and hands in 12-hour shifts. During this time, the heavy cylinder above them would slowly sink

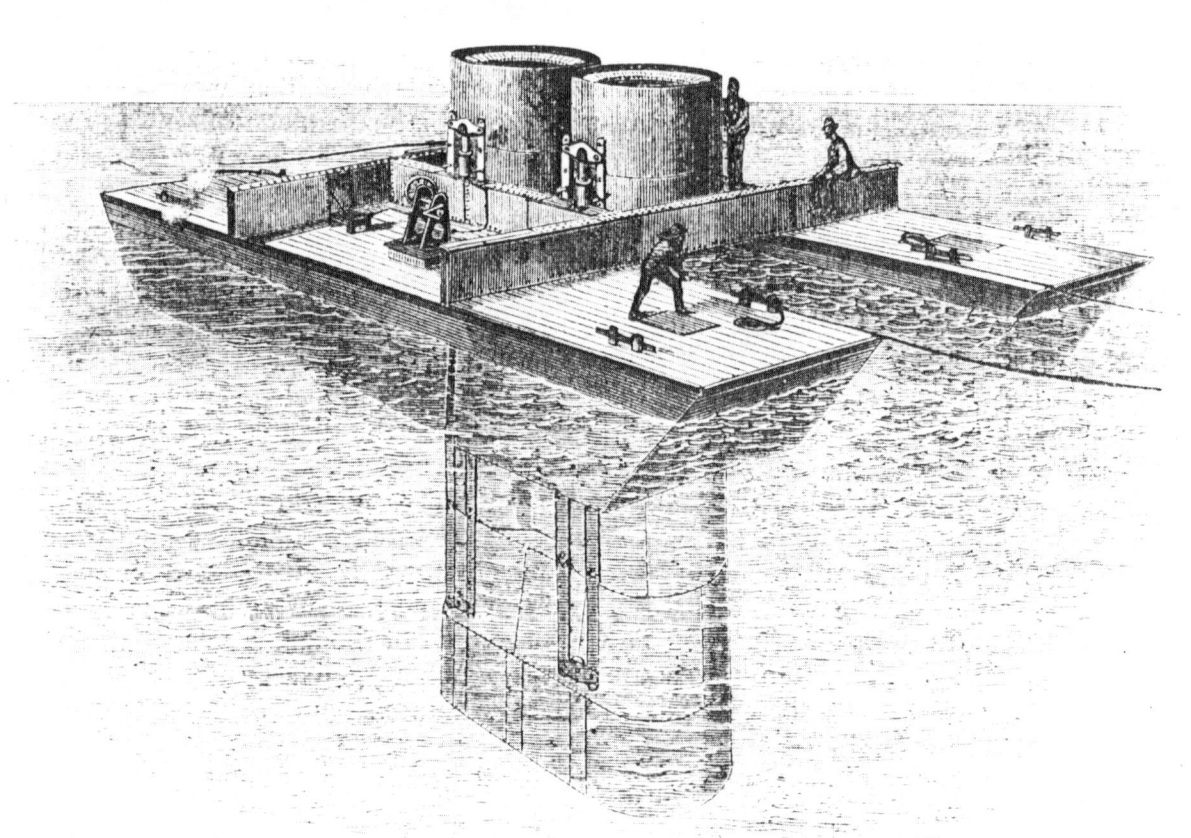

14
Double-caissons floating on pontoons, being towed to the piers [7].

14
Schwimmende Doppelsenkkästen auf Pontons für die Pfeilerfundamente [7].

through the mud and clay till it finally rested firmly on the rocky ground. Then the inner side of the cylinder would be lined with brick, reducing the diameter to 4 ft (1.22 m), with just a small access tube for escape. Finally, the interior would be filled with concrete.
After some initial difficulties with the sinking of single caissons (as soon as one hit upon rocks embedded in the sandy soil, the caissons would simply tilt over), now two caissons were joined upon one oval wrought-iron bedplate 13×23 ft (3.96×7.00 m), forming a double-pier foundation. On top of this, a cement and brick pier was erected up to 5 ft (1.52 m) above the high water mark, whence the cast-iron piers would come to stand (Fig. 14).

der schwere Zylinder über ihnen langsam ab, durch Schlamm und Tonschichten, bis er schließlich fest am Felsgrund aufsaß. Dann wurde die Innenwand mit Ziegeln ausgekleidet, wodurch sich der Zylinder-Durchmesser auf 4 Fuß (1,22 m) verringerte, nur mit einer engen Einstiegsröhre als Fluchtweg. Zuletzt goß man das Innere mit Beton aus.
Nach anfänglichen Schwierigkeiten beim Versenken von einzeln stehenden Senkkästen (sobald man auf im Sandboden eingeschwemmte Findlinge stieß, kippten die Kästen einfach um) wurden jetzt jeweils 2 Senkkästen mit Hilfe einer ovalen schmiedeeisernen Grundplatte von 13 × 23 Fuß (3,96 × 7,00 m) zu einem Doppelpfeilerfundament vereinigt. Darauf errichtete man Zement- und Backsteinpfeiler bis 5 Fuß (1,52 m) über die Hochwassermarke, worauf dann erst die gußeisernen Pfeiler zu stehen kamen (Abb. 14).

15
Period illustration of the Tay Bridge as seen from the north, somewhat further to the west than in *Fig. 3* [19, p. 211].

15
Zeitgenössische Darstellung der Tay-Brücke, Nordansicht von einem Standpunkt etwas weiter westlich als in *Abb. 3* [19, S. 211].

Piers

The construction of the numerous piers was far from uniform. Thus the first 14 piers from the south were still erected conventionally as massive masonry piers. Starting with the 15th pier and due to increasing difficulties encountered in foundations (more about this under "First Setbacks"), the new method of open iron piers on masonry footings started to be used – also in order to save weight. Along the northern shore, most of the piers consisted of only 3 trussed cast-iron columns of 16-inch (40.6 cm) or 20-inch (50.8 cm) diameter and 1-inch thickness, depending on the span. These 3 columns were arranged in a straight line at the intermediate piers, the third column facing the current being inclined at an angle of 1:3 *(Fig. 15)*. Only the main piers (every third or fourth one) were doubled-up for better stability of the bridge in its longitudinal axis, and braced like yokes. – Particularly in this northern section of the bridge near Dundee, the jarring mix of different styles of piers is quite obvious, reminding us once more of the words of Herr Havestadt about "...a conglomerate of various pier and bridge systems borrowed from smallish means..." (see Fig. 15 and especially Fig. 21).

First Setbacks

On August 8, 1874, a caisson imploded, probably because of faulty casting. There were 7 deaths, after 14 other caissons had been sunk previously without any trouble. And in 1875, de Bergue died, when every new pier erected by him seemed to push him closer to the ground. For he had discovered that the report about the soil conditions at the riverbed had been far too optimistic. Already during the sinking of the 10th pier, the rocky soil was found to be considerably deeper than anticipated, at places even down at 40 ft (12.19 m) [8,

Pfeiler

Im Einzelnen war dann der Aufbau der zahlreichen Zwischenstützen keineswegs einheitlich. So wurden die ersten 14 Pfeiler von Süden noch konventionell als massive Mauerwerkspfeiler ausgeführt. Erst ab dem 15. Pfeiler ging man wegen zunehmender Schwierigkeiten bei der Gründung (mehr darüber unter «Erste Rückschläge») zu der neuen Methode der aufgelösten Eisenpfeiler auf gemauerten Sockeln über – nicht zuletzt auch, um das Pfeilergewicht zu verringern. Im Viaduktbereich am Nordufer bestand die Mehrzahl der Stützen sogar nur aus je 3 verstrebten gußeisernen Säulen von 16 Inch (40,6 cm) oder 20 Inch (50,8 cm) Durchmesser, mit 1 Inch (2,54 cm) Wandstärke, je nach Spannweite. Diese 3 Säulen standen bei den Zwischenstützen in gerader Linie hintereinander, wobei die dritte, stromaufwärts gerichtete Säule im Verhältnis 1:3 geneigt war *(Abb. 15)*. Nur die Hauptstützen (jeder dritte oder vierte Pfeiler) wurden zwecks besserer Stabilität in Längsrichtung der Brücke verdoppelt und als Bockgerüst verstrebt. – Besonders in diesem nördlichen Brückenverlauf vor Dundee wird das unruhige Vielerlei der Pfeilerausführungen deutlich und läßt uns der Worte des Herrn Havestadt vom «...aus kleinen Verhältnissen entlehnten Konglomerat verschiedener Pfeiler- und Brückensysteme» gedenken (vergl. *Abb. 15* und besonders *Abb. 21*).

Erste Rückschläge

Am 28. 8. 1874 implodiert ein Senkkasten, wahrscheinlich wegen schlechtem Guß. Es gibt 7 Tote, nachdem vorher schon 14 Kästen anstandslos versenkt worden waren. Und 1875 stirbt de Bergue, als jeder neue Pfeiler, den er aufstellte, ihn mehr und mehr zu Boden zu drücken schien. Denn man hatte entdeckt, daß der Bericht über die Bodenbeschaffenheit am Flußgrund viel

p. 27]. The old soundings had shown quite different figures. According to these, bedrock was to be no deeper than 18 to 20 ft (5.49–6.10 m) under the bottom slick. At the 26th pier, the situation became truly critical, as now the central shipping channel had been reached.

New soundings showed the rocky soil dropping sharply from both sides, forming a deep channel in the middle of the river. One sounding even went down as far as 157 ft (47.85 m) under the riverbed without hitting any trace of rock. A layer of densely-packed gravel over mighty strata of clay had fooled those who made the soundings earlier on. On the evening of the same day, de Bergue took to bed, knowing that it had been his very own men who had committed this error nine years ago. Two days later he was dead – due to a heart ailment which his doctor is said to have ascertained already months ago. Hammond claimed, alternatively, that de Bergue had already in 1873 "provided convincing proof that he was insane" [8, p. 27]. Actually, de Bergue had been gravely affected, aside from many other setbacks, by the accidental deaths of the seven workmen in 1874. He felt personally responsible for the fate of those men.

His death was a great blow to the whole project – coinciding with the moment when the very ground beneath the builders' feet was literally caving in. The contractors declared right away their inability to continue this work after the death of their superior. The news spread through Dundee like wildfire and was the talk of stockmarkets in Manchester and London. The shareholders of NBR were shocked – twice 100,000 Pounds of their money seemed to have been thrown into the water. But the general public had anticipated the collapse, as usual. There was a limit to everything: for bridges, there were rivers enough in the world, but firths should be left in peace. The more powerful shareholders hoped that this would be the end of this story at last and that nobody would attempt to rescusitate this unlucky endeavour again [4, p. 440].

Once again though, it is Bouch who succeeds in stopping the impending closure of a work half-done. In a fiery speech to the directors of NBR and the old bridge building company, he speaks of the national disgrace to leave so grand and necessary a task in a half-finished state; he speaks of the endurance and tenacity of the Anglo-Saxon race whose representatives – he is saying this with a curteous bow – have rallied here around him in this crisis. It is a masterly speech of persuasion.

Afterwards, his first act is to let go of half of his workforce. Only four weeks later, a new contractor – the

zu optimistisch gewesen war. Schon beim Versenken des zehnten Pfeilerfundamentes fand sich der Felsgrund wesentlich tiefer als erwartet, stellenweise erst bei 40 Fuß (12,19 m) [8, S. 27]. Die alten Bohrungen hatten ganz andere Werte ergeben: Danach sollte der Fels nicht tiefer als 18 – 20 Fuß (5,49 – 6,10 m) unter dem Grundschlamm liegen. Beim 26. Pfeiler wurde die Sache bedrohlich, denn nun war die mittlere Fahrrinne erreicht.

Neue Bohrlöcher zeigten, daß der Felsgrund von beiden Seiten plötzlich scharf abfällt und im gesamten Flußbett eine tiefe Mittelrinne bildet. Eine Sondierung wurde sogar bis auf 157 Fuß (47,85 m) unter das Flußbett getrieben, ohne auf irgendwelchen Fels zu stoßen. Eine Schicht zusammengebackener Kiesel über mächtigen Tonlagern hatte seinerzeit die Bohrleute getäuscht. Am Abend desselben Tages legte sich de Bergue zu Bett – denn seine eigenen Leute hatten vor 9 Jahren den Fehler bei den Bohrungen gemacht. Zwei Tage später war de Bergue tot – aufgrund eines Herzleidens, wie sein Arzt sagte, das er schon vor Monaten festgestellt habe. Nach Hammond hatte sich jedoch bereits 1873 zweifelsfrei erwiesen, daß de Bergue verrückt geworden war – ohne daß hierzu besondere Angaben gemacht werden [8, S. 27]. Tatsache ist, daß de Bergue neben den vielen Rückschlägen das Unglück mit den 7 Arbeitern vom Jahre 1874 sehr mitgenommen hatte. Er fühlte sich persönlich verantwortlich für diese Männer.

Sein Tod war ein schwerer Schlag für das ganze Unternehmen – er fällt mit dem Augenblick zusammen, als den Erbauern sozusagen der Boden unter den Füßen verschwindet! Die Baufirma erklärt sofort, daß sie nach dem Tode ihres Chefs außerstande sei, den Brückenbau weiterzuführen. Die Nachricht geht in Dundee wie ein Lauffeuer um, und ist in 24 Stunden das Börsengespräch von Manchester und London. Die Aktionäre der NBR ringen die Hände – zweimal 100 000 Pfund Sterling ihres Geldes scheinen nutzlos ins Wasser geworfen zu sein. Das übrige Publikum hat wie gewöhnlich den Zusammenbruch vorausgesehen: Es habe alles seine Grenzen; für Brücken seien Flüsse genug auf der Welt; Meeresarme solle man in Ruhe lassen. Die wohlhabendsten Aktionäre hoffen, daß die Geschichte wenigstens damit ein Ende haben und niemand versuchen würde, das unglückliche Unternehmen wieder auf die Beine zu stellen [4, S. 440].

Und wieder ist es Bouch, der die drohende Einstellung des halbfertigen Werkes verhindern kann. Mit einer zündenden Rede vor den versammelten Direktoren der NBR und der alten Brückenbaugesellschaft spricht er von der nationalen Schande, ein so großartiges und notwendiges Werk halbfertig zu verlassen, spricht von

firm of Hopkins, Gilkes & Co – is found that is prepared to finish the job, for now 250,000 Pounds. This firm had been the second-lowest bidder at the time. They keep on Albert Groethe as resident engineer and manager, and establish a foundry at Wormit. A half a year has now been lost, though at first the second phase has a good start; the problems associated with a certain foreman called Fergus Ferguson at Wormit, who will prove himself a most unreliable supplier of cast iron in the years to come, are not yet apparent.

In 1876, new large caissons are sunk in the central channel. Here, the invention of a young draughtsman – the son of one of the seven dead caisson workers – is of great help and even being patented. With a pump and vacuum hose, the excavation material, a mixture of water, sand and gravel, can now be lifted "with a horrid roar into so-called suction boats at the surface", saving a lot of work [4, p. 445][7]. Neither Hopkins nor Bouch have placed much faith into this contraption before. Finally, a certain Gerrit Willem Camphuis from Holland arrives just in time to master the difficulties of foundations in bottomless silt – without any previous experience in bridge building, but with a well-trained team of pile drivers [8, p. 27]. He constructs artificial platforms on piles in soil of sand and clay which, however, will no longer support the large brick monoliths that had been planned originally for the piers.

Further Alterations

In order to further reduce the number of footings necessary for the piers, Bouch alters his design for the central spans: He reduces the 14 large spans to 13, thereby increasing their length from 200 to 245 ft (60.96 to 74.68 m) for eleven girders, and to 227 ft (69.19 m) for two girders. Thus he saves one whole pier, while increasing the central portion of the bridge from 2,800 ft to 3,149 ft (853.44 to 959.82 m), equaling almost 1 km. In any different type of bridge, such an alteration would have changed the whole plan; yet for such a simple additive bridge of so many equal or similar parts, this did not seem to pose a problem *(Fig. 11)*.

The foundations for the tall piers are made to fit the new soil conditions in the riverbed, using over-sized footings. Instead of the original double-caissons of 9 ft (2.74 m) diameter each, new giant caissons of 31 ft (9.45 m) diameter and 20 ft (6.09 m) height are being used now. Totally pre-assembled at the shore and al-

7 The same procedure will later be used to a major extent in building the Forth Bridge.

der Ausdauer und Zähigkeit der angelsächsischen Rasse, deren Vertreter – er sagt dies mit einer galanten Verbeugung – sich ja in dieser Krise hier um ihn geschart hätten, usw. Seine Rede ist ein Meisterstück.

Danach entläßt er zunächst einmal die halbe Arbeiterschaft, und schon in vier Wochen ist ein neuer Unternehmer – die Firma *Hopkins, Gilkes & Co* – gefunden, die bereit ist, den Bau für jetzt 250 000 Pfund Sterling fertigzustellen. Sie war seinerzeit der zweitbilligste Bieter gewesen und übernimmt nun auch den bewährten Albert Groethe als verantwortlichen Ingenieur und Manager auf der Baustelle. Dazu wird die Eisengießerei in Wormit errichtet. Dennoch geht dem Projekt ein halbes Jahr verloren. Aber zunächst hat die zweite Bauphase einen guten Start – wäre da nicht ein gewisser Werkmeister namens Fergus Ferguson in Wormit, der sich als ein höchst unbefriedigender Lieferant von Gußeisen erweisen soll.

1876 werden neue große Senkkästen in der Mittelrinne versenkt. Hier ist die Erfindung eines jungen Zeichners – es ist der Sohn eines der 7 ertrunkenen Senkkastenarbeiter – eine große Hilfe und wird sogar patentiert: Mit einer Pumpe und einem Vakuumschlauch befördert man den Aushub, ein Gemisch aus Waser, Sand und Steinen, «mit furchtbarem Gebrüll in sogenannte Saugboote an die Oberfläche», was viel Arbeit erspart [4, S. 445][7]. Weder Hopkins noch Bouch wollten zunächst an diese Sache glauben. Schließlich kommt noch ein gewisser Gerrit Willem Camphuis aus Holland wie gerufen, der ohne jede Erfahrung im Brückenbau, aber mit einer geübten Truppe von Pfahlbauern die Schwierigkeiten bei der Fundierung im bodenlosen Schlick meistert [8, S. 27]. Er schafft künstliche Plattformen auf Pfahlgründungen im Sand- und Tonboden, die dann allerdings nicht mehr die großen Ziegel-Monolithe der Pfeiler tragen können, wie ursprünglich geplant.

Erneute Änderungen

Um die Zahl der notwendigen Pfeilerfundamente zu verringern, ändert Bouch seinen Entwurf für den hochliegenden Mittelteil: Die 14 großen Fachwerkträger werden auf 13 reduziert, dafür wird deren Länge von je 200 Fuß (60,96 m) bei 11 Trägern auf 245 Fuß (74,68 m) und bei 2 Trägern auf 227 Fuß (69,19 m) vergrößert. Damit spart man einen Pfeiler und verlängert gleichzeitig den Mittelteil der Brücke von insgesamt 2800 Fuß (853,44 m) auf 3 149 Fuß (959,82 m) – also auf fast einen Kilometer. Eine solche Änderung hätte bei

7 Dasselbe Verfahren wird später in großem Stil beim Bau der Firth-of-Forth-Brücke angewandt.

ready lined inside with brick, they are then floated to the site, sunk to the ground and filled with concrete. According to the original plans, a circular, almost equally large body of masonry would have to be placed on top of this; but despite the large base, there was now hardly enough room for the eight columns planned for each pier. Therefore, Bouch and the contractor decided, against great resistance by the responsible designer, *to reduce the number of iron columns from 8 to 6 at every pier.* The result is a hexagonal masonry pier of 27×16 ft (8.23×4.88 m), pre-assembled at the shore and lowered in the same way. On top of this, masonry is extended till above the high-water level, topped off by four layers of granite blocks *(Fig. 16)*. As shall become evident after the accident, only the blocks of the uppermost layer have been dowelled [19, p. 213].

On top of these piers, the 6 cast-iron columns are braced in two sets of three, placed in opposition in plan *(Figs. 17 and 31)*. The tall piers of the central portion consist of 7 sections, on the average 11 ft (3.35 m) in length, and braced accordingly in 7 sections each; the exact length of these castings depends on the respective heights of the piers rising towards the middle of the Firth. However, *Fig. 35* looking South and *Fig. 50* looking North show only 6 or 5 sections over the entire height of the piers. Because the tall central piers have all been destroyed in the disaster, we are looking here at the last ones of the smaller piers left standing, with the ends of the shorter spans above. Only in *Fig. 35,* can the broken-off lower portion of one of the tall piers still be made out.

Despite diameters of 15 and 18 inches (38 cm and 46 cm) for the two outer columns, with a wall thickness of 1¼ inch (32 mm), this hexagonal arrangement further weakens the piers, making them vulnerable to lateral loads and vibrations. These are precisely those piers which have already exceeded the others in height by 16 ft (4.88 m); they reach a total height of 86 ft (26.21 m) in order to support the High Girders in the centre. It is a very risky game indeed, considering those extraordinarily slender and airy piers!

jeder anderen Brückenform den gesamten Plan umgeworfen – bei einer so einfachen, additiven Brücke von lauter gleichen oder ähnlichen Teilen aber schien das kein Problem zu sein *(Abb. 11)*.

Die Fundamente für die hohen Pfeilertürme werden der anderen Bodenbeschaffenheit im Flußbett angepaßt und mit übergroßen Pfeilerfüßen versehen: Anstatt der ursprünglichen Doppelsenkkästen von je 9 Fuß (2,74 m) Durchmesser kommen jetzt Riesensenkkästen von 31 Fuß (9,45 m) Durchmesser und 20 Fuß (6,09 m) Höhe zum Einsatz. Zur Gänze am Ufer vorgefertigt und bereits mit ihrem Ziegelmantel ausgekleidet, werden sie zum Pfeilerpunkt geschwommen, abgesenkt und mit Beton verfüllt. Darauf wäre nun planmäßig ein kreisrunder, fast ebensogroßer Mauerwerkskörper zu stehen gekommen – der aber trotz großer Grundfläche für die 8 pro Pfeiler vorgesehenen Säulen kaum Platz geboten hätte: Also setzen Bouch und der Unternehmer gegen den großen Widerstand des verantwortlichen Statikers durch, *daß die gußeisernen Säulen von 8 auf 6 Stück je Pfeiler reduziert werden.* Damit ergibt sich ein sechseckiger Mauerwerkskörper von 27 × 16 Fuß (8,23 × 4,88 m), der in gleicher Weise am Ufer vorgefertigt und abgesenkt wird. Darauf kommt das Ziegelmauerwerk bis über die Flutmarke, mit einer vierlagigen Abdeckung aus Granitblöcken *(Abb. 16)*. Wie sich nach dem Einsturz herausstellen soll, wurde aber nur die oberste Lage dieser Granitblöcke verdübelt [19, S. 213].

Auf diesen Pfeilersockeln faßt man die 6 gußeisernen Säulen jeweils in 2 Dreiergruppen zusammen, welche sich im Grundriß paarweise gegenüberstehen und miteinander verstrebt werden (vergl. *Abb. 17 und 31*). Dabei sind die Hochpfeiler des Mittelteils aus sieben Gußstücken von durchschnittlich 11 Fuß (3,35 m) Länge zusammengesetzt und dementsprechend in je 7 Gefachen miteinander verstrebt; die genaue Länge dieser Gußstücke hängt von der jeweiligen Pfeilerhöhe ab, die ja zur Flußmitte hin ansteigt. *Abb. 35* mit Blickrichtung Süd und *Abb. 50* mit Blickrichtung Nord zeigen allerdings nur 6 bzw. 5 Gefache über die ganze Höhe der Pfeiler: Da die hohen Mittelpfeiler alle beim Einsturz zerstört worden waren, blicken wir jeweils auf die letzten der kürzeren Pfeiler, die mit den Enden der kleineren Fachwerkträger noch stehengeblieben und niedriger waren als die Hochpfeiler des Mittelteiles. Nur in *Abb. 35* ist noch der untere Rest eines gebrochenen Hochpfeilers zu sehen.

Trotz eines Durchmessers von 15 Inch (38 cm) bzw. 18 Inch (46 cm) für die beiden äußeren Säulen, und einer Wandstärke von 1¼ Inch (32 mm), werden mit dieser Sechseck-Anordnung die Pfeiler noch weiter geschwächt und anfällig gegen seitliche Belastung oder

16
Hexagonal masonry pier on top of circular foundation, entirely pre-assembled at the shore and floated to the site [6].

16
Sechseckiger Mauerwerkskörper über kreisrundem Pfeilerfundament, zur Gänze am Ufer vorgefertigt und eingeschwommen [6].

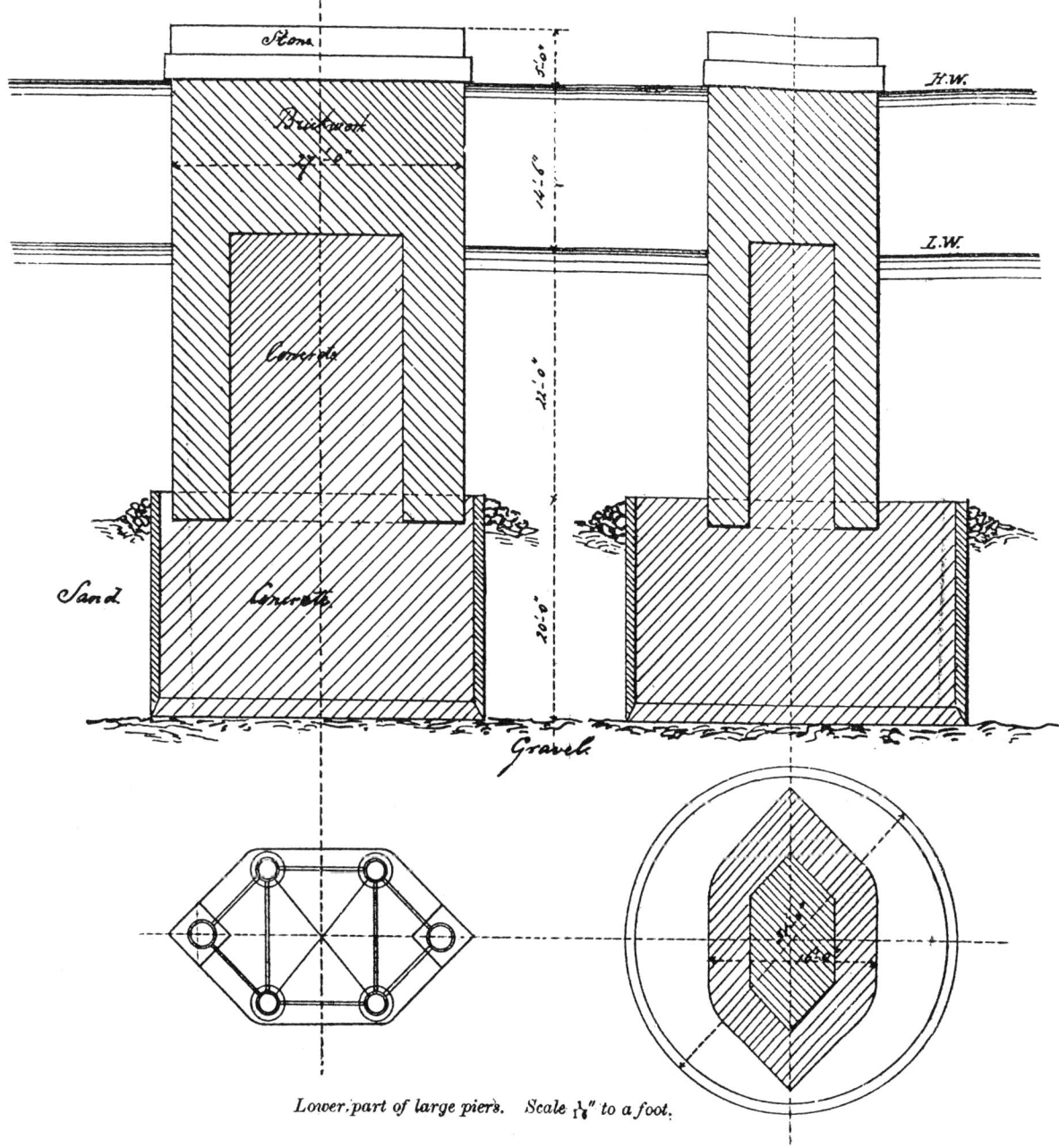

Schwingungen – also gerade diejenigen Pfeiler, die ohnehin schon um 16 Fuß (4,88 m) höher sein müssen als die anderen, nämlich insgesamt 86 Fuß (26,21 m), um die Hochträger im Mittelteil zu halten: Es ist ein riskantes Spiel mit diesen ungemein schlanken und luftigen Pfeilern!

17
Longitudinal and transverse elevation of the tall piers supporting the High Girders. Over a hexagonal base, 6 braced cast-iron columns rise in 7 sections up to the required height of 86 ft (26.21 m) above high-water level [6].

17
Längs- und Queransicht der Hochpfeiler unter den Mittelträgern: Auf 6-eckigem Grundriß erheben sich 6 verstrebte Säulen aus Gußeisen in 7 Gefachen bis auf eine Höhe von 86 Fuß (26,21 m) über Hochwasser-Niveau [6].

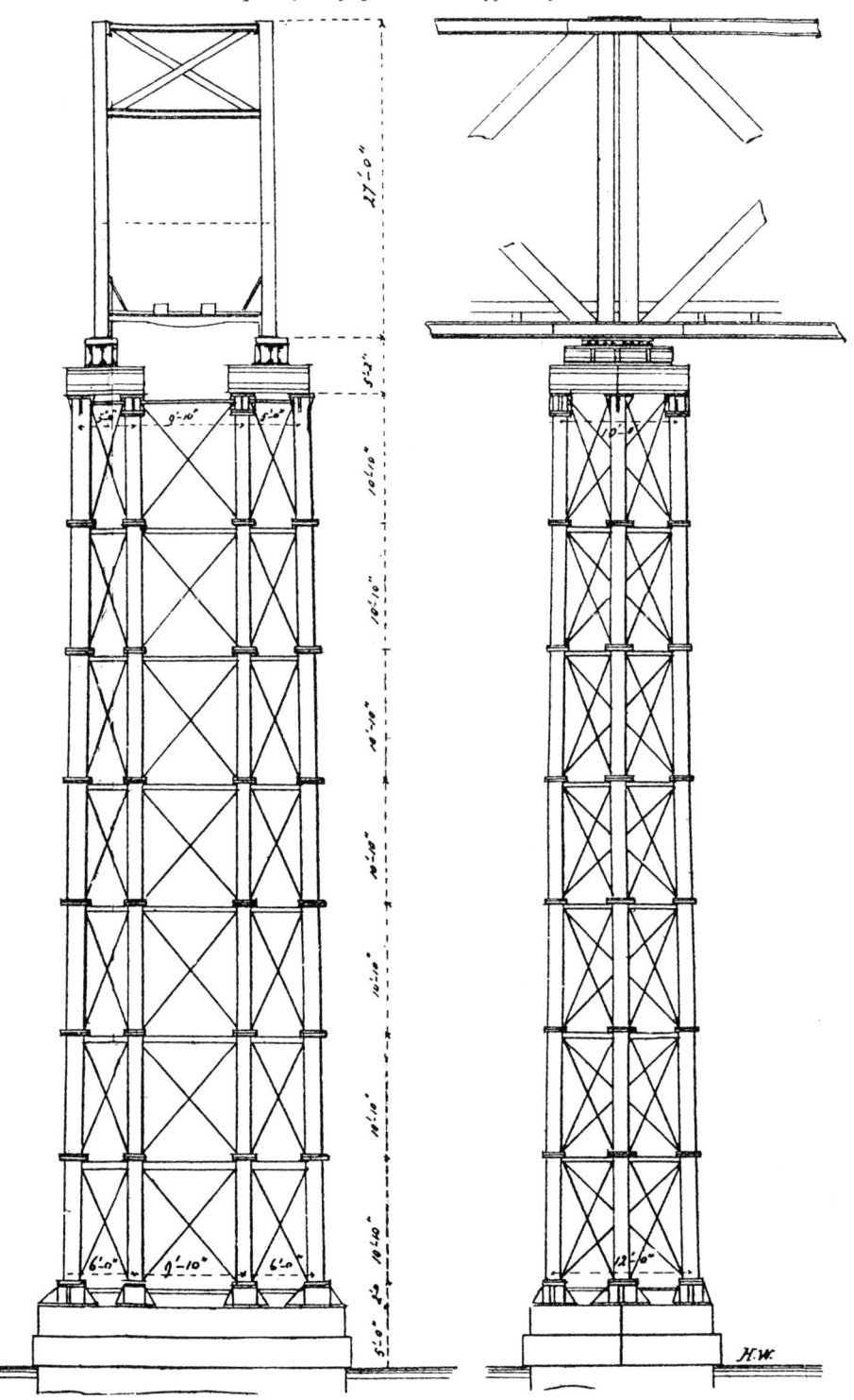

Iron part of large piers. Scale $\tfrac{1}{16}''$ to a foot.

"...my cast-iron piers are standing there as if to grow into heaven. I admit, it is a sight one has to get used to. Sometimes it grips my heart, as I stare from my boat at this girder-work up there in the blue firmament. One imagines to see the whole horrible mass of iron to totter at times. Naturally it is the movement of the boat, or it is the clouds flying by that deceive the eye. Still, it's just as well that nobody gets to see this."

"...the job is organized now like a flying factory. My piers stand their ground. They mean a tremendous saving compared to the masonry work of the Menai Bridge. Still, they appear somewhat extravagant reaching up 86 ft (26.23 m) above water level, the required height for the rail bed in the center of the river."

Horrible mass of iron and
Comparing the Menai Bridge

(Excerpts from Eyth [4, pp. 432, 436] referring to the giant piers of Robert Stephenson's Britannia Bridge – see *Fig. 13*).

The tubular sections of the columns are braced via cast-on lugs with the wrought-iron horizontal and diagonal struts. For this, iron cones or pegs are used in regulating the tight fit of the tension rods – a method to be questioned at the inquest later on. More over, the top ends of the 6 cast-iron columns are not held in place by one common head plate; rather, 3 tops each are held together by a triangular cap *(Fig. 17)*, forming two separate supports for the two main girders. Here, a single joint head plate would surely have improved the lateral stability of the piers, corresponding to the prevailing wind direction along the River Tay.
Work goes on, and 5 years after the cornerstone has been set, a journalist describes progress by no means as rapid as the more sanguine shareholders would have imagined – in the apparent expectation to see the edifice rise up like a fabric in fairyland [8, p. 27]. By April 1876, only 76 of 88 spans are completed, still missing 18 girders to fill the gaps. By the autumn of 1876, two large electric arc-lamps with parabolic mirrors are installed at the job site to speed up the work at night, – "each lamp with an illuminating power equal to 1,000 candles, the current generated by two of Gramme's electro-magnetic machines driven by an engine of 4 horse power" [21, p. 122]. People were overwhelmed by this spectacle, which was to be so commonplace to their grandchildren. On weekends they took the ferry to Newport and the train on to Wormit, just to have a look at those exciting works from close up.

Girders

By February 1877, the last of the smaller girders is bolted into place on its pier – "...girders resting now on slender stilts, drawing two unbroken dark lines from South and North toward the center of the Firth.... They are being pumped upward into the sky and equipped with columnar legs growing from their bodies while rising" [4, p. 446]. During this intensive work phase, 500 men are employed on both shores, supported by 4 steamers, 12 barges and countless smaller vessels. Everything is organized so well and the men are so accustomed to their job that a serious setback would almost seem unthinkable: The end is at sight.

The large girders of wrought iron are truly giant trusses. Assembled on shore as 245-ft-long (74.68 m) hollow cages, they measure 18 ft (5.49 m) across and 36 ft (10.97 m) in height, in accordance with the size of a train car. Each box-girder requires 4 weeks of labour, uses 18,000 rivets and weighs 190 tons. The problem though is not this pre-assembly, despite its novelty, but the transport of these "frightfully long rectangular tubes" [4, p. 448] – even after the most careful planning beforehand. Since the central shipping channel of the Firth of Tay was not to be obstructed by permanent scaffolding, once again the same ingenious method is used which had already been tested with so much success during the construction of Robert Stephenson's Britannia Bridge for the first time. Eyth describes this ever exciting spectacle with much imagination, using the "cosmic clockwork" analogy of the tides. The principle was quite simple as the girders, resting pre-assembled on their stagings, were left to free themselves floating on barges placed underneath; once towed to the site *(Fig. 18)*, they would again be left to set themselves into place atop the piers as the tide was going out. Thus the lifting and lowering of the giant masses was left entirely to the forces of nature; only horizontal movement was carried out by steam tugs. At the piers, the girders began to extend, so to speak, their own legs from underneath, saving a scaffolding altogether. First, the lower sections of the 4 central columns were erected, using them as supports for pushing the girders up slowly, step-by-step with hydraulic presses, until all sections were assembled and the lifting process was finished. Simultaneously the outer heavier columns could be extended independently *(Fig. 19)* [19, p. 213].

Fachwerkträger

Im Februar 1877 wird der letzte der kleineren Fachwerkträger auf seinen Pfeilerköpfen festgebolzt – «...Gitterbalken, die jetzt, auf zierlichen Stelzen stehend von Süden und Norden her gegen die Mitte der Bucht zwei ununterbrochene dunkle Linien ziehen... Sie werden ins Blaue hinaufgepumpt und mit ihren Säulenbeinen versehen, die ihnen während des Aufsteigens unter dem Leib wachsen» [4, S. 446]. In dieser intensivsten Arbeitsphase sind 500 Arbeiter auf beiden Ufern im Einsatz, unterstützt von 4 Dampfern, 12 Barken und unzähligen kleineren Booten. Alles ist nun so gut organisiert und die Leute sind dermaßen eingearbeitet, daß eine ernsthafte Störung kaum noch denkbar erscheint: Das Ende ist in Sicht.

Die großen Fachwerkträger aus Schmiedeeisen sind gewaltige Gewerke: Sie werden als 245 Fuß (74,68 m) lange Hohlkästen am Ufer fertig zusammengefügt, sind 18 Fuß (5,49 m) breit und 36 Fuß (10,97 m) hoch, entsprechend dem Querschnitt eines Eisenbahnzuges. Jeder dieser Doppelträger benötigt 4 Wochen Bauzeit, verbraucht 18 000 Nieten und wiegt 190 Tonnen. Das Problem ist aber nicht diese Vormontage, bei aller Neuartigkeit, sondern der Transport dieser «entsetzlich langen viereckigen Röhren» [4, S. 448] – auch wenn er vorher noch so gut überlegt worden ist. Da ja die Fahrrinne des Firth of Tay nicht durch feste Gerüste unterbrochen werden durfte, bediente man sich wieder jener ingeniösen Methode, die bereits beim Bau von Robert Stephensons «Britannia Brücke» zum erstenmal erfolgreich erprobt worden war. Sehr anschaulich schildert Eyth diesen immer wieder aufregenden Vorgang mit Hilfe des «Weltalluhrwerkes» von Ebbe und Flut. Im Grunde war es ja sehr einfach, denn man überließ es den auf der Arbeitsbühne vormontierten Trägern, sich bei Flut mittels untergeschobener Barken freizuschwimmen, schleppte sie dann zur Einbauöffnung *(Abb. 18)* und ließ sie dort bei weichender Flut wieder auf die Pfeilersockel absinken: Damit blieb das Heben und Senken dieser gewaltigen Massen vollkommen den Naturkräften überlassen; nur die horizontale Bewegung geschah mit Hilfe der Dampfschlepper. Auf den Pfeilersockeln ließ man die Träger gewissermaßen ihre eigenen Standbeine bauen und sparte so ein Gerüst: Zuerst stellte man die untersten Schaftlängen der 4 mittleren Säu-

von dem Schauspiel, das für ihre Enkel so alltäglich werden sollte. An Wochenenden nahmen sie die Fähre nach Newport und von dort den Zug nach Wormit, um diese aufregenden Arbeiten aus nächster Nähe zu beobachten.

18
One of the High Girders on its staging, waiting to be towed on barges to the site [7].

18
Einer der großen Mittelträger auf der Arbeitsbühne, vor dem Abtransport auf Barken zur Einbaustelle [7].

len auf und drückte daran die Träger mittels hydraulischer Pressen langsam und hubweise nach oben, bis schließlich alle Schaftlängen zusammengesetzt und der Hubvorgang beendet war. Währenddessen konnte man die äußeren stärkeren Säulen unabhängig weiterbauen *(Abb. 19)* [19, S. 213]

"Tomorrow the first of the 250 ft girders will reach its full height. The hydraulic lifting devices now work perfectly. We make headway with that giant box-girder by 25 ft daily, reaching its dizzy heights within four days. From close by the thing looks ghastly, from afar, watching from the shore, it looks like some magic spell, like happening in a dream – secretly, silently, as if out of its own. There isn't a sound to be heard and one loses all sense of distance at the size of those masses. Higher and higher the thing rises to rest like hovering in mid-air, as if iron had lost all weight. In such moments one can hardly surpress a foolish sense of pride – nor is one inclined to do so, to tell you the truth. With all the many secret worries going along, moments of such illusion shouldn't be spoiled. Ever since the ground caved in from under our feet one year ago, I won't trust the following day any more."

Morgen kommt der erste der zweihundertfünfzig Fuß langen Gitterbalken auf seine volle Höhe. Die hydraulischen Hebevorrichtungen arbeiten jetzt musterhaft. Wir kommen mit den Riesenkasten täglich um fünfundzwanzig Fuß weiter, so daß er in vier Tagen seine schwindlige Höhe erreicht. In der Nähe sieht die Sache gruselig aus, in der Ferne, vom Ufer gesehen, wie ein Zauber, wie etwas, das im Traum geschieht; heimlich, still, wie von selbst. Man hört keinen Laut und hat, bei der Größe der Massen, alles Gefühl für Entfernungen verloren. Höher und höher steigt das Ding und ruht, wie schwebend, in der Luft, als ob Eisen kein Gewicht mehr hätte. Das sind Augenblicke, in denen man ein dummstolzes Gefühl nicht ganz unterdrücken kann — noch mag, um Dir's ehrlich zu sagen. Man hat so viele geheime Sorgen nebenher, daß man sich Augenblicke solcher Illusionen nicht verderben darf. Seitdem uns vor einem Jahr der Erdboden unter den Füßen verschwand, traue ich dem morgigen Tag nicht mehr.

Spell of dreams: Iron without weight
[4, pp. 445, 446]

Traum-Zauber: Eisen ohne Gewicht
[4, S. 445, 446]

19
Lifting of the girders unto the piers with the help of hydraulic presses at the southern shore of the Tay Bridge, October 5, 1872. In the right foreground, a completed girder still resting on its staging can be seen [26, p. 85].

19
Hebung der Brückenträger auf den Pfeilern mit Hilfe hydraulischer Pressen am Südende der Tay-Brücke, 5. Oktober 1872. Rechts vorne ist ein fertigmontierter Träger auf der Arbeitsbühne zu erkennen [26, S. 85].

High Girder Going Down

On February 27, 1877, there is another accident with one of the High Girders, luckily without loss of life. During an upcoming storm – in a routine tugging manoeuvre now being executed for the 63rd time – the 190-ton colossus is towed on its barges almost up to the piers, when it suddenly sinks into the tide!
The tugging manoeuvre used to be handled regularly by two tugs towing both barges with their heavy load in between, from the inland side until almost up to the bridge; there the countercurrent of the tide coming from the east would slow them down to such a degree that the girder-ends could now be lowered unto the piers which reached just above the high tide level. As well-trained as this team of workmen, sailors and tug pilots was, they knew no commandos for turning back; such a case had never been encountered before and would have been rather complicated anyway. So despite the upcoming storm and high waves, it was impossible now to stop the operation. In this situation where the tidal current coincided with an easterly gale quite rare in this region, there was no way but to proceed as usual and to take the risk – with the result that one of the large girders sank accidentally for the first

Mittelträger versinkt

Am 27. Februar 1877 kommt es mit einem der großen Mittelträger zu einem weiteren Unfall, der glücklicherweise ohne Todesopfer abgeht. Bei aufziehendem Sturm – und beim 63. Mal dieses Routine-Manövers – wird der 190 Tonnen schwere Koloß auf seinen Barken bis kurz vor die Brückenpfeiler geschleppt, um dann plötzlich in den Fluten zu versinken!
Das Schleppmanöver bestand darin, daß bei Flut zwei Schlepper die beiden Barken mit ihrer großen Last dazwischen von der stromaufwärtigen Seite bis hart vor die Brücke schleppten; dort ließen sie sich vom Gegenstrom der Flut, die von Osten kam, soweit abbremsen, bis man schließlich die Träger-Enden auf den beiden Pfeilerplattformen einfach absetzen konnte, die ja bei Flut eben noch über Wasser ragten. Das gut eingespielte Team von Bauarbeitern, Schiffern und Schlepperführern kannte aber keine «Rückwärts»-Kommandos – dieser Fall war nie geprobt worden und wäre obendrein sehr kompliziert gewesen, so daß man trotz des aufkommenden Sturmes und hoher Wellen nicht mehr umkehren konnte! In diesem Falle, wo der Flutstrom und ein in dieser Gegend seltener Oststurm aufeinandertrafen, blieb also gar nichts

Excerpts

from Max Eyth: "Behind plow and bench-vice" [4, pp. 448–452]

– collected here for a sample description of the loss of one girder
"Cosmic clockwork": Lifting forces of the moon
Hovering girders
Tugboat-manoeuvre in the storm

Auszüge

aus Max Eyth [4, S. 448 – 452]

– zusammengestellt als Probeseite zum Sturz eines Trägers
«Weltall-Uhrwerk»: Hebekräfte des Mondes
Schwebende Gitterbalken
Schleppmanöver im Sturm

Als ich ans Ufer kam, waren die Barken schon unter dem Gitterbalken festgelegt, die mächtigen Schwellen und Holzklötze aber, die ihn tragen mußten, noch einen halben Fuß unter dem Eisenwerk. Langsam, fast unmerklich, stiegen sie empor wie der Zeiger eines Uhrwerks. Es war das Weltalluhrwerk, das man hier laufen sehen konnte. Die Schiffer der Boote, ein Dutzend Arbeiter und die Führer des Trupps warteten plaudernd auf das Steigen des Wassers, das in rauschenden Wellen vom Meer heraufkam. Das Wetter hatte sich plötzlich geändert. Ein frischer Wind jagte in leichten Stößen über die Wasserfläche, und im Osten stieg eine Mauer schwerer Wolken auf, die scharf gegen den blauen Himmel abstach.

Man warf den zwei kleinen Schraubendampfern schon die Schlepptaue zu. Alle Aufmerksamkeit war auf unsre Barke selbst gerichtet, die jetzt die Unterkante des Balkens berührte. Hier wurde noch ein mächtiger Holzteil untergeschlagen, dort mit hastiger Anstrengung eine sich verschiebende Schwelle zurechtgerückt. Man hörte da und dort ein leises Knistern, ein dumpfes Knarren. Das Boot drückte jetzt gewaltig von unten. Die Bohlen der Holzbrücke, denen die Last allmählich abgenommen wurde, stöhnten auf. Es wurde ihnen mit jeder Minute leichter. Jetzt fiel ein Holzklotz, auf dem ein Teil des Gitterbalkens geruht hatte, polternd aus seiner Lage, jetzt am andern Ende ein zweiter. Die zwei Schleppdampfer zogen die ausgestreckten Taue sanft an, um ihre Länge genau zu regeln. Jetzt endlich regte sich der mächtige Gitterbalken mit einem leisen Ruck, und plötzlich schwebte er einen Zoll hoch über dem Lager, auf dem er entstanden war. Zwanzig Arbeiter warfen sich Unterlagen über den Haufen, auf denen er entlang der Brücke geruht hatte. Man wartete noch zehn Minuten, dann schwankte er sechs Zoll über jedem festen Punkt der Plattform. Der Vormann kommandierte: „Seile los!", und die zwei Dampfer zogen langsam und vorsichtig die Schlepptaue an.

Majestätisch segelte das wunderliche Doppelfahrzeug in den Strom hinaus: die zwei Barken mit dem zweihundertfünfzig Fuß langen Riesenbalken, der sie verband. Der Holzstaden, von dem wir abtrieben, befindet sich oberhalb der Brücke. Da die Strömung während der steigenden Flut landeinwärts zieht, läßt man die Barken in dieser Richtung hinauftreiben, um sie weiter oben über die Bucht wegzuschleppen. Dann erst wird der Balken langsam gedreht und in einer Stellung parallel zur Brücke dieser entgegengeschleppt. Das gewohnte, wenn auch noch immer etwas unbehagliche Manöver gelang wie früher immer, doch bemerkten wir jetzt erst, wie unruhig der Strom war. Je weiter wir gegen die Mitte der Bucht kamen, um so höher wurden die Wellen. Der blaue Himmel war verschwunden, ein pfeifender Wind kam vom Meer her, und da und dort zeigten sich die weißen Schaumkämme einer regelrechten See. Man fuhr natürlich immer mit der größten Vorsicht, so daß die Fahrt gewöhnlich dreißig bis sechsunddreißig Minuten dauerte. Die Arbeiter betrachteten sie als eine ihrer Lustbarkeiten und saßen gewöhnlich plaudernd auf dem Rand der Barken. Heute wurde einer nach dem andern still und sah nachdenklich über die windbewegte Fläche, auf der weiter unten schon stürmische Regengüsse hinfegten.

Von Zeit zu Zeit spritzte jetzt etwas Wasser über Bord, denn die Boote gingen mit ihrer gewaltigen Last ziemlich tief. Manchmal traf eine Welle die Bootseite mit einem lauten, harten Schlag. Dann ging ein Zittern durch unsern Gitterbalken, von einem Boot zum andern, wie wenn man eine Saite berührt. Dazu heulte jetzt der Wind hörbar und brachte dicke Nebelwolken den Fluß herauf. Die Stimmung wurde unbehaglich.

Unser Fahrzeug hatte noch nicht die Mitte des Stroms erreicht, der uns mit Gewalt von der Brücke ab nach oben trieb, was übrigens ganz im Plan des Manövers lag.

Der Gitterbalken stand jetzt parallel mit der Brücke, quer über den Fluß. Beide Dampfer zogen mit Macht gegen die wütende Strömung, die uns entgegenbrauste. Am Bug unsrer Boote spritzten die Wellen jetzt beständig über Bord. Die Barken hoben und senkten sich in unruhiger Bewegung, die sich dem Balken mitteilte, der haushoch über seine Unterlagen emporragte. Da sich beide Boote jedoch nicht in gleichem Tempo bewegen wollten, so zitterten und knirschten die Unterlagen bösartig.

Wir näherten uns jetzt den Pfeilerinseln, auf denen ein halbes Dutzend Leute uns erwartete, förmlich eingehüllt in den weißen Gischt einer kleinen Brandung. Die Dampfer waren schon zwischen den Pfeilern durchgefahren und ließen die Schleppseile sinken. Jetzt erst, an den Pfeilerinseln als festen Punkten, sah man, wie unser gewaltiges Zwillingsfahrzeug schwankte und schaukelte. Es war grausig.

Beim besten Willen kann ich nicht genau erzählen, was nun vor sich ging. Es war den Steuerleuten wahrscheinlich nicht möglich gewesen, die genaue Mitte zwischen unsern zwei Pfeilern einzuhalten. Auch hatten die Schlepper uns nicht ganz parallel mit der Brücke herangezogen. Wir mit unserm Balkenende waren noch ein paar Fuß vom Pfeiler entfernt. Da kam ein furchtbarer Stoß von der andern Seite. Fünf, sechs Leute fielen zu Boden. Dann ein zweiter. Die Unterlagsschwellen krachten und drehten sich, die dicken Seile, die den Balken aufrechthielten, knallten entzwei und flogen wie Peitschenschnüre durch die Luft. Am fernen Ende stieg die Bootspitze aus dem Wasser wie ein Pferd, das sich bäumt. Bei uns neigte sich der Gitterbalken nach hinten, langsam, unaufhaltsam; die Unterlagsschwellen stürzten zermalmt in einen Holzstücke speienden Haufen übereinander, und dann war es zehn Sekunden lang ein Zischen und Tosen, ein Klatschen und Schlagen, ein Knirschen und Sausen, in dem man nicht wußte, ob man im Wasser oder auf dem Land, auf den Füßen oder auf dem Kopf stand. Und das Geschrei!

Als ich mich wieder mit einiger Besinnung umsehen konnte, stand ich neben dem alten Kapitän auf der Pfeilerinsel. Unser großer Gitterbalken war spurlos verschwunden, und der reißende Strom jagte drüber weg, da und dort noch ein wenig gurgelnd, als habe ihm der ungewohnte Bissen nicht übel geschmeckt.

Das Erstaunliche ist, daß nicht ein Mann verloren ging. Am andern Ende waren die meisten ins Wasser gesprungen. Da jedoch die Leute auf den Pfeilern zwei Kähne bei sich hatten, wurden sie ohne Schwierigkeit aufgefischt. Auf unsrer Seite gab es einen Beinbruch und ein paar zerbrochene Rippen, wofür wir Gott danken dürfen.

time before the bridge was even finished. The girder was later recovered, only to go down later – almost 2 years and 10 months to the day – for the second and final time.

Trial Run and Opening

At long last, on September 22 of 1877 at 10 o'clock in the morning, the last rail is bolted fast in one of the large girders, between piers Nr. 32 and 33. Immediately thereafter the first locomotive leaves Dundee for a trial run across the bridge, carrying a half-burned garland around its stack and a rosary slung over the valves. For the engineer it is a scary sight, as the handrail and the flooring next to the tracks are still missing. One appears to be suspended mile-high on the narrowest base above the water, while steering the engine right up into the air *(Fig. 20)* [4, p. 456]!
Four days later, the small locomotive "Lochee" is pulling a special train – "The Directors' Special" – packed with elegant ladies and gentlemen across the bridge, all invited by courtesy of the NBR and the Tay Bridge Undertaking. In front of the "Lochee", another engine is piloted by Thomas Bouch himself. Soon, he will be knighted by Queen Victoria for his services.
On November 18, 1877, Major-General Charles Scrope Hutchinson, the Government Inspector of the Board of Trade, arrives from London for three days to inspect the bridge from one end to the other. This done, he needs three months to write his report. For the vertical loading test, he uses three heavy locomotives, running first one, then two coupled together, three etc. up to six across every span. Sending them across with various speeds, he finally reaches 40 m.p.h. and even climbs into one of the High Girders, notebook in hand. Afterwards he repeats the whole procedure while directing a theodolite from the south shore at the top beams of the central spans, in order to observe lateral vibrations. But these were "nothing at all excessive, as my judgement went", he later told the court. In all his tests he hardly paid attention, if at all, to the possible effects of wind pressure. When he was asked about this months later, he replied:

> "The subject never entered into the calculations that I made, and never has done. It has never been, to my knowledge, customary hitherto to take wind pressure into account in calculating parts of bridges of this description." [21, p. 125]

The thought, however, had indeed occurred to him nevertheless: he recommended a speed limit of 25 m.p.h. on the bridge and concluded his report with the remark that he would hope to observe, at the opportunity of another visit, the effects of a strong wind

anderes übrig, als wie gewohnt zu verfahren und die Sache zu riskieren. – Damit war einer der großen Mittelträger zum erstenmal und noch vor Fertigstellung der Brücke in den Fluten versunken. Es gelang später, ihn wieder zu heben – bis er dann, fast auf den Tag genau 2 Jahre und 10 Monate später, zum zweitenmal und endgültig untergehen sollte.

Probelauf und Einweihung

Endlich, am 22. September 1877 um 10.00 Uhr morgens, wird zwischen dem 32. und 33. Pfeiler in einem der langen Mittelträger die letzte Schiene auf die Schwellen gebolzt. Und sogleich fährt von Dundee aus die erste Lokomotive – mit einer halbverbrannten Girlande am Schornstein und einem Rosenkranz über den Ventilen – zur Probe über die Brücke: Für den Fahrer halsbrecherisch anzusehen, da die Bohlung an der Seite der Gleise und das Handgeländer noch fehlen: Man scheint auf schmalster Fläche turmhoch über dem Wasser zu hängen und die Maschine direkt in die Luft hineinzuziehen *(Abb. 20)* [4, S. 456]!
Vier Tage später führt die kleine Lokomotive «Lochee» einen Sonderzug – «the Directors' Special» – voll feiner Damen und Herren über die Brücke, alle eingeladen auf Kosten der NBR und der Brückenbaugesellschaft. Vorne fährt noch eine Pilotmaschine mit Thomas Bouch am Steuer voraus: Er sollte alsbald von Queen Victoria für seine Verdienste geadelt werden.
Am 18. November 1877 kommt Generalmajor Charles Scrope Hutchinson, der Regierungsinspektor vom *Board of Trade* (Industrie- und Handelskammer) aus London und inspiziert drei Tage lang die Brücke von einem Ende zum anderen. Anschließend braucht er drei Monate, um seinen Prüfbericht zu schreiben. Für die vertikale Lastprüfung benutzt er 6 schwere Lokomotiven und läßt zuerst eine, dann zusammengekuppelt zwei, drei usw. bis zuletzt sechs über jeden Träger laufen. Er schickt sie mit verschiedenen Geschwindigkeiten hinüber, schließlich sogar mit 40 Meilen/Stunde (64 km/h), und klettert auch einmal in einen der Hochträger, mit Notizbuch in der Hand. Dann läßt er noch einmal alles wiederholen, während er vom Südufer aus einen Theodoliten auf die Obergurte der Mittelträger richtet, um seitliche Schwingungen zu beobachten. Diese waren, wie er später vor Gericht erzählte, «keineswegs übermäßig, in meiner Beurteilung». Bei all seinen Tests beachtete er kaum, wenn überhaupt, die mögliche Wirkung von Winddruck. Als man ihn Monate später danach befragte, erklärte er:

> «Das war nie Gegenstand meiner Berechnungen gewesen. Es war auch bisher, meines Wis-

20
Trial runs at the southern end of the Tay Bridge, during the autumn of 1877 [26, p. 86].

20
Probeläufe am Südende der Tay-Brücke im Herbst 1877 [26, S. 86].

when a train was crossing the bridge. But before he had occasion to make this second visit, those effects would be demonstrated most dramatically.

The official opening celebration was scheduled to take place a few months later; it had been decided to run – during the winter months – only goods-trains across the bridge so the public might get used to the matter. Only on March 31, 1878, after months of testing, does the official opening for passenger service take place. To the sounds of Handel's Victory March "See, See, the Conquering Hero Comes", the Lochee once again sets out with carriages of high visitors in its tow [25, p. 230]. – The common people had to wait till the next day. They arrived in swarms at Tay Bridge Station, excited to embark on the 6:25 morning train for their first journey ever. Aside from some anger about the fare having been raised from 6 to 9 pence, there was

sens, nicht üblich, den Winddruck bei der Berechnung von Brückenteilen dieses Typs zu berücksichtigen.» [21, S. 125]

Allerdings hatte ihn der Gedanke daran wohl befallen: Er empfahl eine Geschwindigkeitsbegrenzung von 25 Meilen/Stunde (40 km/h) und bemerkte abschließend, er hoffe, bei der Gelegenheit eines erneuten Besuches die Wirkung eines starken Windes zu beobachten, wenn ein Zug über die Brücke fährt. Bevor er aber diesen Besuch wahrmachte, waren selbige Wirkungen auf dramatische Weise bewiesen worden.

Die eigentliche Eröffnung sollte erst in einigen Monaten feierlich stattfinden – man wollte während des Winters zunächst nur Güterzüge über die Brücke leiten, damit sich auch das Publikum an die Sache gewöhne. Erst am 31. Mai 1878, nach monatelangen Testfahrten, wird die offizielle Einweihung für den Per-

nothing but praise for this marvel of engineering. Now even ordinary passengers had a chance to travel the distance from Edinburgh to Dundee without detour via Perth, in just 2 hours and 22 minutes. Thus the NBR beat the time of their old rival, the CR, by over one hour.

In the months to follow, the NBR experienced an unheard of prosperity thanks to this bridge. It carried 84% of the traffic from Dundee to Edinburgh and 59% between Edinburgh and Aberdeen. Traffic from Dundee to Fife doubled and the coal pits in Fife could now be exploited to an extent that would have been unthinkable with the old ferry system. By the end of 1878, goods traffic to Dundee had risen by 40%, and there was no longer any doubt that the NBR had gotten a firm grip on railway business in northern Scotland. It possessed now over 500 locomotives, 1,000 carriages and almost 30,000 goods wagons. Per year, the NBR now transported 15 million passengers and earned more than 2 million Pounds in revenue.

Judgement: "Early-American"

For a year now, this powerful structure tracing a dark straight line through the Scottish landscape is considered the pride and triumph of its age, but "... beautiful it isn't, that famous bridge; one mischievous critic even dubbed its style as being 'early-American'" [4, p. 468]. This comparison is not even that much out of place, as the multitude of identical parts indeed is reminiscent of North American bridges without, however, approaching their pragmatic sophistication.

And there a genuine American drops in for a visit: Ulysses Simpson Grant, victorious General of the American Civil War and former President of the United States, had already come on September 1, 1877, by special train from Edinburgh to look at this sight [21, p. 123]. His commentary is quintessential: "It's a very long bridge." – Also Prince Leopold, Queen Victoria's youngest son, comes to remark to Thomas Bouch, how much he admired not just "the elegance of the structure, but its solid substantiality" *(Fig. 21)*.

Indeed, the finished bridge with its widely sweeping curve before Dundee was a pleasing sight. In profile it even performed a few subtle leaps: First it descended from the southern shore over the first three piers, ran level for another three, "then climbed like a tired wave to Pier 29" [21, p. 123], where the central girders began; from there it ran level to Pier 36, began descending steadily to the northern end. It then swung in a nice curve past the beautiful new esplanade *(Fig. 22)* to finally enter the even more beautiful new station; Tay Bridge Station presented itself proudly with extensive

sonenverkehr nachgeholt: Zu den Klängen von Händels Siegesmarsch *«See, See, the Conquering Hero Comes»* erscheint wieder die *Lochee* mit den Wagen hoher Gäste im Gefolge [25, S. 230]. Das gemeine Volk mußte bis anderntags warten: Die Leute kamen in Scharen zur Tay Bridge Station, voller Spannung, den 6.25 Uhr Morgenzug für die erste Fahrt zu besteigen. Außer einigem Ärger über die Erhöhung des angekündigten Fahrpreises von 6 auf 9 Pence gab es nichts als Lob für dieses Wunderwerk der Ingenieure. Jetzt war es auch Passagieren möglich, die Strecke von Edinburgh nach Dundee ohne den Umweg über Perth in nur 2 Stunden und 22 Minuten zu bewältigen. Damit schlug die *NBR* ihre alte Konkurrentin, die *CR* um mehr als eine Stunde.

In den folgenden Monaten erlebte die *NBR* mit dieser Brücke einen ungeahnten Aufschwung: Sie beförderte nun 84% des Verkehrs von Dundee nach Edinburgh und 59% zwischen Edinburgh und Aberdeen. Der Verkehr von Dundee nach Fife verdoppelte sich, und die Kohlegruben in Fife konnten jetzt in einem Maße erschlossen werden, wie es mit dem alten Fährensystem unmöglich war. Ende 1878 war der Güterverkehr nach Dundee um 40% gestiegen und es gab keinen Zweifel mehr, daß die *NBR* das Eisenbahngeschäft in Nordschottland fest in der Hand hatte. Sie besaß jetzt über 500 Lokomotiven, 1 000 Personenwagen und fast 30 000 Güterwagen. Pro Jahr beförderte sie 15 Millionen Passagiere und hatte über 2 Millionen Pfund Sterling Einnahmen.

Urteil «Frühamerikanisch»

Seit Jahresfrist wird nun dieses mächtige Bauwerk, das eine dunkle starre Linie durch die schottische Landschaft zieht, als der Stolz und Triumph der Zeit gepriesen. Aber «...schön ist sie nicht, die berühmte Brücke; ein boshafter Kritiker hatte für ihren Stil die Bezeichnung ‹frühamerikanisch› erfunden» [4, S. 468]. So abwegig ist der Vergleich gar nicht, denn mit ihrer Vielzahl identischer Bauteile ähnelt sie eher den nordamerikanischen Brückenbauten, ohne jedoch deren pragmatische Raffinesse zu erreichen.

Und es kommt denn auch ein echter Amerikaner zu Besuch: Ulysses Simpson Grant, siegreicher General des amerikanischen Bürgerkrieges und damals Expräsident der USA, war bereits am 1. 9. 1877, kurz vor Fertigstellung, mit einem Sonderzug aus Edinburgh angereist, um die Sehenswürdigkeit zu besichtigen [21, S. 123]. Sein Kommentar sagte alles: «Es ist eine sehr lange Brücke.» – Auch Prinz Leopold, Königin Viktorias jüngster Sohn, war schon dagewesen und bemerkte zu Thomas Bouch, wie sehr er «nicht nur die

21
Probably the finest photograph preserved of the Tay Bridge, as seen from the northern bank, by the Scottish photographer George Washington Wilson. The haughty elegance of this structure, the fragile piers and girders, the slender profile, the obvious minimum of masses barely securing their own stand, aided by the sweeping course across the firth, so inviting to watch a train rolling along with a long smokey trail – all that explains quite well the admiration of the time for this structure – "the most stupendous bridge ever created". – In the boat in the middle foreground, Mr. Bell, one of the five men in the maintenance staff, can be seen at work, checking a pier for trouble [3].

21
Wahrscheinlich die beste erhaltene Aufnahme der Tay-Brücke vom Nordufer, aus dem Atelier des schottischen Fotografen George Washington Wilson: Die spröde Eleganz der Brücke, die grazilen Stützen und Träger, das schlanke Profil, das augenfällige Minimum an Masse, gerade noch an der Grenze der Standsicherheit, dazu der Schwung der Linienführung, so einladend zur Betrachtung eines dahinrollenden Eisenbahnzuges mit langer Rauchfahne – all das erklärt die Bewunderung der Zeit für dieses Bauwerk – «the most stupendous bridge ever created». – Im Boot in der Mitte vorne prüft Mr. Bell, einer der 5 Leute vom Bauunterhalt, einen Brückenpfeiler [3].

22
Dundee's newly constructed Esplanade of 1875, as seen around 1903. In the background, the "new" Tay Bridge, as rebuilt by Barlow [2, p. 67].

22
Die 1875 neu angelegte Esplanade von Dundee, ca. 1903, dahinter die «neue» Tay-Brücke Barlow's [2, S. 67].

public toilets, three classes of waiting lounges for ladies and three refreshment rooms of different classes.

In the end, just one visitor was still needed to make the creators truly happy: Queen Victoria had lent her name to a whole era and could at last be persuaded, in June of 1879, to take a detour via Dundee on her return trip from Balmoral to Windsor Castle. And so she came one afternoon with the Royal Train, to cross for the first – and presumably only – time the new Tay Bridge.

No better than others is Max Eyth able to escape the spell of this structure; his almost poetic description of the spectacle may stand here in lieu of so many others:

Eleganz des Bauwerkes, sondern auch dessen solide Gediegenheit und Substanz» bewunderte *(Abb. 21)*.

In der Tat war die fertige Brücke mit ihrer weitausholenden Kurve vor Dundee nicht übel anzusehen. Im Höhenprofil vollführte sie sogar ein paar spröde Sprünge: Zunächst fiel sie vom südlichen Hochufer über die ersten drei Pfeiler etwas ab, verlief weitere drei Felder eben, erhob sich dann aber wieder «wie eine müde Welle» [21, S. 123] bis hinauf zu Pfeiler 29, wo die Mittelträger begannen; von dort verlief sie eben bis Pfeiler 36, um dann stetig bis zum Nordende abzufallen; dort kurvte sie dann in elegantem Schwung an der schönen neuen Esplanade vorbei *(Abb. 22)* um schließlich in einen noch schöneren neuen Bahnhof einzumünden: Tay Bridge Station präsentierte sich stolz mit ausgedehnten Toilettenanlagen, drei Klassen von Wartesälen für Damen und drei Erfrischungsräumen verschiedener Klassen.

Schließlich fehlte nur noch ein Gast, um die Bauherren vollends glücklich zu machen, und der kam im Juni 1879: *Queen Victoria* hatte einer ganzen Epoche ihren Namen gegeben und konnte nun endlich dazu bewegt werden, auf ihrem Rückweg vom Hochlandschloß Balmoral nach Windsor Castle den Umweg über Dundee zu nehmen. Und so kam sie eines Nachmittags mit dem königlichen Zug und fuhr höchstselbst zum ersten – und wohl einzigen – Mal über die neue Tay-Brücke. – Ebensowenig kann sich der Zeitzeuge Max Eyth der Wirkung dieses Bauwerks ent-

Central portion without gravity
Mittelteil ohne Schwerkraft

Firth of Tay with flying sailing vessels [4, pp. 467, 468]
Firth of Tay mit fliegenden Segelschiffen [4, S. 467, 468]

("Enno-Bay" = Firth of Tay, "Pebbleton" = Dundee)
(«Ennobucht» = Firth of Tay, «Pebbleton» = Dundee)

"Large shadows and sun spots were sweeping across the wide landscape, livening up the majestic waterway of Enno Bay miraculously, spread out some 70 ft below. On the opposite shore, hardly visible in the shade of the hills, the houses of Pebbleton huddled, and along the far beach a string of hamlets and towns. In the background towards the North, the solemn peaks of the Scottish Highlands were rising. Towards the far West, heavy clouds were piling up, and the sun seemed to have to set shortly into the golden mass. Across the glimmering and agitating water, a dozen sailing vessels were flying towards the sea. Here and there a steamer could be seen, towing a brigg or schooner upstream with sails drawn. But everything stood back at this spot in view of the grand structure... Its dizzy height above the water, the incredible length gave the structure its own character – and even in structures, the characteristic often is worth more than the beautiful. Here, iron and stone had turned to resoluteness, willpower and sense of life. At the Northern end, far in the misty distance, the bridge performed a wide curve, still far from the shore, toward the West; so a long line of slender piers could be seen clearly, whereas most of them were invisible from our viewpoint looking longitudinally along the bridge. All the more it seemed as if the giant girders were actually hanging up in the air. Particularly the central portion, rising well above the other girders for a length of one kilometer, gave the impression as if the laws of gravity had lost all meaning in such giant edifices. The railway crossing the firth carried just one track. On both sides of the rails there was a narrow asphalted footpath, protected by an iron railing at the water side. But between the rails and sleepers, one could still look down through the girders into the green water, watching the iron-work supporting the wooden sleepers. The current flowing through some 80 ft below drew our attention in a curious fashion, not exactly helping our sense of security."

ziehen – seine fast schon dichterische Beschreibung des Schauspiels mag hier stellvertretend stehen für viele andere:

> Große Schatten und Sonnenflecke flogen über die weite Landschaft und belebten in wunderlicher Weise die mächtige Wasserfläche der Ennobucht, die sich etwa siebzig Fuß unter uns dehnte. Am andern Ufer, kaum sichtbar im Schatten der Hügel, lagen die Häuser von Pebbleton und am entfernten Strande hin eine Reihe von Dörfern und Städtchen. Im Hintergrund gegen Norden ragten die ruhigen Gipfel des schottischen Hochlands empor. Im fernen Westen türmten sich schwere Wolken auf, und die Sonne schien in kurzer Zeit in der vergoldeten Masse versinken zu müssen. Auf dem flimmernden, lebhaft bewegten Wasser flog ein Dutzend Segelschiffe der See zu. Da und dort sah man einen Dampfer, der eine Brigg oder einen Schoner mit gerefften Segeln heraufschleppte. Aber alles trat an dieser Stelle zurück vor dem mächtigen Bauwerk.
>
> ...die schwindelnde Höhe über dem Wasserspiegel, die riesige Länge gaben dem Bauwerk seinen eignen Charakter, und auch in Bauwerken ist das Charakteristische oft mehr wert als das Schöne. Hier war in Eisen und Stein Entschlossenheit, Wille, Lebenszweck. Am Nordende, in dunstiger Ferne, machte die Brücke noch weit vom Ufer ihren gewaltigen Bogen gegen Westen, so daß eine lange Reihe ihrer schlanken Pfeiler deutlich hervortrat, während weitaus die Mehrzahl von unserm Standpunkte aus, in der Längenrichtung der Brücke, nicht gesehen werden konnte. Um so mehr schien es, als ob die riesigen Gitterbalken förmlich in der Luft hingen. Namentlich der mittlere Teil, der in der Länge von einem Kilometer hoch über die andern Partien hervorragte, machte den Eindruck, als ob die Gesetze der Schwere bei so gewaltigen Bauten keine Geltung mehr hätten. Die die Bucht überschreitende Bahn war nur eingleisig. Auf beiden Seiten der Schienen war ein schmaler asphaltierter Fußsteg, der nach der Wasserseite hin durch ein Eisengeländer geschützt war. Zwischen den Schienen und Schwellen jedoch konnte man noch immer durch die Gitterbalken ins grüne Wasser hinuntersehen und das Eisenwerk betrachten, auf dem die hölzernen Schwellen lagerten. Die achtzig Fuß unter uns durchziehende Strömung, die den Blick in wunderlicher Weise mitzog, trug nicht zum Gefühl der Sicherheit bei

Building Costs

The total costs came to 320,000 Pounds Sterling in the end (according to Hammond [8] to 350,000 Pounds). This would amount to a cost overrun of 48 %, based on the contract sum of 217,000 Pounds, as agreed upon 6 years previously. But in view of the magnitude and novelty of this undertaking, with all the unaccountable difficulties experienced with the foundations, the change of the contractors and the many other problems, this is by no means an exceptional amount – even less so if compared to cost overruns today for major projects. One could almost take this to be a "normal" result – had there not been some major

Baukosten

Die Baukosten beliefen sich schließlich auf 320 000 Pfund Sterling (nach Hammond [8] auf 350 000 Pfund). Das entspricht einer Kostenüberschreitung von 48 %, legt man die vor 6 Jahren vereinbarte Bausumme von 217 000 Pfund zugrunde. Angesichts der Größe und Neuartigkeit dieses Unternehmens mit all den unvorhersehbaren Schwierigkeiten bei der Gründung, dem Wechsel der Baufirma und vielen sonstigen Widrigkeiten ist das aber keineswegs ein ungewöhnlich hoher Betrag – schon gar nicht im Vergleich zu Kostenüberschreitungen heutiger Großprojekte. Fast könnte man sagen, das sei ein «normales» Ergeb-

and irresponsible "savings" obtained in the wrong places.

Already in Victorian Britain of the 1870's, railway construction no longer appeared as the heroic adventure it was in the pioneering days of Stephenson and Brunel. But this great bridge was really something special and by far the longest bridge of the world. Never before had such masses of material been used for a single bridge: during its 6 years of construction 600 workmen had used 3,700 tons of cast iron, 3,500 tons of wrought iron, 87,000 cubic ft of timber, 15,000 barrels of cement, 10 million bricks and over 2 million rivets. Now, a major part of this volume was resting on 85 spidery piers.

Precursors

Sir Thomas Bouch and his engineers had already left the job site, when the problems began in earnest. First, the railway company had appointed a most conscientious old man and former assistant to Bouch, Mr. Henry Noble, as resident Bridge Inspector. He was supposed to check for screws, bolts and rivets that may have loosened. But Mr. Noble possessed neither the qualifications nor the practical experience to prepare him for the heavy responsibility of maintaining the world's longest bridge, since all his life he had dealt with concrete and brick buildings, never with iron structures. Although he was entrusted with a crew of 7 men and the use of the steamer "Tay Bridge", he obviously was overtaxed. Still, already by September of 1878, hardly four months after the official opening, he discovered that the wind ties had gotten loose at the cast-iron lugs of the piers. Calculation of these ties proved subsequently that they provided only about half the strength necessary to withstand the maximum lateral wind pressure with a train on the bridge [8, p. 29]. Henry Noble, while taking soundings from his boat near the piers — one may call them "hearing-tests" — clearly heard the ties rattling whenever a train was passing through. The cast-iron lugs on the piers, intended to hold the bolts of the wind ties in place, were defective due to poor workmanship and faulty supervision at the foundry in Wormit. Since the bolts kept falling out again and again, one took to driving them back in by force every time — an act which eventually came to be regarded as so routine a measure of maintenance, that it was no longer worthy of report.

By the autumn of 1879, the structure had already lost much of its novelty. The nation's imagination was now captivated by yet another miracle still to come: "Except in matter of length, the Tay Bridge will be a mere

nis — wäre da nicht obendrein an falscher Stelle, wie sich zeigen sollte, noch ganz erheblich und in unverantwortlicher Weise «gespart» worden.

Im viktorianischen England der 1870er Jahre erschien der Eisenbahnbau schon nicht mehr als das heroische Abenteuer, das er noch in den Pioniertagen von Stephenson und Brunel dargestellt hatte. Aber diese großartige Brücke war nun wirklich etwas ganz Besonderes — und bei Weitem die längste der Welt: Noch nie zuvor waren derartige Mengen von Material für eine einzige Brücke verbaut worden: In 6 Baujahren hatten 600 Arbeiter 3 700 Tonnen Gußeisen, 3 500 Tonnen Schmiedeeisen, 87 000 Kubikfuß Holz, 15 000 Faß Zement, 10 Millionen Ziegel und über 2 Millionen Nieten verbaut. Und ein großer Teil dieses Volumens ruhte auf 85 spindeligen Pfeilern.

Vorboten

Sir Thomas Bouch und seine Ingenieure hatten schon nichts mehr mit dem Bau zu tun, als die Probleme erst so richtig begannen. Zunächst hatte die Bahngesellschaft einen äußerst gewissenhaften alten Mann und früheren Mitarbeiter Bouch's, Mr. Henry Noble, zum Brückeninspektor ernannt. Er sollte nach Schrauben, Bolzen und Nieten sehen, die sich etwa gelöst hatten. Mr. Noble besaß aber weder die Qualifikation noch die praktische Erfahrung, ihn für die schwere Verantwortung des Unterhaltes der längsten Brücke der Welt zu befähigen; denn er hatte sein Leben lang mit Beton- und Ziegelbauten, aber nie mit Eisenkonstruktionen zu tun gehabt. Obwohl man ihn mit einer Mannschaft von 7 Leuten und dem Dampfer *«Tay Bridge»* ausgestattet hatte, war er offensichtlich überfordert. Immerhin entdeckte er schon im September 1878, also noch keine 4 Monate nach der offiziellen Eröffnung, daß sich die Windverbände an den gußeisernen Haltezapfen der Pfeilertürme gelockert hatten. Bei der späteren Nachrechnung dieser Verbände zeigte sich, daß sie nur etwa die Hälfte der notwendigen Festigkeit besaßen, um dem Maximaldruck eines Seitenwindes mit einem Zug auf der Brücke standzuhalten [8, S. 29]. Henry Noble stellte von seinem Boot nahe den Pfeilern Untersuchungen an — man könnte sagen, Hörversuche — und vernahm deutlich das Klappern der Verstrebungen, sobald ein Zug durchkam. Die fehlerhaften gußeisernen Ösen an den Pfeilern, welche die Bolzen der Windverbände halten sollten, waren das Ergebnis schlechter Werkarbeit und mangelnder Aufsicht in der Gießerei von Wormit. Weil diese Bolzen immer wieder herausfielen, hatte man sich angewöhnt, sie jedesmal mit Gewalt wieder einzuschlagen — bis man das anscheinend als eine Art routinemäßi-

trifle compared to the Forth Bridge", wrote the press. Although familiarity need not mean contempt, some criticism had been voiced nevertheless; this was, in part, even more disquieting than one was prepared to accept: General Hutchinson had recommended a speed limit of 25 m.p.h. (40 km/h) for the bridge, but quite a few travellers were convinced that this limit was being dangerously exceeded (Fig. 23). Soon one could observe respectable gentlemen riding in first-class compartments with watch in hand, timing the trains from the ferries or from the shore. They concluded and later testified before the court that speeds of 30, 35 and sometimes even 42 m.p.h. (48, 56 and 67 km/h) had been reached – by drivers who were racing the ferries, as everyone knew!

The ex-Provost of Dundee so suffered from "mental discomfort" on the bridge that he gave up train rides altogether, taking to the ferries instead. He complained about the excessive speeds to the stationmaster and to Henry Noble. But Noble did nothing and the stationmaster did little more than to advise his drivers to "go cannily", because the Lord Provost had complained. One driver reported how he had chuckled and taken "the whole thing for perfect nonsense". But also other passengers, especially the ladies, were frightened by the obvious rattling of the girders and by "a prancing motion" of the trains [21, pp. 126, 127].

Painters working in the girders told of ever more bolts falling out along the bridge and that they were rusting through at the lower piers at an alarming rate. Also the vibrations under passing trains were reported to be most uncomfortable to the men.

All these faults and complaints were obviously well known for months – and the astonishing neglect in maintaining this bridge was surely a central cause for the disaster. Already during construction, the matter of the faulty casting of the piers had been no secret to the engineers. Because the castings supplied by Hopkins & Co were worse than those by the old Hugenot de Bergue, quite a few columns had to be returned to the foundry. But the full extent of this shoddy workmanship came to light only at the public inquiry.

Load Assumptions and Guesswork

It should be noted that the engineers in charge, including Bouch himself, were far from certain about valid load assumptions, even in the planning phase. In those days, there were no British Standard Specifications (comparable to the German DIN-Standards) in existence which one could have consulted for the permissible wind loads, giving some orientation for cal-

gen Bauunterhaltes hinnahm und solche Vorfälle nicht mehr extra meldete.

Mit dem Herbst 1879 hatte das Bauwerk schon viel von seiner Neuartigkeit verloren. Die Vorstellungskraft der Nation berauschte sich jetzt an einem anderen Wunderwerk, das erst noch kommen sollte: «Außer in ihrer Länge wird die Tay-Brücke eine Kleinigkeit sein im Vergleich zur Forth-Brücke», schrieb die Presse. Obgleich die Gewöhnung noch nicht Geringschätzung bedeuten mußte, war doch Kritik laut geworden, teilweise noch beunruhigender, als man zunächst wahrhaben wollte: General Hutchinson hatte eine Geschwindigkeitsbegrenzung von 25 Meilen/Stunde (40 km/h) für die Brücke empfohlen, aber viele Reisende waren überzeugt, daß diese Grenze gefährlich überschritten wurde (Abb. 23). Alsbald sah man achtbare Männer mit der Taschenuhr in der Hand in ihren 1. Klasse-Abteilen sitzen, oder sie machten Zeitmessungen der Züge von den Fähren oder vom Ufer aus. Sie folgerten, und bezeugten das später vor Gericht, daß Geschwindigkeiten von 30, 35 und manchmal sogar 42 Meilen/Stunde (48, 56 und 67 km/h) erreicht wurden, und zwar von Zugführern, die – wie jedermann wußte – sich Wettläufe mit den Fähren lieferten!

Der Ex-Bürgermeister von Dundee litt an «geistiger Unpäßlichkeit» auf der Brücke, so daß er die Bahnfahrten aufgab und sich lieber den Booten anvertraute. Er beklagte sich über die hohen Geschwindigkeiten beim Stationsvorsteher und bei Henry Noble. Aber dieser unternahm nichts und der Stationsvorsteher wenig mehr, als daß er seine Zugführer anwies «es schlau anzustellen», weil sich der Bürgermeister beschwert hätte. Ein Lokführer berichtete, wie er darüber geschmunzelt und «das Ganze für baren Unsinn gehalten habe». Aber auch andere Fahrgäste, insbesondere die Damen, waren voller Angst über das offensichtliche Rütteln der Träger und «eine sich aufbäumende Bewegung» der Züge [21, S. 126, 127].

Anstreicher auf den Trägern erzählten von immer mehr herausgefallenen Bolzen entlang des Schienenweges, und daß diese an den unteren Stützen erschreckend schnell durchrosteten. Auch wären den Malern bei der Arbeit die Vibrationen von durchfahrenden Zügen jedesmal höchst unbehaglich gewesen.

All diese Schäden und Beschwerden waren offensichtlich schon seit Monaten bekannt – und die erstaunliche Sorglosigkeit beim Unterhalt der Brücke war sicher eine ganz wesentliche Unglücksursache. Schließlich war ja schon während der Bauzeit die Sache mit dem mangelhaften Guß der Pfeilerrohre den Ingenieuren nicht verborgen geblieben: Die Gußstücke, welche Hopkins & Co lieferten, waren schlechter als die des alten Hugenotten de Bergue; so man-

NORTH BRITISH RAILWAY COMPANY

M
No. 789.

EDINBURGH, 28th May 1878.

Notice to Enginemen, Guards, Signalmen, and others.

Opening of Tay Bridge for General Traffic.

In anticipation of the Opening for General Traffic, the following Regulations for the working of the Tay Bridge will be brought into force at 12 o'clock Noon on

WEDNESDAY, 29th MAY, 1878.

The Tay Bridge will be worked as one Section, under the combined systems of Block Telegraph and Train Staff and Train Ticket, as follows:—

I. By Block Telegraph.

All Trains and Engines must be signalled in strict accordance with the Revised General Instructions for Working Trains over Single Lines by Tyers' Telegraph Signal—dated 18th August 1873.

SPECIAL NOTE.—This Section is specially excepted from the operation of Clause 11 of the General Instructions above referred to, and therefore a Second Train or Engine *must not* be allowed to follow before the first Train or Engine has been signalled Clear.

II. By Train Staff and Train Ticket.

The Section will be worked by Train Staff and Train Ticket, in accordance with the Rules for working Single Lines of Railway by Train Staff and Train Ticket system, as contained in the Company's Book of Rules and Regulations, at pages 57 to 63 inclusive.

The Colour of Train Staff and Train Tickets will be Dark Blue, and the No. of the Train Staff is No. 4.

The Signalmen at Tay Bridge South and Tay Bridge North Cabins will be the custodiers of the Train Staff and Train Tickets. Before giving the Driver of a Train or Engine the Train Staff or Train Ticket, the Signalman at the one Cabin must be careful to signal on the Train by Block Telegraph to the other Cabin; and until such Signal has been given, and properly acknowledged, by the needle being turned to "Train on Line," the Signalman must not part with either the Train Ticket or Train Staff.

In the event of the failure of the Block Telegraph Instruments, an interval of 15 minutes must be maintained between all trains until the Block Telegraph is again in operation; and great caution must be exercised by all concerned.

Speed of Trains and Engines passing over Tay Bridge.

The Speed of Trains and Engines, when passing over the Tay Bridge, must never exceed the following limits:—

		Speed must not exceed
Passenger Trains and Light Engines	...	25 Miles an Hour.
Goods, Cattle, and Mineral Trains	...	20 " "
Ballast Trains	...	15 " "

Simultaneously with the commencement of the above arrangements, the mode of working the Tay Bridge by Red Cap Pilotman presently in operation will be discontinued.

J. WALKER,
General Manager.

(1000)

Acknowledgment to be signed, cut off, and returned to General Superintendent's Office, Edinburgh, by first Train.

Date ____________ 1878.

I hereby acknowledge to have received copies of Notice, M-No. 789, dated 28th May 1878, in reference to "Opening of Tay Bridge for General Traffic," to which I shall attend.

Signature ____________

Occupation ____________

Station ____________

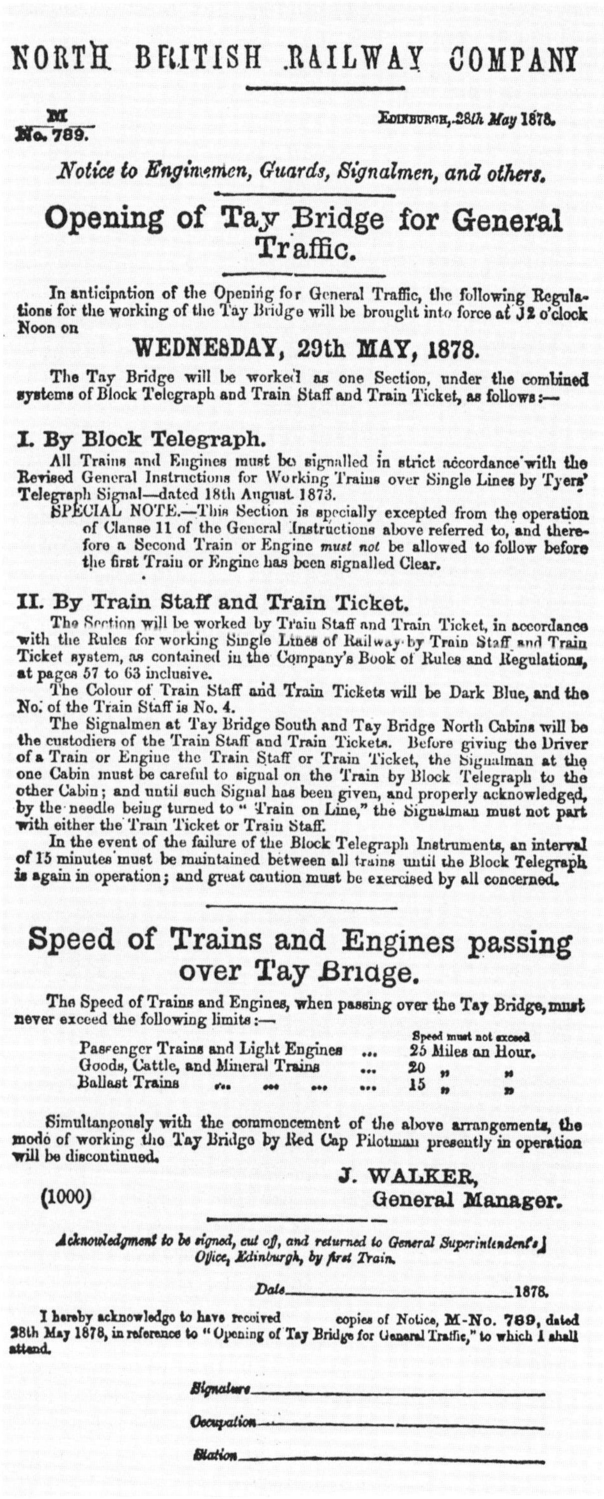

23
Official opening of Tay Bridge to general traffic, stipulating a speed limit of 25 m.p.h. (40 km/h) [26, p. 77].

23
Offizielle Freigabe der Tay-Brücke mit Geschwindigkeitsbegrenzung auf 25 Meilen/Stunde (40 km/h) [26, S. 77].

Denn mit scheinbar kleinen Annahmen bei zweifelhaften Punkten der Kalkulationen läßt sich fast alles ausrechnen, was man haben will. Es war nicht die mathematische Gewißheit, die ich ihm entgegenhalten konnte. Festigkeitskoeffizienten unsrer heutigen Materialien, Winddruckfragen — alles ist so unsicher, daß man mit zehnfacher oder zwanzigfacher oder dreifacher Sicherheit rechnen kann, je nach der Stimmung, ohne sehr fehlzugehen. Jedenfalls läßt sich nicht beweisen, daß man fehlgegangen ist. Ich fühlte nur, wie mich eine geheime Angst packte, die ich mit allem Rechnen nicht los wurde.

Lastannahmen [4, S. 444]

Aber schließlich beruht doch alles mögliche auf Annahmen, auf Theorien, die noch kein Mensch völlig durchschaut und die vielleicht in zehn Jahren wie ein Kartenhaus zusammenfallen. Ein Holzbalken mit seinen Fasern ist noch verhältnismäßig menschlich verstehbar. Aber weißt du, wie es einem Block Gußeisen zumute ist, ehe er bricht, wie und warum in seinem Innern die Kristalle aneinander hängen; ob ein hohles Rohr, das du biegst, auf der einen Seite zuerst reißt oder auf der andern vorher zusammenknickt, ehe es in Stücken am Boden liegt? Wie viel ich über Kohäsion nachgedacht habe, damals und später — namentlich später —, daß mir übel wurde von den ewig kreisenden Gedanken

Kohäsion [4, S. 472]

Ich war in den ersten Jahren nicht ängstlich und hatte keine Ursache dazu. Wenn die Ausführung sorgfältig überwacht wurde und alles streng nach den Plänen durchgeführt werden konnte, so durfte ich so ruhig sein als Bruce und alle Welt. Daß ich aufpaßte, als ob mein Leben daran hing, kann ich beschwören. Aber als die Senkkästen abgeändert und meine Pfeiler statt aus acht nur noch aus sechs Säulen aufgebaut werden mußten, fing ich wieder an zu rechnen. Es war aus mit meiner Ruhe.

Angst [4, S. 474]

culation in bridge construction. Instead, findings gained by experience were being used, which were themselves derived from "cases", frequently from accidents that had occurred in the past. The "Royal Charter Case" had shown that the strongest wind pressure ever to be expected in this country was about 21 lbs/sq ft [8, p. 31].

However, during critical phases of constructing the Tay Bridge, the load assumptions then in use – namely the actual calculations supporting the whole project – were being questioned repeatedly. And the question as to what was actually meant by "cohesion" was something no one yet knew how to answer in scientific terms. The following excerpts indicate the engineers' tenuous grasp of these matters:

Load assumptions [4, p. 444]	"... Because by seemingly small assumptions at doubtful points of calculation, almost anything can be calculated as desired. It wasn't the mathematical certitude which I could hold up to him. Coefficients of strength for today's materials, questions of wind pressure – everything is so uncertain, you may calculate with a tenfold or twentyfold safety-margin or even threefold, according to mood, without failing very much. In any case, it can not be proven that you went wrong. I only sensed some secret fear gripping me which I couldn't get rid of with all my calculations.
Cohesion [4, p. 472]	... But in the end, everything conceivable is based on assumptions, on theories not fully understood by anybody – perhaps in ten years they will all collapse like a set of cards. A wooden beam with its fibres still is relatively conceivable in human terms. But do you know how a block of cast iron really feels inside before it fails, how and why its crystals deep inside keep clinging unto each other? Whether a hollow tube you bend, first fractures, or buckles instead on the opposite side, before it lies broken to pieces on the ground? How much I have thought about cohesion, then and later on – especially later – till I got all sick from those ever-turning thoughts.
Fear [4, p. 474]	... During those first years, I was not fearful, nor had I cause for it. If only the execution was supervised carefully and everything was carried out strictly according to plan, I could rest as peacefully as Bruce (= Bouch) and the rest of the world. I can swear to having watched out, as if my life depended on that. But when the caissons were altered and my piers had to be built out of only six instead of eight columns, I started my calculations all over again. My peace of mind was gone."

ches Stück Säule mußte deshalb in die Gießerei zurückgehen. Das ganze Ausmaß dieser Pfuscharbeit wurde allerdings erst bei der öffentlichen Untersuchung deutlich.

Lastannahmen und «Guesswork»

Schließlich soll festgehalten werden, daß sich die leitenden Ingenieure inkl. Bouch schon während der Planung keineswegs sicher waren über die gültigen Lastannahmen. Es gab zu dieser Zeit noch keine *British Standard Specifications* – vergleichbar den DIN-Normen – welche etwa den zulässigen Winddruck im Brückenbau festhalten und Orientierungshilfen bei den Berechnungen hätten geben können. Stattdessen richtete man sich nach Erfahrungswerten, die ihrerseits wiederum von *«cases»*, also eingetretenen (Unglücks-)Fällen abgeleitet waren: Hatte doch erst der *«Royal Charter Case»* gezeigt, daß der stärkste Winddruck, den man je in diesem Land erwarten konnte, bei 21 Pfund/Quadratfuß lag [8, S. 31].

In kritischen Bauphasen der Tay-Brücke wurden allerdings die damals gebräuchlichen Lastannahmen – also die eigentlichen Berechnungsgrundlagen, auf denen das ganze Werk beruhte – immer wieder in Frage gestellt. Und die Frage, was überhaupt «Kohäsion» sei, wußte man wissenschaftlich noch nicht zu beantworten.

Der 28. Dezember 1879

Die englischen Quellen über die entscheidenden Stunden auf dem Firth of Tay sind reich an Personen und *«human interest stories»* aller Art. Allein schon die Erzählweise, etwa in Prebble's «High Girders» [20] oder in Thomas' «Tay Bridge Disaster» [26] ist ebenso anschaulich packend, wie sie für deutsche Leser manchmal «unwissenschaftlich» erscheinen mag. Will man aber überhaupt etwas von der Athmosphäre jenes Tages vermitteln – und damit vom harten und doch so gemächlich-selbstgewissen Lebensstil im victorianischen Schottland – ist es notwendig, wenigstens in diesem Abschnitt die wichtigsten Personen zu Worte kommen zu lassen. Die folgende Übersicht versucht, schon wegen der vielen ähnlich-lautenden Namen, etwas Übersicht in das historische Geschehen zu bringen. Dabei geht es hier um die unmittelbar an den Ereignissen Beteiligten – hauptsächlich Eisenbahner und Hafenleute, sowie um die wenigen Augenzeugen des Brückeneinsturzes:

December 28, 1879

British sources describing those decisive hours at the Firth of Tay are full of personal accounts and human-interest stories of all kinds. Even the style of reporting, such as in Prebble's *"High Girders"* [20] or in Thomas' *"Tay Bridge Disaster"* [26] is as captivating, as it may occasionally seem "un-scientific" to German readers. Yet if one wants to capture anything at all of the mood of that day – including the hard yet leisurely self-assured life-style of Victorian Scotland – it is necessary to include the tales of the major persons involved, at least in this section of the book. The following chart is an attempt to sort out the many (often similar) names, and bring some order into the historic account. Thus, the list focuses on those people directly involved – mostly railway and harbour personnel, as well as the few actual eye-witnesses of the disaster.

Personages present on the night of December 28, 1879, in the sequence of their appearance:

1) *Mr. Wright,* Captain of the Tay ferry *Dundee* (paddle-wheeled steamer)
2) *William Robertson,* Captain and harbourmaster at Dundee
3) *James Roberts,* foreman of the engine shed at Dundee Station
4) *James Smith,* stationmaster at Dundee Station
5) *Alexander Kennedy,* driver of the local train to Newport
6) *Robert Shand,* guard on the local train to Newport
7) *John Black,* employee at the parcels office of CR
8) *John Buik,* fitter with the CR
9) *David McBeth,* guard aboard the *Edinburgh*
10) *Donald Murray,* mail guard aboard the van of the *Edinburgh*
11) *David Mitchell,* driver of engine Nr. 224
12) *John Marshall,* stoker of engine Nr. 224
13) *Mr. Linskill,* the first-class passenger who still got off on time
14) *William Robertson,* carriage-inspector at Leuchars Station
15) *Thomas Robertson,* stationmaster at Leuchars Station
16) *Robert Morris,* stationmaster at St. Fort Station
17) *Alexander Ingles,* porter at St. Fort Station
18) *William Friend,* ticket collector at St. Fort Station
19) *George Ness,* engine cleaner ⎫
20) *David Johnston,* railway-man ⎬ as passengers aboard the *Edinburgh*
21) *David Scott,* railway-man ⎭
22) *Thomas Barclay,* signalman at the southern signal box at Wormit
23) *George Clark,* wine merchant in Dundee
24) *William Clark,* eyewitness Nr. 1 and brother of George Clark
25) *James Black Lawson,* eyewitness Nr. 2, Dundee
26) *Mr. Smart,* eyewitness Nr. 3, Dundee
27) *Alexander Maxwell,* eyewitness Nr. 4, son of Mr. Maxwell, Dundee

Handelnde Personen am Abend des 28. 12. 1879 in der Reihenfolge ihres Erscheinens:

1) *Mr. Wright,* Kapitän der Tay-Fähre «Dundee» (Raddampfer)
2) *William Robertson,* Kapitän und Hafenmeister von Dundee
3) *James Roberts,* Werkmeister des Lokomotivschuppens am Bahnhof von Dundee
4) *James Smith,* Stationsvorsteher am Bahnhof von Dundee
5) *Alexander Kennedy,* Zugführer des Lokalzuges nach Newport
6) *Robert Shand,* Schaffner des Lokalzuges   "   "
7) *John Black,* Angestellter am Paketschalter der CR
8) *John Buik,* Monteur bei der CR
9) *David McBeth,* Schaffner des «Edinburgh»-Zuges
10) *Donald Murray,* Postmann im Gepäckwagen des «Edinburgh»
11) *David Mitchell,* Lokführer der Unglücksmaschine Nr. 224
12) *John Marshall,* Heizer auf   "   "   "
13) *Mr. Linskill,* jener 1. Klasse-Passagier, der noch rechtzeitig ausstieg
14) *William Robertson,* Waggon-Inspektor in Leuchars Station
15) *Thomas Robertson,* Stationsvorsteher von   "   "
16) *Robert Morris,* Stationsvorsteher von St. Fort Station
17) *Alexander Ingles,* Träger auf   "   "   "
18) *William Friend,* Ticket-Einsammler in   "   "
19) *George Ness,* Lokomotivputzer ⎫
20) *David Johnston,* Eisenbahner ⎬ Als Passagiere auf dem Unglückszug
21) *David Scott,* Eisenbahner ⎭
22) *Thomas Barclay,* Signalmann auf der Südkabine der Tay-Brücke bei Wormit
23) *George Clark,* Weinhändler in Dundee
24) *William Clark,* Augenzeuge Nr. 1 und Bruder von George Clark
25) *James Black Lawson,* Augenzeuge Nr. 2, Dundee
26) *Mr. Smart,* Augenzeuge Nr. 3, Dundee
27) *Alexander Maxwell,* Augenzeuge Nr. 4, Sohn von Mr. Maxwell, Dundee
28) *William Millar,* Augenzeuge Nr. 5, Freund von Alexander Maxwell
29) *Peter Barron,* Augenzeuge Nr. 6, Waggon-Inspektor bei der CR
30) *John Wallace,* Augenzeuge Nr. 7, Fernhall und Edinburgh
31) *Henry Sommerville,* Signalmann auf der Nordkabine der Tay-Brücke, Dundee
32) *Provost Brownlie,* Bürgermeister von Dundee
33) *Ex-Provost Robertson,* Augenzeuge Nr. 8 und Altbürgermeister von Newport
34) *John Watt,* Augenzeuge Nr. 9 und Streckenmeister auf dem Abschnitt Leuchars-Tay-Bridge
35) *W. B. Thomson,* Augenzeuge Nr. 10, Ingenieur und Schiffbauer in Dundee

Der Sonntag des 28. Dezember 1879 begann so friedlich wie jeder andere Dezembertag am Tay vor Dundee. Das Barometer der *Dundee Meteorological*

28) *William Millar*, eyewitness Nr. 5, friend of Alexander Maxwell
29) *Peter Barron*, eyewitness Nr. 6, carriage inspector with CR
30) *John Wallace*, eyewitness Nr. 7, Fernhall and Edinburgh
31) *Henry Sommerville*, signal-man at the northern signal box near Dundee
32) *Provost Brownlie* of Dundee
33) *Ex-Provost Robertson* of Newport and eyewitness Nr. 8
34) *John Watt*, eyewitness Nr. 9 and foreman surfaceman responsible for the section from Leuchars to Tay Bridge
35) *W. B. Thomson*, eyewitness Nr. 10, engineer and shipbuilder in Dundee

The Sunday of December 28, 1879, began as peacefully as any other December day along the Tay near Dundee. The barometer of the Dundee Meteorological Society was standing at normal and nothing indicated anything unusual. When Captain Wright took his ferry *Dundee* across to Newport at 1:15 p.m., the weather was fair and the water calm. Only with dusk settling in during the 4:15 crossing did the wind freshen up. Captain Wright returned his paddlewheeler on time at 4:45 p.m. from Newport, to set out again at 5:15 p.m.

This was the great age of amateur meteorologists, and quite a few residents along the Tay that night had been recording the type and force of this exceptional storm. At his home in Scotscraig, a hill to the east of Newport, a retired Admiral had watched a small lunar eclipse with his telescope. Fierce rain drove him back into the house, and with the upcoming storm, the Admiral grew very concerned for all the men at sea that night. He suddenly noticed a sharp drop of the barometer and asked his servants whether they had touched it. By 5 p.m. the wind turned westward bringing along sleety weather, and the old salt estimated the wind at 75 to 78 miles (120–125 km/h). Aboard the training ship *Mars*, the captain also had noticed the alarming drop of the barometer and he measured a gale force of 10 to 11, almost the maximum. By 5:30 p.m. roofing tiles and chimney pots were clattering into the streets of Dundee.

According to Captain Wright, "the river was now getting up very fast". When the ferry finally arrived again in Dundee shortly past 6 p.m. the citizens knew that they were headed for one of those sudden and fierce storms along the Tay. The harbourmaster, Captain William Robertson, had gone down to his office shortly before 6 p.m. to check on everything *(Fig. 24)*. From his window he could barely make out the dark silhouette of the bridge, its navigation lights strung out like a thin line of pearls across the Firth. A crawling band of lights at rail level indicated the passage of the 5:50 local from Newport to Dundee.

Society stand normal und nichts deutete auf Ungewöhnliches hin. Als Kapitän Wright um 13.15 Uhr seine Fähre «Dundee» nach Newport übersetzte, war das Wetter gut und das Wasser ruhig. Erst mit einsetzender Dämmerung auf der 16.15er Überfahrt frischte es auf. Kapitän Wright brachte seinen Raddampfer um 16.45 Uhr planmäßig von Newport zurück, um dann noch einmal um 17.15 Uhr überzusetzen.

Es war das große Zeitalter der Amateur-Meteorologen, und so mancher Anwohner des Tay hatte an jenem Abend die Art und Geschwindigkeit dieses außergewöhnlichen Sturmes aufgezeichnet. Zuhause auf *Scotscraig*, einem Hügel östlich von Newport, hatte ein pensionierter Admiral mit seinem Fernrohr eine kleine Mond-Eklipse ausgemacht. Heftiger Regen hatte ihn dann ins Haus getrieben und mit aufziehendem Sturm wurde der Admiral sehr besorgt um alle Männer, die in einer solchen Nacht auf See sein mußten. Er bemerkte, wie plötzlich das Barometer scharf gefallen war – so tief, daß er seine Dienstboten fragte, ob sie es berührt hätten. Um 17.00 Uhr drehte der Wind auf West, fegte Graupelschauer einher und der alte Seebär schätzte die Geschwindigkeit auf 75 – 78 Meilen/Stunde (120–125 km/h). An Bord des Schulschiffes «Mars» hatte der Kapitän ebenfalls das alarmierende Fallen des Barometers bemerkt und bereits eine Windstärke von 10 bis 11, also fast das Maximum, gemessen. Ab 17.30 Uhr schepperten in Dundee Dachpfannen und Kaminköpfe auf die Straßen.

In den Worten von Kapitän Wright «erhob sich der Fluß jetzt sehr rasch». Als die Fähre endlich kurz nach 18.00 Uhr wieder in Dundee festmachte, wußten die Bürger, daß ihnen einer jener plötzlichen und heftigen Tay-Stürme ins Haus stand. Der Hafenmeister, Kapitän William Robertson, war noch kurz vor 18.00 Uhr zum Hafenamt *(Abb. 24)* gegangen, um nach dem Rechten zu sehen. Von seinem Fenster konnte er gerade noch die dunklen Umrisse der Brücke ausmachen – ihre Navigationslichter waren wie eine dünne Perlenschnur über den Firth gespannt; ein kriechendes Lichterband auf Schienenhöhe zeigte die Passage des 17.50er Lokalzuges von Newport nach Dundee an.

Weiter westlich am Lokomotivschuppen der *NBR*, beim Nordende der Brücke, wurde der Sturm bedrohlich. Der dortige Werkmeister James Roberts fürchtete schon, der Wind würde die großen Tore des Schuppens hinwegfegen, und schickte seine Leute, alles zu verbarrikadieren. Als aber der Sturm drei vollbeladene Kohlewaggons aus einem Nebengleis heraus und *bergauf* zum 400 Yard (366 m) entfernten Gütergelände trieb, hielt er den Zeitpunkt für gekommen, den Stationsvorsteher in dessen Haus aufzusuchen. Der neue Tay-Brücken-Bahnhof lag in einer Gleis-

Die Brück' am Tay

(28. Dezember 1879)

> When shall we three meet again?
> *Macbeth*

„Wann treffen wir drei wieder zusamm?"
   „Um die siebente Stund', am Brückendamm."
     „Am Mittelpfeiler."
          „Ich lösche die Flamm."
„Ich mit."

   „Ich komme vom Norden her."
„Und ich vom Süden."
        „Und ich vom Meer."

„Hei, das gibt einen Ringelreihn,
Und die Brücke muß in den Grund hinein."

„Und der Zug, der in die Brücke tritt
Um die siebente Stund'?"
        „Ei, der muß mit."
„Muß mit."

   „Tand, Tand,
ist das Gebilde von Menschenhand!"

Auf der *Norder*seite, das Brückenhaus –
Alle Fenster sehen nach Süden aus,
Und die Brücknersleut' ohne Rast und Ruh
Und in Bangen sehen nach Süden zu,
Sehen und warten, ob nicht ein Licht
Übers Wasser hin „Ich komme" spricht,
„Ich komme, trotz Nacht und Sturmesflug,
Ich, der Edinburger Zug."

Und der Brückner jetzt: „Ich seh' einen Schein
Am anderen Ufer, Das muß er sein.
Nun, Mutter, weg mit dem bangen Traum,
Unser Johnie kommt und will seinen Baum,
Und was noch am Baume von Lichtern ist,
Zünd' alles an wie zum heiligen Christ,
Der will heuer *zweimal* mit uns sein, –
Und in elf Minuten ist er herein."

Und es war der Zug. Am *Süder*thurm
Keucht er vorbei jetzt gegen den Sturm,
Und Johnie spricht: „Die Brücke noch!
Aber was thut es, wir zwingen es doch.
Ein fester Kessel, ein doppelter Dampf,
Die bleiben Sieger in solchem Kampf.
Und wie's auch rast und ringt und rennt
Wir kriegen es unter: das Element.

Und unser Stolz ist unsre Brück';
Ich lache, denk' ich an früher zurück,
An all den Jammer und all die Noth
Mit dem elend alten Schifferboot;
Wie manche liebe Christfestnacht
Hab ich im Fährhaus zugebracht
Und sah unsrer Fenster lichten Schein
Und zählte, und konnte nicht drüben sein."

Auf der Norderseite, das Brückenhaus –
Alle Fenster sehen nach Süden aus,
Und die Brücknersleut' ohne Rast und Ruh
Und in Bangen sehen nach Süden zu;
Denn wüthender wurde der Winde Spiel,
Und jetzt, als ob Feuer vom Himmel fiel',
Erglüht es in niederschießender Pracht
Überm Wasser unten . . . Und wieder ist Nacht.

„Wann treffen wir drei wieder zusamm?"
   „Um Mitternacht, am Bergeskamm."
     „Auf dem hohen Moor, am Erlenstamm."

„Ich komme."
   „Ich mit."
      „Ich nenn' euch die Zahl."
„Und ich die Namen."
     „Und ich die Qual."

„Hei!,
   Wie Splitter brach es entzwei."
     „Tand, Tand,
Ist das Gebilde von Menschenhand."

Theodor Fontane

24

The harbour of Dundee around 1878. In the centre, the office of Captain William Robertson, showing the harbourmaster's viewing platform on top [2, p. 55].

24

Der Hafen von Dundee um 1878. In der Bildmitte das Hafenamt von Kapitän William Robertson, mit der Beobachtungsstation des Hafenmeisters im Obergeschoß [2, S. 55].

Further to the west near the engine shed of the NBR, at the bridge's northern end, the storm became threatening indeed. The shed foreman there, James Roberts, feared the wind might blow off the large gates of the shed, so he had his men barricade everything. But when the gale started pushing three fully-loaded coal wagons out of a siding uphill to the goods yard some 400 yards (366 m) away, he decided the moment had come to seek out the stationmaster at his house.

The new Tay Bridge Station was situated at a curve in the rails running through a tunnel at the east end which passed underneath the dock yards. But the west side was wide open, resembling a funnel in which the wind entered unhindered. Entrapped in this cave-like space, the accumulating force of the wind burst through the glass roof. When stationmaster James Smith arrived at the scene, already a large section of the roof had been demolished with the rest ready to fly off at any moment. The platform and the steps leading up to Union Street Bridge were littered with broken glass. Smith ordered the station gates closed to prevent accidents.

Once more the 5:50 local was to cross the bridge back to Newport. The driver that night was Alexander Kennedy, a substitute, with Robert Shand as the guard. The train had left Newport on time, with two CR employees aboard the van, John Black of the parcels office and John Buik, a fitter. As the train picked up speed across the bridge, Black and Buik peered through the

kurve, an deren Ost-Ende die Bahn einen Tunnel unter dem Dock-Gelände durchfuhr. Die Westseite aber lag weit offen, so daß die Station einer Höhle glich, in die der Wind ungehindert einfiel: Da er aus diesem Trichter nirgendwo herauskonnte, sprengte er sich seinen Weg durch das Glasdach. Als Stationsvorsteher James Smith am Ort erscheint, ist schon ein erheblicher Teil des Daches zerstört und der ganze Rest droht jeden Augenblick wegzufliegen. Der Bahnsteig und die Treppen hinauf zur Union-Street-Brücke sind mit Glasscherben übersät; Smith läßt den Bahnhofseingang schließen, um möglichen Verletzungen vorzubeugen.

Nocheinmal sollte der Lokalzug über die Brücke gehen, zurück nach Newport um 17.50 Uhr. Der Zugführer an jenem Abend war Alexander Kennedy, ein Ersatzmann, mit Robert Shand als Schaffner. Der Zug hatte Newport zuvor pünktlich verlassen, mit zwei CR-Angestellten im Gepäckwagen: John Black vom Paketschalter, und John Buik, ein Monteur. Während der Zug auf der Brücke Fahrt aufnahm, spähten Black und Buik durch das «dookit»-Fenster an der Ostseite des Gepäckwagens. Dort hatten sie einen guten Blick voraus und spürten, wie das Gefährt von den heftigen Windstößen hin und her geworfen wurde. Feuer flog von den Rädern des Wagens voraus und sprang dann den ganzen Zug entlang. Buik wollte Alarm schlagen, aber Shand hatte schon frühere Nachtstürme auf der Brücke erlebt und wußte, daß die Funken von den Rad-

"dookit" window at the eastern side of the van. From here they had a good view ahead and felt how the train was being thrown to and fro by the fierce gales. Sparks started flying from the wheels of the carriage ahead, – finally jumping alongside the whole train. Buik was about to call the alarm, but Shand had witnessed night storms before on that bridge; he knew the sparks came from the flanges of the wheels being pushed sideways against the rails by the storm. As a precaution he gave a few turns at the handbrake and directed a red warning light ahead from the eastern window. But the driver saw nothing and only sensed a slowing down, as always when the train had to go against strong wind. "So I let her out a bit more", as he later told his mates at the platform of Dundee. Then suddenly the men in the van felt the mighty thrusts of the storm battering against the train.

Whenever the worst blows hit, the van would rock from side to side and seem to lean over. Many observers that night described the storm as one continuous roar, accentuated in short intervals by violent blasts. The storm was pushing and pressing against the bridge, pulling and straining at the piers; the utmost margin of safety had been reached. Indeed, the 5:50 local was the last train to pass the bridge – barely by a hair's breadth. As the guard recounted afterwards, not even 500 Pounds could have enticed him to go and try another crossing that night.

The Final Hours

At this time, the "express" from Edinburgh was still travelling far in the South of Fife. Actually, this was the same train which went as a mail train on Sundays at 1:20 p.m. from Dundee to Burntisland; from there it started its return trip only at 5:20 p.m. back to Dundee, usually arriving shortly before 7:30 at night. But the railwaymen used to call this train the *Edinburgh* as it was in fact a connecting train, even though it originated only from Burntisland. The connecting train from Edinburgh left Waverly Station at 4:15 p.m., called at Abbeyhill, Leith Walk and Trinity, getting into Granton on the Forth by 4:35 p.m. There the paddlewheel ferry *William Muir* took on the passengers, leaving for Burntisland at 4:43 p.m. The crossing normally took 34 minutes, allowing for a change-over time of just 3 minutes, till departure of the connecting train from Burntisland at 5:20 p.m.; this was tight timing, but in those days, before the Tay Bridge was built, it still meant a total travelling time from Edinburgh to Dundee of 3 hours and 12 minutes *(Fig. 25)*!

The so-called *Edinburgh,* ready for departure at Burntisland, had 5 carriages and a van. Behind the loco-

flanschen kamen, wenn der Wind die Waggons seitlich gegen die Schienen preßte. Er kurbelte vorsichtshalber die Handbremse etwas an und richtete eine rote Warnlampe vom Ostfenster nach vorne. Aber der Lokführer sah nichts, spürte nur die Verzögerung, wie jedesmal, wenn der Zug gegen starken Wind anging. «Also habe ich noch eins draufgegeben», wie er später seinen Kollegen am Bahnsteig von Dundee erzählte. Aber dann spürten die Männer im Gepäckwagen die gewaltigen Windstöße wie Schläge gegen den Zug. Der Wagen schaukelte von Seite zu Seite und schien sich jedesmal überzulehnen, wenn die schlimmsten Stöße trafen. Viele Beobachter dieser Nacht beschrieben den Sturm als ein fortdauerndes Brüllen, akzentuiert in kurzen Abständen von grausamen Windstößen. Der Sturm drückte und preßte auf die Brücke, zerrte und rüttelte an den Pfeilern: Die absolute Grenze der Standsicherheit war erreicht. Der 17.50er «*Local*» hatte als letzter Zug – und wirklich nur um Haaresbreite – die Brücke gerade noch überqueren können. Der Schaffner meinte hinterher, keine 500 Pfund würden ihn dazu verleiten, in dieser Nacht nocheinmal hinüberzufahren.

Die letzten Stunden

Um diese Zeit befand sich der «*Express*» aus Edinburgh noch weit im Süden der Grafschaft Fife. In Wahrheit handelte es sich hier um denselben Zug, der sonntags als Postzug um 13.20 Uhr von Dundee nach Burntisland fuhr; von dort trat er dann um 17.20 Uhr seine Rückfahrt nach Dundee an und pflegte kurz vor 19.30 Uhr einzutreffen. Aber die Eisenbahner nannten diesen Zug immer den «*Edinburgh*», da er zwar seine Reise erst in Burntisland aufnahm, aber eigentlich ein Anschlußzug war: Der vorgeschaltete Verbindungszug von Edinburgh verließ Waverly Station um 16.15 Uhr, hielt dann in Abbeyhill, Leith Walk und Trinity, bis er um 16.35 Uhr in Granton am Firth of Forth eintraf. Dort nahm das Fährschiff «William Muir», ein Raddampfer, die Passagiere auf und legte um 16.43 Uhr in Richtung Burntisland ab. Die Überfahrt dauerte normalerweise 34 Minuten – was bei der planmäßigen Abfahrt des Anschlußzuges von Burntisland um 17.20 Uhr eine Umsteigezeit von 3 Minuten erlaubte: Das war knapp kalkuliert, bedeutete aber damals, vor dem Bau der Forth-Brücke, immer noch eine Gesamtfahrzeit von Edinburgh nach Dundee von 3 Stunden und 12 Minuten *(Abb. 25)*!

Der sogenannte «*Edinburgh*» also, welcher in Burntisland bereitstand, hatte 5 Waggons und einen Gepäckwagen. Hinter der Lokomotive kam erst ein 3. Klasse-Wagen mit 5 Abteilen, gefolgt von einem 1. Klasse-

25

The railway network of the North British Railway, showing the line from Edinburgh to Dundee [15, p. 10].

25

Das Eisenbahnnetz der North British Railway mit der Strecke Edinburgh-Dundee [15, S. 10].

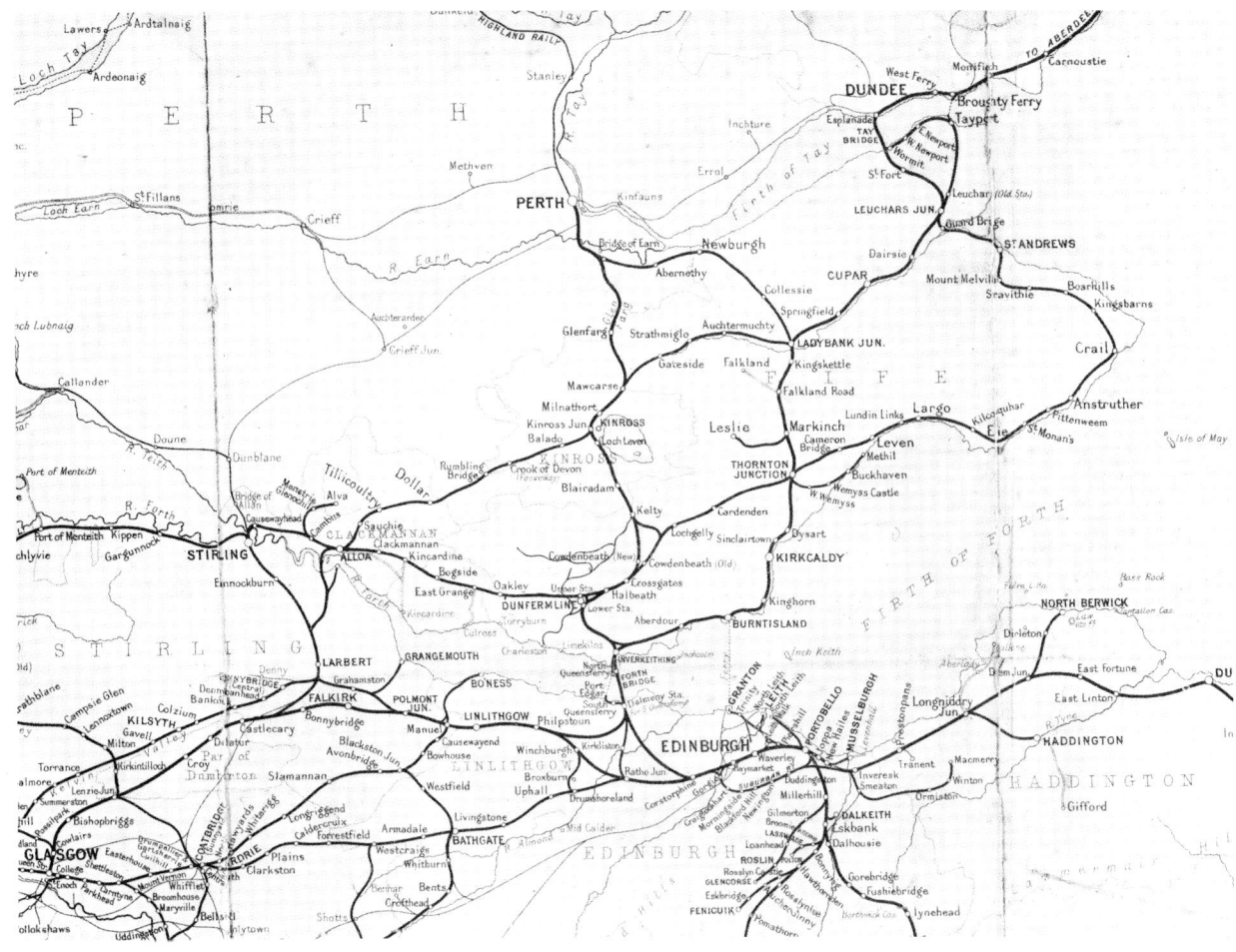

motive there was first a third-class carriage with 5 compartments, then a first-class carriage, also with 5 compartments but longer by 10 ft (3.05 m). With over 14 tons of weight, this was by far the heaviest carriage and the only one on 6 wheels. Then came two more third-class carriages and one second-class carriage. The second class was at the time about to be abandoned by the NBR and the carriage of that night was old with only 4 compartments; its approximately 5 tons made it the lightest vehicle of the train. In those days, railway carriages were still assembled in light-weight construction and were not very sturdy in accidents.

The van at the rear was in the hands of guard David McBeth and mail guard Donald Murray. The olive-green locomotive Nr. 224, one of the new Wheatley bogies of the type 4-4-0 *(Fig. 26)* had been built in Cowlairs in 1871. It was driven by David Mitchell, with the stoker John Marshall at his side. The train

Wagen, ebenfalls mit 5 Abteilen, aber um 10 Fuß (3,05 m) länger; mit über 14 Tonnen Gewicht war dies bei Weitem der schwerste Waggon und der einzige auf 6 Rädern. Es folgten 2 weitere 3. Klasse-Wagen und noch ein 2. Klasse-Wagen; die 2. Klasse war aber damals bei der *NBR* schon fast ausgemustert und der Wagen dieser Nacht war alt, mit nur 4 Abteilen; seine ca. 5 Tonnen machten ihn zum leichtesten Gefährt des Zuges: Zu jener Zeit waren Eisenbahnwaggons noch in der Leichtbauweise von «carriages» zusammengefügt und wenig haltbar bei Unfällen. Der Gepäckwagen am Ende stand in der Obhut des Schaffners David McBeth und des Postmannes Donald Murray. Die olivgrüne Lokomotive Nr. 224, eine der neuen Wheatley-Drehschemel-Lokomotiven vom «Typ 4-4-0» *(Abb. 26),* war 1871 in Cowlairs gebaut worden. Sie wurde geführt von David Mitchell, mit dem Heizer John Marshall an seiner Seite. Der Zug

had a total length of 225 ft and a total weight of a little over 114 tons.

After the loading of baggage and mailbags, and once the passengers had changed from the ferry to the train, it started on its last journey at 5:27 p.m., being 7 minutes late. This "express" was to stop at all 14 stations then existing between Burntisland and Dundee; it therefore was neither an express train, nor was it a real "Edinburgher". On its leisurely trip along the coast it first called at Kinghorn, Kirkcaldy, Sinclairtown and Dysart. At Thornton Junction, 29 miles (46 km) to the South of Dundee, the train was only slightly behind schedule. There followed stops at Markinch, Falkland Road, Kingskettle and Ladybank Junction, where passengers from the Perth line embarked.

The slow journey through Fife went on without the growing storm hampering the course significantly. It was almost 7 p.m. when the train arrived at Leuchars Junction, from where in those days a trunk-line went on to St. Andrews. There was no connection on Sundays, however, so one passenger bound for St. Andrews, a Mr. Linskill, had ordered a coach to pick him up at Leuchars. He had been one of only two first-class passengers on the "express". At first, no coach was anywhere in sight. Since Leuchars Station stood isolated in the clearing above the village, and with the storm growing fiercer all the time, Mr. Linskill climbed back into his compartment in order to spend the night in Dundee instead. Above the noise of the wind, the hammering sounds of carriage inspector William Robertson rang out as he tapped the wheels, walking along the train. The stationmaster was about to give his signal "Go", when he took one last glance down the road. Just then he saw the lights of a coach coming up – it was Mr. Linskill's vehicle. So the stationmaster hurried to the first-class carriage to help the St. Andrews passenger out with his baggage. Linskill would be the last person to leave the train alive.

The last stop before the bridge at 7:05 p.m. was St. Fort Station, where the tickets of Dundee passengers were collected. Stationmaster Robert Morris and porter Alexander Ingles helped the ticket collector William Friend with his work – in such a night they wanted to get the process over with in a hurry. When they hastened down the train opening doors, they noticed how unusually full it was for a Sunday night since New Year's Eve was approaching. Most of the passengers were nameless faces appearing briefly under the carriage lamps; also children were among them. But the St. Fort men did recognise a few friends travelling as passengers, among them George Ness, an engine cleaner who had married the daughter of John Brand,

hatte eine Gesamtlänge von 225 Fuß (68,58 m) und ein Gesamtgewicht von etwas über 114 Tonnen.

Nach dem Verladen von Gepäck, Postsäcken und dem Umsteigen der Passagiere von der Fähre auf den Zug, begann dieser um 17.27 Uhr, mit 7 Minuten Verspätung, seine letzte Reise. 14 Stationen gab es zwischen Burntisland und Dundee, und dieser «Express» sollte an allen Stationen haltmachen: Er war also weder ein Expreßzug noch ein echter «Edinburgher». In gemächlicher Fahrt entlang der Küste hielt er zunächst in Kinghorn, Kirkcaldy, Sinclairtown und Dysart. In Thornton Junction, 29 Meilen (46 km) südlich von Dundee, war er nur noch wenig hinter der fahrplanmäßigen Zeit zurück. Es folgten die Stationen Markinch, Falkland Road, Kingskettle und Ladybank Junction, wo Reisende von der Perth-Linie hinzukamen.

Die geruhsame Fahrt durch Fife verlief, ohne daß der hereinbrechende Sturm den Weg nennenswert beeinflußte. Es war fast 19.00 Uhr, als der Zug in Leuchars Junction hielt, von wo (damals noch) eine Nebenlinie nach St. Andrews führte. Sonntags gab es aber keine Verbindung, weshalb ein Mr. Linskill – er war einer von nur zwei 1. Klasse-Passagieren – sich eine Kutsche nach Leuchars bestellt hatte. Eine solche war aber weit und breit nicht zu sehen und da Leuchars Station isoliert auf offenem Feld oberhalb des Dorfes stand, auch der Sturm immer heftiger wurde, erklomm Mr. Linskill wieder sein Abteil in der Absicht, die Nacht in Dundee zu verbringen. Über dem Lärm des Sturmes konnte man die Hammerschläge des Waggon-Insepkteurs William Robertson hören, wie er den Zug abschritt und gegen die Räder schlug. Auf sein Zeichen zur Abfahrt warf der Stationsvorsteher, Thomas Robertson, einen letzten Blick die Straße hinunter und sah gerade die Lichter einer Droschke heraufkommen: Es war Mr. Linskill's Wagen. Der Stationsvorsteher rannte zum 1. Klasse-Waggon und half dem St. Andrews-Passagier mit seinem Gepäck aus dem Abteil: Er sollte der letzte sein, welcher den Zug lebend verließ.

Der letzte Halt vor der Brücke um 19.05 Uhr war St. Fort Station, wo die Tickets der Dundee-Reisenden eingesammelt wurden. Dort halfen der Stationsvorsteher Robert Morris und der Träger Alexander Ingles dem Fahrkartenschaffner William Friend bei der Arbeit – denn in einer solchen Nacht wollten sie die Sache rasch hinter sich bringen. Als sie den Zug entlanghasteten und die Abteiltüren aufrissen, sahen sie, wie ungewöhnlich voll für einen Sonntag der Zug an diesem Abend war – denn das Neujahrsfest stand ja bevor. Meist waren es nur namenlose Gesichter, die kurz im Lichte der Waggons auftauchten, auch Kinder waren darunter. Die St. Fort-Leute erkannten aber einige Kollegen, die als Passagiere reisten, darunter George

26
The locomotive Nr. 224 of the type 4-4-0 in better times [26, p. 104].

26
Die Lokomotive Nr. 224 vom Typ 4-4-0 in besseren Tagen [26, S. 104].

the driver on the Newport line. Ness had been promoted to fireman just that week. Friend also got brief glimpses of his mates Donald Murray, David Johnston and David Scott, all men quite familiar to him. With the clang of the last carriage door, Morris gave the green light and the train rolled on towards the bridge, just two miles ahead.

The narrow valley harbouring the rail line beyond St. Fort gave some shelter (see map, Fig. 27). But as soon as the train reached the shore near Wormit with the approaches to the bridge, the full force of the gale came roaring down. John Marshall, the stoker, leaning over the side of his cabin, spotted the signal box in the curve before the bridge (Fig. 28) and next to it signalman Thomas Barclay holding up the baton for the bridge. On the one-track line across the bridge, only the driver carrying the baton had the right of way. The stoker grabbed the baton, calling out a greeting in that instant as the engine was abreast with the signal box. Now the train was curving towards the bridge, its broadside fully exposed to the storm. The moon was covered by clouds, but the night was clear. Through the portholes of the cabin – to the left the driver, to the right the stoker at his stand – the one-track line could be seen ahead, leading thin and straight up to the high deck of the bridge. Over a mile beyond, the lights of Dundee were spackling: So near and yet unreachably far this community of travellers, thrown together by chance, was now suspended above the water, left to the mercy of the elements. The bridge was trembling, as a hollow rumbling and hissing of

Ness, einen Lokomotivputzer, der die Tochter von John Brand, dem Zugführer der Newport-Linie, geheiratet hatte. Ness war erst in dieser Woche zum Heizer befördert worden. Friend erhaschte auch einen flüchtigen Blick seiner Kollegen Donald Murray, David Johnston und David Scott, alles Männer, die er gut kannte. Mit dem Schlag der letzten Abteiltür gab Morris grünes Licht und der Zug rollte der Brücke entgegen – 2 Meilen voraus.

Das enge Tal, durch das die Strecke hinter St. Fort führt, bot einigen Schutz (vergl. Situationskarte, Abb. 27). Aber sobald der Zug vor Wormit das Flußufer mit dem Brückenanfahrtsweg erreichte, traf ihn die volle Gewalt des Orkans. John Marshall, der Heizer, lehnte sich über die Seite der Kabine, sah das Signalhäuschen an der Biegung vor der Brücke stehen (Abb. 28) und daneben den Signalmann Thomas Barclay, wie er den Stafettenstab für die Brücke hochhielt. Denn auf der eingleisigen Brückenstrecke hatte jeweils nur der Zugführer im Besitze des Stabes freie Fahrt. Der Heizer schnappte sich den Stab und schrie noch einen Gruß hinaus in jenem kurzen Moment, als Nr. 224 gleichauf mit dem Signalhaus war. So kurvte der Zug auf die Brücke zu, seine Breitseite gegen den Sturm gewandt. Der Mond war hinter Wolken verborgen, die Nacht aber klar. Durch die Bullaugen im Führerhaus – links der Lokführer, rechts der Heizer auf seinem Stand – ließ sich die eingleisige Strecke voraus überblicken, wie sie schmal und gerade auf das hohe Brückendeck zuführte. Über eine Meile weiter vorne funkelten die Lichter von Dundee: Nahe und doch unerreichbar ferne hing jetzt die Zufallsgemeinschaft der Reisenden über dem Wasser, dem Spiel der Elemente ausgeliefert. Die Brücke bebte, während wie gewöhnlich ein hohles Rumpeln und das Zischen der vorbeilaufenden Fachwerkstreben die Einfahrt in die Hochträger signalisierten: Es waren die letzten Augenblicke.

Wie man später im Zeitvergleich feststellte, hatte der Orkan jetzt, ab 19.00 Uhr, seine vollste Stärke von 145 km/h erreicht, was einer Windstärke von 10 bis 11 auf der Beaufort-Skala entspricht. Mit großer Wahrscheinlichkeit hatten die rhythmischen Stöße des Sturmes begonnen, eine heftige Schaukelbewegung zwischen den beiden Dreiecksverbänden der Hochpfeiler auszulösen. Die Haltezapfen sprangen vermutlich als erste von den Säulen weg, die Verstrebungen schwangen frei und die Bolzen scherten von den Flanschen der Säulenverbindungen. Das zusätzliche Gewicht des Zuges war zuviel für das schwankende Bauwerk. Als die Lokomotive das Nordende des 5. Mittelträgers fast erreicht hatte, begannen die Pfeiler über ihren unteren Verbindungen einzuknicken. Die Hochträger neigten

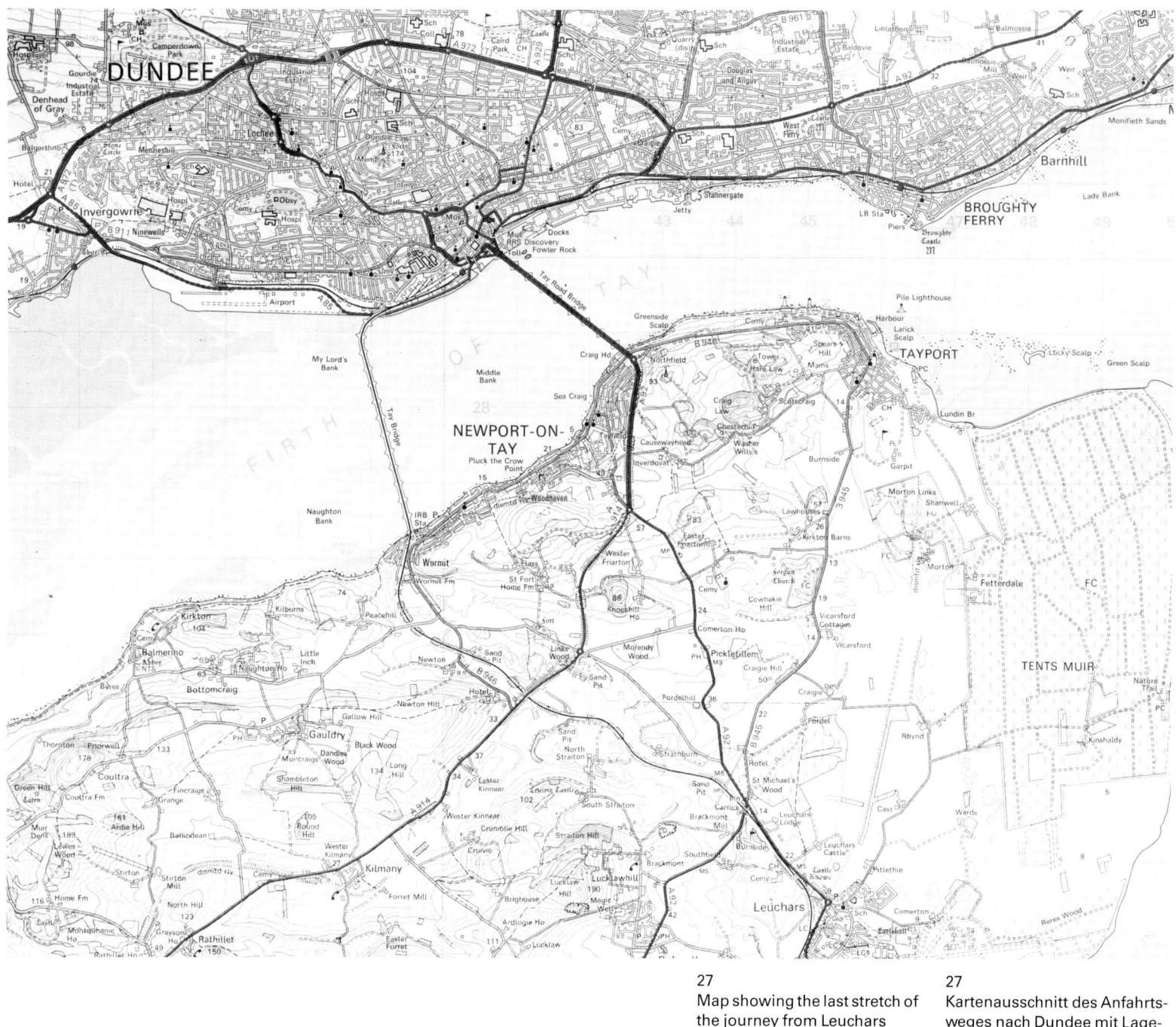

27
Map showing the last stretch of the journey from Leuchars Junction to Dundee, with location of the Tay Bridge [16].

27
Kartenausschnitt des Anfahrtsweges nach Dundee mit Lageplan der Tay-Brücke [16].

28
The southern curve leading into the Tay Bridge, in a photograph by G. W. Wilson during the summer of 1879. At the far right, the trunk line going off to Newport (abandoned today) can be seen, and positioned inside the fork; the old signal box of Thomas Barclay to the right. In front, the transition from the double line near Wormit to the then single line across the bridge [3, p. 16].

28
Die Südkurve der Tay-Brücke in einer Aufnahme von G. W. Wilson im Sommer 1879. Ganz rechts erkennt man die soeben fertiggestellte Abzweigung nach Tayport (heute aufgelassen), in der Gabelung davor die Signalkabine von Thomas Barclay. Vorne ist der Übergang der zweigleisigen Strecke von Wormit auf die eingleisige Brücke erkennbar [3, S. 16].

sich nach Osten, zuerst langsam, dann immer schneller. Im letzten Moment brachen die Träger weg von den zusammenfallenden Pfeilern und stürzten gegen die Wasseroberfläche, mit dem Zug in ihrer Mitte. Obwohl sich die Träger seitlich übergelegt hatten, blieben die Waggons während des Sturzes aufrecht stehen. Als die Träger auf der Oberfläche prallten, brach das Wasser durch die offenen Verstrebungen und drückte die leichten Wagen nach oben, so daß sie mit den Dächern gegen die westseitige Trägerwand stießen. Innerhalb weniger Augenblicke stürzten so alle 13 Hochträger mit Hunderten von Tonnen Eisen samt dem Zug in den Tay: Die Passagiere hatten keine Chance.

passing trusses signalled, as usual, the train's entry into the High Girders. These were the last moments. As would later be proven by time checks, the storm had now, by 7 p.m., reached its fullest force of 90 m.p.h. (145 km/h) corresponding to Force 10 to 11 on the Beaufort Scale. Most probably, the rhythmic blasts of the storm had triggered a fierce rocking motion, between the two sets of triangular bracings of the tall iron piers. Most likely, the holding lugs tore off the columns first, then the bracings swang loose and the bolts sheared from the flanges of the columns. The added weight of the train was too much for the swaying structure. When the engine had almost reached the northern end of the fifth high girder, the iron piers began to buckle upon their lower joints. The High Girders tilted over to the east, slowly at first, then faster. In the last moment, they broke loose from the collapsing piers and hurtled toward the surface of the water, still holding the train in their midst. Although the girders had turned over unto their side, the carriages remained standing upright during the fall. When the girders hit the surface, the water broke through the open bracings and pushed the light carriages upward, so they crushed with their roofs against the western wall of the girders. Thus, within a few seconds, all 13 of the High Girders with hundreds of tons of iron crashed into the Tay, holding the train inside. The passengers never had a chance.

Eyewitnesses

It is one of the peculiarities of this disaster that, while there were several eyewitnesses, almost none of them were able to describe exactly afterwards how the collapse had really taken place. It seemed as if most witnesses in those decisive moments were either distracted or otherwise hindered from actually comprehending this event as a disaster – or they were not willing to believe, in fact, what they saw. Far too unreal must have appeared this sight in the exceptional circumstances of that night. For the citizens of Dundee, it had become a pleasant pastime to gather behind the windows of dark rooms and to watch the night trains with their fairy-like lights crawling across the spidery bridge. The wide Firth with its elevated vantage points at both banks of the Tay offered ample opportunity for that. And as difficult as it had been for Dundee to get used to the bridge at first, so now its citizenry had grown proud, in those 2 years and 3 months, of this structure. But on this 28th of December, the citizens were either at evening mass or busy barricading their houses against the storm. At the moment when the bridge tumbled into the water, the

Augenzeugen

Es ist eine der Besonderheiten dieser Katastrophe, daß es zwar mehrere Augenzeugen gab, daß aber hinterher fast keiner von ihnen genau beschreiben konnte, wie der Einsturz wirklich vonstatten gegangen war. Ja es schien, daß die meisten Zeugen in den entscheidenden Sekunden entweder abgelenkt oder sonstwie daran gehindert waren, den Vorgang wirklich als einen Einsturz wahrzunehmen – bzw. nicht glauben wollten, was sie sahen. Allzu unwirklich muß das Geschehen in der Ausnahmesituation jener Nacht gewirkt haben. Dabei war es für die Bürger von Dundee ein beliebter Zeitvertreib geworden, sich an den Fenstern dunkler Räume einzufinden und die Nachtzüge mit ihren feenhaften Lichtern auf dem Weg über die spindelige Brücke zu verfolgen. Die Weite der Bucht mit ihren erhöhten Aussichtspunkten zu beiden Seiten des Firth of Tay bot dazu reichlich Gelegenheit. Und so lange es zunächst für Dundee gedauert hatte, sich an die Brücke zu gewöhnen, so stolz war man in den 2 Jahren und 3 Monaten ihrer Existenz auf dieses Bauwerk geworden. Aber an jenem 28. Dezember waren die Bürger beim Abendgottesdienst oder damit beschäftigt, ihre Häuser gegen den Sturm dicht zu machen. Im Augenblick, als die Brücke ins Wasser stürzte, war die Gewalt des Orkans solcherart, daß kaum eine der 140 000 Seelen der Stadt das Krachen des Eisengerippes hörte. Nur eine Handvoll Zeugen an beiden Ufern sah wirklich die Brücke fallen.

Zeuge 1: Das Haus von George Clark, einem Weinhändler, stand bei Magdalen Green am Nordufer, nicht weit von der nördlichen Signalkabine, mit bestem Blick auf die Brücke – und zwar gegenüber von Magdalen Point. Die Sichtentfernung bis zum Nordende der Brücke betrug 140 Yard (128 m) *(Abb. 29)*. Wenn man aus den Vorderfenstern des Hauses blickte, erschien die Brücke etwa 10 Grad nach rechts, so daß man ihre Ostflanke in spitzem Winkel entlangsehen konnte. – Clark's Bruder William war gerade zu Besuch und um 19.00 Uhr hielten beide Männer Ausschau nach dem Zug, wenngleich von benachbarten Zimmern. George Clark hatte das chinesische Meer besegelt und wußte, was ein Orkan war. Er hatte noch erwähnt, wie sehr ihn diese Nacht an einen Hurrikan erinnerte. Nun war er besonders daran interessiert, dessen Wirkung auf den Zug zu beobachten. Als die Lichter zu flackern begannen, wußte er, daß der Zug in die Hochträger eingefahren war. Gerade da wandte er den Kopf ab, hörte aber fast zeitgleich seinen Bruder vom nächsten Zimmer ausrufen: «Schau das Feuer – der Zug ist über die Brücke!» George Clark wandte

29
Magdalen Green around 1905. In former times, this open green extended from Dundee's city edge down to the beach along the Tay. In 1845, the construction of the Dundee & Perth Railway (behind the music pavilion at the shoreline to the right) interrupted this natural link of the green with the seaside. Today, the motorway A 85 dissects this area [2, p. 78].

29
Magdalen Green um 1905: Früher reichte diese offene Grünfläche vom Stadtrand Dundee's bis zum Badestrand am Tay hinunter. 1845 wurde mit dem Bau der Dundee & Perth Railway (hinter dem Musikpavillon am Ufer rechts) die natürliche Verbindung zwischen «Green» und Wasser unterbrochen. Heute führt die Schnellstraße A 85 durch dieses Gebiet [2, S. 78].

force of the gale was such that almost none of the 140,000 inhabitants noticed the crashing of the iron structure. Only a handful of eyewitnesses on either shore actually saw the bridge fall.

Eyewitness Nr. 1: The house of George Clark, a wine merchant, was situated at Magdalen Green in Dundee, not far from the northern signal box, offering an excellent view of the bridge from opposite Magdalen Point. The viewing distance to the bridge's near end was 140 yards (128 m) (Fig. 29). When one looked towards the bridge from the front windows of the house, it appeared about 10° to the right, such that its eastern side could be seen at an oblique angle. Clark's brother William was visiting, and around 7 p.m. both of them were looking out for the train, although from adjoining rooms. George Clark had sailed the China Seas and recognised a hurricane when he saw one. He had mentioned how much this night reminded him of a cyclone. Now he was especially anxious to observe its effect on the train. When the lights began to flicker, he understood that the train was entering the High Girders. At that moment he turned around, hearing his brother call out from the next room almost simultaneously: "Look at the fire – the train is over the bridge!« George Clark quickly turned to the window again, but there was nothing to be seen; the bridge was just a faint shadow in the dark, without lights. William Clark recounted how he had followed the course of the train, up to a point which he thought to be the third pier of the central girders, when suddenly

sich rasch wieder dem Fenster zu, aber es war nichts mehr zu sehen: Die Brücke war nur noch ein Schatten in der Finsternis, ohne Lichter. William Clark erzählte, wie er den Weg des Zuges verfolgt habe bis zu einem Punkt, den er für den 3. Pfeiler der Mittelträger hielt, als auf einmal drei Lichtblitze erschienen, rasch hintereinander, und in schiefem Winkel vom Schienendeck auf der Ostseite der Brücke ins Wasser zu fallen schienen.

Zeugen 2 und 3: James Black Lawson wohnte etwa 300 Yard (274 m) westlich von George Clark's Haus, knapp westlich der Brücke. Um 19.00 Uhr ging er aus, um die Wirkung des Sturmes auf die Brücke zu beobachten. Er schritt in östlicher Richtung bis zu einem Punkt nahe dem Hause Clark und traf dort einen Bekannten, einen Mr. Smart. Die beiden Männer spähten zur Brücke hinüber nach einem Anzeichen des Zuges. Der Wind peitschte die Wasseroberfläche zu Schaum und Teppiche von Gischt fegten über das Nordende der Brücke. Alsbald erschienen die Lichter des Zuges, aber wegen des steilen Blickwinkels schienen sie Lawson und Smart nur sehr langsam voranzukommen. Schwankend näherten sich die Lichter der Brückenmitte, als sie urplötzlich verschwanden und etwas, das wie ein Feuerball aussah, von der Brücke ins Wasser fiel. «Da ist der Zug im Wasser!» schrie Smart. Und Lawson erzählte später der «Times»: «Es war wie ein kometenhafter Ausbruch wilder Funken, von der Lokomotive gewaltsam in die Finsternis geschleudert. In einer langen Spur war der Feuerstrahl zu sehen, bis zu

three brilliant flashes of light appeared in quick succession; they seemed to fall at an angle into the water from the rail-deck at the eastern side of the bridge.

Eyewitnesses Nr. 2 and 3: James Black Lawson lived some 300 yards (274 m) to the west of George Clark's home, slightly west of the bridge. At 7 p.m. he went outside to observe the effect of the storm on the bridge. He walked in an easterly direction to a spot near Clark's house and met an aquaintance there, a Mr. Smart. Both of them peered across to the bridge for any sign of the train. The wind was beating the surface of the water into foam and carpets of spray were engulfing the bridge's north end. Soon the lights of the train did appear, but due to the steep angle of vision they seemed to make headway very slowly. The swaying lights approached the center of the bridge when they suddenly disappeared and something that looked like a ball of fire fell from the bridge into the water. "There is the train in the river", shouted Smart. Later, Lawson was to tell to the *Times:* "It was like a comet's outburst of wild sparks, being thrown forcefully from the engine into the darkness. In a long trail the fiery streak could be seen until it vanished down into the stormy sea. Then there was total darkness on the bridge" [25, p. 231]. Both men ran across the CR yards to Buckingham Point, seeking shelter underneath the bank-side girders. They believed they had seen a steam fountain out there near the center spans, but then concluded that it may well have been rising spray (compare also eyewitnesses Nr. 8 and 10!).

Eyewitnesses Nr. 4 and 5: Not far from Magdalen Green stood the house of the Maxwells. Young Alexander Maxwell had invited his friends, among them William Millar, to watch the train crossing the long bridge – always a fascinating spectacle. Eventually, shortly past 7 p.m., the lights of the train appeared as it turned unto the bridge from the far side, like an uncertain glimmer in the distance. Alexander noticed the signal lamp flicker to the north end of the central spans, and he thought the lamp might have set the casing afire. When he turned his eyes again to the approaching train as it entered the high girders – he noticed one, two, three big flashes of fire, lasting just long enough to allow one to recognise the iron structure. But he got the impression that the flashes occurred in front of the train, and he reaffirmed this later on. William Millar thought the flashes originated from the train itself, and he joked that the stoker must be in a great hurry to get home since he was already extinguishing his fire by throwing the coals into the water. It was difficult to make out any details – ragged clouds were racing

seinem Verlöschen unten in der stürmischen See. Dann herrschte auf der Brücke völlige Finsternis.» [25, S. 231] Die Zeugen rannten über das *CR*-Gelände und zum Buckingham Point, wo sie Schutz unter einem der landseitigen Brückenträger suchten. Sie meinten, eine Dampf-Fontäne draußen bei den Hochträgern zu sehen, aber kamen überein, daß es wohl auch aufsprühende Gischt gewesen sein konnte (vergl. hierzu Zeugen 8 und 10!).

Zeugen 4 und 5: Nicht weit entfernt und ebenfalls bei Magdalen Green stand das Haus der Maxwells. Der junge Alexander Maxwell hatte seine Freunde, darunter auch William Millar, eingeladen; gemeinsam wollten sie den Zug über die lange Brücke fahren sehen – immer wieder ein besonderes Schauspiel. Endlich, kurz nach 19.00 Uhr, sahen sie von jenseits die Lichter des Zuges auf die Brücke einbiegen, wie ein fernes ungewisses Glimmern. Alexander fiel auf, daß dabei die Signallampe nördlich (also diesseits) der Hochträger flackerte und dachte schon, daß die Lampe vielleicht ihr Gehäuse zum Brennen gebracht hatte. Doch dann wandte er den Blick wieder auf den näherkommenden Zug, wie er in die Mittelträger einfuhr – und da gewahrte er eins, zwei, drei große Lichtblitze: Sie dauerten gerade lange genug, das eiserne Gitterwerk zu erkennen. Er hatte aber den Eindruck, daß sich die drei Blitze *vor* dem Zug ereigneten, und blieb auch später dabei. William Millar glaubte, daß die Blitze vom Zug selbst kamen, und scherzte noch, daß der Heizer in großer Eile sein mußte, heimzukommen, denn er würde bereits sein Feuer löschen und die Kohlen in den Fluß werfen. Es war schwer, Genaueres zu erkennen, denn Wolkenfetzen jagten über den Mond, bis die Szene wieder klar im fahlen Licht erschien. Und da sah er einen weißen Strahl vor der Brücke aufflammen. «Die lassen jetzt Dampf ab», rief er – und dann, mit veränderter Stimme: «Die Brücke ist runter!» Doch keiner wollte ihm glauben und der alte Mr. Maxwell, der hinzugetreten war, verlangte nach dem Fernrohr. Der Mond kam für einige Minuten heraus und Mr. Maxwell sah jetzt deutlich eine lange Lücke von 1000 Yard (914 m) in der Brücke und die schwarzen Pfeilerstümpfe aus dem Wasser ragen.

Zeuge 6: Peter Barron war Waggon-Inspekteur bei *CR* und wohnte in Balgay Lodge an Blackness Road. Als gegen 19.00 Uhr ein Kaminkopf von seinem Hause fiel, kam er nachzusehen, und schloß die Läden. Er überquerte die Straße, um nach der Brücke und dem Zug auszuschauen. Der Sturm war jetzt so stark geworden, daß Barron sich an zwei Torpfosten festklammern mußte, um nicht weggeweht zu werden. Der Firth war

30
This vantage of the Tay Bridge may be close to that of eyewitness Peter Barron. The bridge was rebuilt after the disaster at the same spot by H. Barlow – with the central portion in altered form, as it appears today.

30
Das heutige Blickfeld auf die Tay-Brücke entspricht in etwa dem Standort des Augenzeugen Peter Barron. Die Brücke wurde nach dem Einsturz an gleicher Stelle von H. Barlow wiederaufgebaut – der Mittelteil bietet sich dem Betrachter heute in dieser Form.

across the moon – until the scene cleared up again in the pale light. And then he observed a white steamy beacon rising up in front of the bridge. "They are blowing steam now," he called out – and then with an altered voice, "The bridge is down!" But nobody wanted to believe him; old Mr. Maxwell came up and asked for his telescope. The moon came out for a few minutes and Mr. Maxwell saw clearly now a long gap of 1,000 yards (914 m) in the bridge. He also saw the dark stumps of the piers jutting up above the water.

Eyewitness Nr. 6: Peter Barron, a carriage inspector with CR, lived in Balgay Lodge on Blackness Road. When a chimney pot fell from his roof, he went outside to check and close the shutters. He crossed the street to look at the bridge and the train. The storm had now become so strong that Barron was forced to get a firm grip on two gateposts, so as not to be blown away. The river was dark, but the end of the bridge could be seen 200 yards (183 m) to the west. Since Barron was watching it at an angle from high ground, he had a better vantage point than the other eyewitnesses *(Fig. 30)*. Unlike their rather imprecise observations, his were far clearer – or at least he claimed later on to have seen all quite clearly: He watched the train crawl onward from the far end toward the central section; then he thought he saw part of the bridge – what seemed to him the first or second of the central spans – fall into the river. Nervously he rubbed his eyes and when he looked again after a second or two, he saw a second portion of the bridge fall; simultaneously with the fall of the southernmost span he noticed a light flashing out in the water only to disappear immediately thereafter. For a long time he had his eyes fixed at that point. When the moon came out, Barron spotted against the glittering silver of the Firth a long gap where the central spans were supposed to be. He also noticed the large stumps of the piers, clearly lined up from end to end.

Eyewitness Nr. 7: John Wallace of Edinburgh was an advanced 88 years of age in 1956, but blessed with a long memory. In a newspaper interview[8] that year he recalled vividly how he stood as an eleven-year-old boy with his father at the garden gate of their home at Fernhall. His father was coachman to the family who had an estate on the shores of the Tay. After the two had taken care of the horses, as they did every night, John Wallace was watching the train snake along the coastline and then swing in an arc toward the bridge.

8 *Scottish Sunday Express,* Nov. 25, 1956: "The Long Memory of Mr. Wallace"

dunkel, aber man konnte den Anfang der Brücke 200 Yard (183 m) nach Westen ausmachen. Da Barron die Brücke in schiefem Winkel und von weiter oben beobachtete, hatte er ein besseres Blickfeld als die anderen Augenzeugen *(Abb. 30)*. Nach deren eher vagen Beobachtungen sah er genauer, was geschah – oder sagte zumindest, es genau gesehen zu haben: Er sah den Zug von jenseits gegen die Mittelträger herankriechen; dann vermeinte er, einen Teil der Brücke, was ihm wie der 1. oder 2. Mittelträger schien, ins Wasser fallen zu sehen. Voller Aufregung rieb er sich die Augen, und als er wieder hinblickte, ein, zwei Sekunden später, sah er einen zweiten Teil der Brücke fallen; gleichzeitig mit dem Fall des südlichsten Hochträgers hatte er ein Licht aufblitzen und gleich wieder verlöschen gesehen. Lange hielt er seine Augen auf diesen Punkt gebannt. Der Mond kam heraus und gegen das schimmernde Silber des Wassers gewahrte Barron eine lange Lücke, da wo die Mittelträger hätten sein sollen. Dafür erkannte er die großen Pfeilerfüße, deutlich aufgereiht von einem Ende zum anderen.

Zeuge 7: Der im Jahre 1956 bereits 88jährige John Wallace aus Edinburgh hatte ein langes Gedächtnis. In einem Zeitungsinterview[8] erinnerte er sich lebhaft, damals als Elfjähriger mit seinem Vater am Gartentor ihres Hauses in Fernhall gestanden zu haben. Sein Vater war dort Kutscher bei einer Familie, die ihren Ansitz an den Ufern des Tay hatte. Als die beiden wie jeden Abend die Pferde versorgt hatten, sah John Wallace den Zug sich an der Küste entlangschlängeln und dann im Bogen auf die Brücke herausfahren. Die winzigen Lichter krochen unstet bis zur Mitte – dann war es nur noch dunkel, gefolgt von einzelnen Lichtblitzen, und über das Sturmgeheul hinweg erscholl das Krachen berstender Eisenmassen. Erst am nächsten Tag kam mit dem Postboten die endgültige Nachricht der Katastrophe; auf seinem Schulweg entlang der Küste bemerkte Wallace herangespülte Wrackteile des Zuges. Später verfolgte er die Hebung der Träger und des Zuges; ganze Waggons kamen hoch und wurden am Ufer abgestellt. Als Kinder spielten sie in den Wagen, wo die unglücklichen Passagiere gesessen waren.

Inzwischen wartete Stationsvorsteher James Smith voller Ungeduld an der Tay Bridge Station auf die Ankunft des «Edinburgh». Ein Blick auf die Signale zeigte ihm, daß sie für den Zug ausgelöst worden waren, d. h. er mußte die Signalkabine am Südende der Brücke passiert haben. Als 10 Minuten herum waren und noch immer kein Zug erschien, wurde Smith unruhig.

8 Scottish Sunday Express, 25. 11. 1956: «The Long Memory of Mr. Wallace».

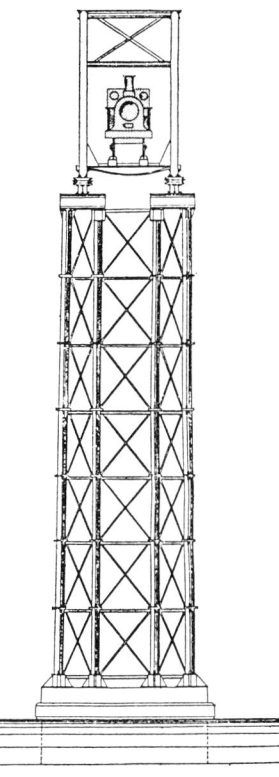

31
This cross-section of the Tay Bridge illustrates well the very slender profile of Bouch's design, showing the position of the locomotive within the High Girders [26, p. 88].

31
Dieser Querschnitt der Tay-Brücke mit Position einer Lokomotive im Mittelträger macht den hohen Schlankheitsgrad der Bouch'schen Konstruktion deutlich [26, S. 88].

The tiny lights edged out fitfully halfway across the bridge – then there was only darkness, followed by several flashes of light. Above the roaring storm came the awesome thunder of bursting iron, as the giant girders toppled. Only the next day did the postman bring definite news of the disaster; on his walk to school along the shore, young Wallace noticed bits of wreckage from the train washed ashore. Later he watched the lifting of the girders and the train; whole carriages came to the surface and were deposited on the shore. As children they played in the compartments where the unlucky passengers had sat.

Meanwhile, stationmaster James Smith was waiting impatiently at Tay Bridge Station for the arrival of the *Edinburgh*. Looking at the signals he knew that they had been cleared for the train, so it must have passed the signal box at the bridge's southern end. But when there was still no sign of the train after 10 minutes had gone by, Smith grew uneasy. Although he did have a telegraph machine at his office, it was not manned on Sundays. On the platform he saw a crowd of railwaymen and citizens waiting anxiously. Upon his question as to whether anybody could work the machine, Shand, the guard of the local to Newport, tried to get through to the southern box. But the machine was dead. Smith reasoned that the wires must have been ripped off by the storm. All he could do now was to wait for the arrival of the train.

By then, George Clark had joined James Lawson at Magdalen Point. After comparing impressions, they became certain that they had witnessed the collapse of the bridge. Clark had gotten his telescope but it was useless in the gushing spray. So they both ran down the esplanade to find out more at the northern signal box. They tried to get the attention of the signalman in his box, sitting high up on the landward side of the bridge, but nobody could hear their calls over the howling gales. When they were already halfway up the steps, Henry Sommerville the signalman, came towards them. But in the horrible roar no talk was possible and the three sought shelter under the bridge. "Where is the train?" Lawson called. "It's been a long time on the bridge", replied Sommerville. That was enough for the two – as fast as they could they ran back to the station. There Smith and Shand had just come out the office after their futile attempts at the telegraph. "The train is in the river", shouted Lawson and told what he had observed near Magdalen Point and heard just now from Sommerville. The stationmaster was not convinced that the train had gone down, but he indicated it might have been halted in time, or may have gone back to the south bank. Smith warned the two not to tell anybody, and ordered his

Er hatte zwar einen Telegraphenapparat in seinem Büro, aber der war sonntags nicht besetzt. Auf dem Bahnsteig sah er eine Schar von Eisenbahnern und Bürgern in angstvoller Erwartung: Auf seine Frage, ob jemand das Gerät bedienen könne, versuchte Shand, der Schaffner des Lokalzuges, zur Südkabine durchzukommen. Aber der Apparat blieb tot. Smith folgerte, daß die Drähte im Sturm gerissen sein mußten; er konnte nichts tun, als die Ankunft des Zuges abzuwarten.

Inzwischen hatte sich George Clark zu Lawson bei Magdalen Point gesellt. Nach dem Vergleich ihrer Eindrücke waren sie sicher, den Einsturz der Brücke mitangesehen zu haben. Clark hatte sein Fernglas geholt, aber es war nutzlos wegen der sprühenden Gischt. So rannten beide die Esplanade entlang zur nördlichen Signalkabine, um vielleicht dort Näheres zu erfahren. Sie versuchten, sich dem Signalmann bemerkbar zu machen, dessen Kabine hoch auf dem landseitigen Teil der Brücke stand; aber ihre Rufe gingen im Sturmgetöse unter. Als sie schon halbwegs die Stufen hinaufgeklettert waren, kam ihnen Henry Sommerville, der Signalmann, entgegen. In dem schrecklichen Lärm war aber kein Gespräch möglich und zu dritt suchten sie Schutz unterm Brückenbogen. «Wo ist der Zug?» schrie Lawson. «Er war schon lange Zeit auf der Brücke gewesen», erwiderte Sommerville. Das war genug für die beiden – so schnell sie konnten, rannten sie zur Station. Dort traten gerade Smith und Shand aus dem Büro, nach ihren vergeblichen Versuchen mit dem Telegraphen. «Der Zug ist im Fluß», schrie Lawson und berichtete, was er am Magdalen Point gesehen und soeben von Sommerville gehört hatte. Der Stationsvorsteher war nicht überzeugt, daß der Zug verloren war, sondern deutete an, er könnte vielleicht rechtzeitig angehalten haben, vielleicht zurück auf die Südseite gefahren sein. Smith warnte die beiden, niemand etwas zu sagen, ließ seine Leute den Bahnhof räumen und alle Tore schließen.

Lawson eilte zum Hafenamt, Clark zum Fährmeister und Smith suchte nach James Roberts. Er fand ihn im Lokomotivschuppen und Roberts war ebenso besorgt wie er. Denn Sommerville war schon dagewesen mit der bösen Nachricht, daß zwei Bürger den Einsturz gemeldet hätten. Inzwischen hatte der «Edinburgh» schon vor 33 Minuten die Südkabine passiert und es gab keine Spur von ihm. An der Nordkabine hatte Sommerville das Signal «Zug fährt in Abschnitt ein» um 19.14 Uhr von Barclay erhalten und somit erwartet, daß der Zug um 19.19 Uhr seine Kabine passieren würde – denn 5 Minuten dauerte die Überfahrt der Brücke. Als um 19.23 Uhr noch immer kein Zug ankam, stieg Sommerville zur obersten Stufe seiner Kabinen-

32
Contemporary illustration from the northern deck of the Tay Bridge after the disaster of December 28, 1879 [17].

32
Zeitgenössische «artist's impression» vom Deck der Tay-Brücke nach dem Einsturz am 28. Dezember 1879 [17].

men to clear the station of people and to have all gates closed.
Lawson hurried to the harbourmaster's office, Clark to see the ferry superintendent and Smith went to look for James Roberts. He found him in the engine shed and Roberts was just as concerned as he was, for Sommerville had already been there with the awful news that two citizens had brought about the collapse. Meanwhile the *Edinburgh* was now 33 minutes past the southern box and there was no sign of her. At the northern box, Sommerville had received the signal "train entering section" at 7:14 p.m. from Barclay and he therefore expected the train to pass his box by 7:19 p.m. since the crossing of the bridge normally took 5 minutes. But when at 7:23 p.m. there was still no trace of the train, Sommerville climbed up to the last step of his signal box to look across the river.

treppe und spähte über den Firth – von dem Standpunkt mußte er die Lichter des Zuges sehen können; aber es war nichts dergleichen auszumachen. Er versuchte, Barclay über Telefon und Telegraphen zu erreichen, aber beides funktionierte nicht – es war einfach unmöglich, mit der Fife-Küste Verbindungen aufzunehmen. So blieb nur noch eine letzte Möglichkeit für Smith und Roberts: Sie mußten selbst hinaus auf die Brücke und nachsehen, was passiert war. Niemand hätte einen solchen Gang in jener Nacht freiwillig unternommen, außer im äußersten Notfall. Also machten sie sich auf den Weg über die Brücke, ungewiß, was sie finden würden. Das Bauwerk lag in völliger Dunkelheit da, es gab keinen Schutz, der Wind jaulte durch die Streben und zerrte an der Kleidung. Tief unten, unsichtbar im Dunkel, hämmerten die Wellen gegen die Pfeilerfüße. Stöße von Gischt spritzten durch

The lights of the train should have been visible from up there, but nothing of the like was to be seen. He tried to reach Barclay by telephone and telegraph, but neither would work – it was simply impossible to make contact with the Fife coast. There was only one last possibility for Smith and Roberts: they had to go out on the bridge themselves to see what had happened. Nobody would have undertaken such a walk voluntarily that night, unless in a real emergency. They started out across the bridge, uncertain as to what they would find. The structure stood in total darkness and there was no shelter; the wind was howling die offenen Träger und durchnäßten die Männer. Langsam und hintereinander setzten sie Schritt vor Schritt, die Hand fest am Geländer. Oft mußten sie anhalten und Atem schöpfen; die Windstöße zwangen sie in die Knie und auf das hölzerne Brückendeck. Die Strapaze war zuviel für Smith; Schwindel überkam ihn und er blieb ans Geländer geklammert zurück, während Roberts sich weiterhangelte.

Kriechend fand sich Roberts auf dem letzten der niedrigen Träger. Jetzt hätte der käfigartige Rahmen des ersten Hochträgers gegen den Himmel erscheinen müssen; aber davon war nichts zu sehen – nur noch Leere.

Max Eyth describes the same scene from the other end of the bridge to the south [4, pp. 488, 489]:

Now the hard-ringing bang of a piece of iron could be heard from afar – now once again. This was strange, unnatural. I stopped and listened, but I only heard the whistling of the wind and the dull humming hiss of the sea under my feet. Onwards!
... We seemed to get closer to the central portion of the bridge. If I had counted correctly, the 26th pier was now behind us. I remembered that from the 27th onwards, the rails were running inside of the high girders, instead of along their upper flanges, as previously. My hopes were growing that everything might still turn out well. Besides, the storm had subsided quickly since the last ten minutes. The black clouds above us showed some rents and light-brown edges. I began to regain my spirits.
... Suddenly one could look quite far in all directions. It felt like standing in the center of a magic bowl, deep down below us the foamy sea in a dusky circle, around us firm and clear the rails, the sleepers, the handrail, and in front of us, suddenly and sharply severed, the end of the bridge jutting out into the empty void.

I proceeded another twenty steps almost without thinking, just following a painful urge driving me on. Then I clutched the railing with both my hands again, looking out into the misty blue, where just two hours ago the giant tunnel-like girders had started. They had disappeared, blown off without a trace.
I looked around with the utmost effort of all my nerves. Far, far in the distance, one could see the bridge again, its end coming up from the North bank of the bay and rising high up from the water, like a slender vertical pole. Between that end and ours, extended an empty gap almost 1 km wide, over which the heaving sea was surging across in untamed force and freedom. Only a row of white dots indicated the course of the former bridge at water level. It was the surf gushing up at the remains of the piers that had vanished. I counted them mechanically, without thinking. Twelve! I knew this was the number of the large piers supporting the higher portion of the bridge. If I was dreaming, then I dreamt it with horrible consistency: So it must have come about. The whole length of the high girders had collapsed.

Die gleiche Szene schildert der Zeitzeuge Max Eyth vom gegenüberliegenden Brückenende im Süden [4, S. 488, 489]

Jetzt hörte man aus weiter Ferne den hartklingenden Schlag eines Eisens — jetzt wieder. Dies war unerklärlich, unnatürlich. Ich hielt an und lauschte, hörte dann aber nur das Pfeifen des Windes und das dumpfe, summende Zischen des Wassers unter meinen Füßen. Weiter!

Wir mußten uns dem mittleren Teil der Brücke nähern. Wenn ich richtig gezählt hatte, lag der sechsundzwanzigste Pfeiler hinter uns. Ich erinnerte mich, daß vom siebenundzwanzigsten an das Bahngleise innerhalb der höher liegenden Gitterbalken läuft, anstatt, wie bisher, auf der oberen Flansche derselben. Meine Hoffnung stieg, daß sich noch alles zum Guten wenden müsse. Auch hatte seit den letzten zehn Minuten der Sturm rasch nachgelassen. Das schwarze Gewölke über uns zeigte Risse und lichtbraune Ränder. Ich fing an aufzuatmen.

Man sah mit einem Male ziemlich weit nach allen Seiten. Es war, als stünde man in der Mitte einer Zauberkugel, tief unter uns in einem dämmerigen Kreis die schaumbedeckte See, um uns bestimmt und klar die Schienen, die Schwellen, das Geländer, vor uns, plötzlich scharf abgeschnitten, das Ende der Brücke, das ins leere Nichts hinausragte.

Ich ging noch zwanzig Schritte vorwärts, fast ohne zu denken, einem qualvollen Drange folgend, der mich weitertrieb. Dann klammerte ich mich wieder mit beiden Händen ans Geländer und sah in das dunstige Blau hinaus, wo noch vor zwei Stunden die riesigen, tunnelartigen Gitterbalken begonnen hatten. Sie waren verschwunden, spurlos weggeblasen.

Ich sah um mich mit der gespanntesten Anstrengung aller Nerven. In weiter, weiter Ferne sah man die Brücke wieder, das Ende, das vom Nordufer der Bucht kam, wie einen schlanken, senkrechten Pfahl, der hoch aus dem Wasser emporragte. Zwischen diesem Ende und dem unsern war eine leere Strecke, fast einen Kilometer breit, über die in ungestörter Kraft und Freiheit das heraufstürmende Meer hinwogte. Nur eine Reihe weißer Punkte bezeichnete über die Wasserfläche weg die Linie der einstigen Brücke. Es war die Brandung, die an den Resten der verschwundenen Pfeiler aufschäumte. Ich zählte sie mechanisch, ohne zu denken. Zwölf! Ich wußte, dies war die Zahl der großen Pfeiler, auf denen der höhere Teil der Brücke geruht hatte. Wenn ich träumte, so träumte ich mit entsetzlicher Folgerichtigkeit. So mußte es gekommen sein. Die ganze Länge der hochliegenden Gitterbalken war eingestürzt.

through the bracings and tore at their clothing. Deep down below, invisible in the dark, the waves were hammering against the piers. Blasts of spray were spattering through the open structure, drenching the men. Slowly and one behind the other they set their steps, holding on tightly to the railing. Frequently they had to stop to catch their breath, as the blasts forced them to their knees on the wooden deck. For Smith it was too much of a trial; dizziness overcame him and he stayed behind clutching the rail, as Roberts went on.

Crawling forward, Roberts found himself along the last of the low girders. Now the cage-like frame of the first high girder should have appeared against the sky; but no such thing was to be seen – only emptiness. He crawled another eight, nine yards toward the end of the girder – and then he saw quite clearly the huge gaping hole *(Fig. 32):* none of the High Girders were left standing, the bridge was abruptly cut off! He noticed the rails torn off – not twisted, almost straight, but bent downwards – and next to them he saw a jet of water gushing forth from the broken end of a water main which used to cross the bridge to Newport; it now shed a long spray of water down into the river. The low tide had set in and the swirling current was disturbed by something submerged in the water. Roberts took all this in, with a single glance. So shocked was he by the spectacle that he shuddered away from the abyss and hastened back from the bridge. "I did not stay many seconds", he later remarked: "Directly I saw it I turned round and came back and had not time to see anything." He found Smith where he had left him and the stationmaster asked anxiously for any sign of the train. Roberts thought he had spotted a red light at rail level on the other end of the gap. Now both of them hoped against all reason that the light might belong to the train, which may perhaps have stopped just short of the gap. When they were off the bridge at last, Roberts headed for the harbourmaster to fetch a boat, while Smith went back to the station to finally telegraph the horrible news to the Edinburgh head office of the NBR, via the post office wire still functioning.

Harbourmaster Robertson had already heard rumors about the disaster and hurried to the harbour office, where a large telescope was directed eastward to keep watch on inbound shipping. He moved the instrument to a westerly window and began to carefully check the bridge. In a slow arc he followed the outline of the bridge from the northern end to a point where it was interrupted by the mast of the sailing vessel *Pladda* – beyond which there was nothing to be seen. In the same way he traced the contour of the bridge from the south shore, until in the centre of the Tay, the silhou-

Roberts kroch bis auf sieben, acht Meter an das Trägerende heran – und da sah er, ganz deutlich, die gähnende Lücke *(Abb. 32):* Keiner von den Mittelträgern war übrig geblieben, die Brücke war jäh zu Ende! Er bemerkte die abgerissenen Schienen – nicht verdreht, fast gerade, aber nach unten gebogen – und daneben einen Wasserstrahl aus dem geborstenen Ende einer Leitung schießen, die über die Brücke nach Newport führte und nun einen langen Sprühregen in den Fluß ergoß. Es war inzwischen Ebbe geworden und die wirbelnde Strömung brach sich da unten an irgendetwas Untergetauchtem im Wasser. All das erfaßte Roberts mit einem einzigen Blick. Er war so entsetzt von dem Schauspiel, daß er vor dem Abgrund schauderte und die Brücke zurückhastete. «Ich bin nicht viele Sekunden geblieben», erzählte er später: «Sobald ich das sah, drehte ich mich um und hatte keine Zeit mehr, noch irgendetwas aufzunehmen.» Er traf Smith, wo er ihn zurückgelassen hatte. Der Stationsvorsteher fragte voller Angst nach irgendeinem Zeichen des Zuges. Roberts meinte, ein rotes Licht etwa auf Schienenhöhe am fernen Ende der Lücke gesehen zu haben. Beide Männer hofften gegen jede Hoffnung, daß das Licht zum Zug gehörte, der vielleicht rechtzeitig vor dem Abgrund gestoppt hatte. Als sie endlich die Brücke hinter sich hatten, ging Roberts zum Hafenmeister, um ein Boot aufzutreiben, während Smith zum Bahnhof zurückkehrte, um über die Postleitung die schreckliche Nachricht an die *NBR*-Hauptverwaltung in Edinburgh zu telegraphieren.

Hafenmeister Robertson hatte bereits Gerüchte über den Einsturz gehört und eilte ins Hafenamt, wo ein großes Teleskop nach Osten postiert war zur Überwachung des einlaufenden Schiffsverkehrs. Er versetzte es an ein Westfenster und begann, die Brücke sorgfältig abzusuchen. In langsamem Bogen folgte er der Brückenlinie vom Nordende bis zu einem Punkt, der vom Masten des Segelschiffes «Pladda» unterbrochen wurde – darüber hinaus war nichts mehr zu sehen. In gleicher Weise zeichnete er den Brückenverlauf von der Südküste aus nach, bis in der Mitte des Tay die dunkle Silhouette abrupt endete. Sogleich wollte er ein Boot zur Unglücksstelle nehmen, aber da der Hafenschlepper auf Strand lag und nicht vor der nächsten Flut in 5 Stunden flott zu machen war, rannte er zum Büro der Tay-Fähre. Dort erfuhr er, daß die «Dundee» soeben abgelegt hatte, um die reguläre Überfahrt nach Newport zu versuchen. Inzwischen war es 20.00 Uhr und er konnte nur auf ihre Rückkehr warten *(Abb. 33).*

Am Bahnhof hatten sich Scharen bestürzter Menschen eingefunden und verlangten Nachricht über Verwandte, die auf dem Zug sein sollten. Kurz nach

33
Following her inauguration in 1875, the paddlewheeler *Dundee* continued crossing the Tay for 42 years, until 1917 [2, p. 17].

33
Die Raddampfer-Fähre *Dundee* besorgte seit ihrer Indienststellung 1875 die Tay-Überfahrt noch 42 Jahre lang, bis 1917 [2, S. 17].

ette ended abruptly. He intended to take a boat out to the scene of the accident, but the harbour tug lay grounded and could not be set afloat before the next flood in five hours, so he ran to the office of the Tay Ferries. But there he was told that the *Dundee* had just left to try a regular crossing to Newport. Meanwhile it was already 8 p.m. and he could only wait for her return *(Fig. 33)*.

At the station, crowds of distraught people had gathered, demanding news of their relatives who were said to be on the train. Then shortly past 9 p.m., a boat from the training ship *Mars* came over. Her sailors had been closer to the place of the accident than anybody else – but the oarsmen themselves wanted to find out more details. Indeed, the deck watch of the *Mars* had clearly seen the train travelling across the bridge, as the lights going by were extended in a long line, crossing the direction of sight in a right angle. But again, all was as if jinxed: he had just seen the lights enter the High Girders, when all at once an exceptionally violent gale from the west had compelled him to turn his back to the structure – most likely, this was the very same blast which brought down the bridge. When the watch looked up the Firth again just seconds later, the lights had all vanished – instead there was a very long gap in the center of the bridge. Aside from the roar of the howling storm, there was nothing to be heard. He gave the alarm signal and with great effort a boat was rowed

21.00 Uhr kam ein Boot zum Schulschiff «Mars» herüber, dessen Besatzung näher als irgendjemand sonst am Unglücksort gewesen war – aber die Bootsleute wollten selbst Genaueres erfahren: Die Deckwache der «Mars» hatte deutlich den Zug auf der Brücke entlangfahren sehen, da die wandernden Lichter in langer Reihe auseinandergezogen waren und die Blickachse rechtwinklig kreuzten. Aber wieder war es wie verhext: Der Posten sah eben noch die Lichter in die Mittelträger einfahren, als ihn plötzlich ein besonders heftiger Windstoß aus Westen zwang, der Brücke kurz seinen Rücken zu drehen – vermutlich war das dieselbe Orkanböe, die den Einsturz auslöste: Als der Posten Sekunden später wieder den Firth hinaufblickte, waren die Lichter verschwunden – dafür gab es eine lange Lücke in der Brückenmitte. Außer dem Geheul des Sturmes war nichts zu hören. Der Schiffer schlug Alarm und mit großer Mühe wurde ein Boot zur Brücke gerudert; aber die Sucher fanden nichts. Das war schlimme Nachricht für die Wartenden. Denn wenn der Zug wirklich auf der Brücke angehalten und ans Fife-Ufer zurückgefahren war, hätte die Deckwache die Lichter sehen müssen. Jetzt mußte man annehmen, daß der Zug untergegangen war. – So schnell die Katastrophe abgelaufen war, so unheimlich war das lange Warten auf Gewißheit, so bruchstückhaft kamen auch Stunden danach die spärlichen Nachrichten. Selbst nach Rückkehr der «Dundee» von Newport gegen 22.00 Uhr wußte der Kapitän nichts

out to the bridge; but the searchers found nothing at all.

This was bitter news for the people waiting. If indeed the train had halted in the nick of time and gone back to the south shore, the watch of the *Mars* would certainly have spotted the lights. Now it had to be assumed that the train had gone down with the bridge. – As fast as the disaster had occurred, so dismal was the long wait for certainty, for scanty news came in only piece-meal fashion over the next hours. Even upon the return of the *Dundee* from Newport around 10 p.m., the captain had nothing further to report. Strong gales and the outgoing tide had forced his ferry to a wide detour across the sea, three times the normal distance; so it had been impossible to get any closer to the scene.

Provost Brownlie of Dundee was himself at the pier and unwilling to give up hope for survivors. He had blankets and provisions sent to the *Dundee* before she set out to sea again – first up to the middle of the Tay and then westward to the big gap, as close as possible. Volunteers rowed a boat under the last standing girders of the north end, to get to the west side of the scene. They avoided the east side in order not to collide with submerged girders. The severed ends of the rails hung high above in mid-air and the water main was still gushing its silvery veil into the pale light – nobody had thought as yet to shut off the water to Newport.

The boat travelled the whole extent of the long gap. From pier 29 to 41 there was no more bridge. The only thing to be seen were the bare masonry bases of the piers. Only from two of them, the remains of torn iron columns were hanging down in an angle (compare *Fig. 1 and Fig. 35*). The surface of the river was strewn with broken planks from the bridge deck, but there was no sign of survivors.

Still, a few ominous traces did appear elsewhere. Downstream at Broughty Ferry, mail bags were being washed up along the beach; the Postmistress there sent a telegram to the Dundee post office, asking what this meant...

Eyewitness Nr. 8: Even along the Fife coast, there had been a few eyewitnesses. Ex-Provost Robertson had been peering out the window of his Newport home shortly past 7 p.m. He was looking out for the train that night as he believed his son to be aboard. He spotted the train setting out unto the bridge, noticed the blinking lights as they entered the central spans. But once again the vision was blocked in the decisive moment, as the neighbour's house obscured his view at that point. To the right, Robertson could see the bridge go

Neues zu berichten. Starke Winde und Strömung hatten die Fähre zu einem weiten Umweg über See, dreimal soweit wie gewöhnlich, gezwungen, so daß man der Brücke nicht näherkommen konnte.

Provost Brownlie, Bürgermeister von Dundee, war selbst am Pier und wollte die Hoffnung auf Überlebende nicht aufgeben. Er ließ Decken und Erfrischungen an Deck der «Dundee» bringen, bevor sie erneut hinausfuhr – zunächst bis zur Mitte des Tay und dann stromaufwärts genau auf die große Lücke zu, so nahe es ging. Freiwillige ruderten ein Boot unter dem letzten der Nord-Träger hindurch, um an die Westseite der Unfallstelle zu gelangen. Sie mieden die Ostseite wegen der möglichen Kollision mit abgestürzten Trägern. Hoch oben hingen die abgebogenen Enden der Schienen in der Luft und die Wasserleitung nach Newport sprühte immer noch ihren Silberschleier in das fahle Mondlicht – keiner hatte noch daran gedacht, das Wasser abzudrehen.

Das Boot folgte der ganzen Länge der Lücke: Von Stütze 29 bis 41 gab es keine Brücke mehr. Das Einzige, was man sehen konnte, waren die nackten Steinstümpfe der Pfeilerfundamente. Nur von zweien hingen noch Reste der eisernen Säulen schräg abgerissen herunter (vergl. *Abb. 1 und 35*). Die Wasseroberfläche war übersät mit Planken vom Brückendeck; aber es gab keine Spur von den Opfern.

Und dennoch waren anderswo schon ein paar ominöse Zeichen aufgetaucht: Stromabwärts bei Broughty Ferry waren auf einmal Postsäcke an Land gespült worden. Die *«Postmistress»* dort schickte ein Telegramm an's Postamt in Dundee, was es denn damit auf sich habe...

Zeuge 8: Auch an der Fife-Küste hatte es einige Zeugen gegeben: Ex-Provost (Altbürgermeister) Robertson blickte kurz nach 19.00 Uhr aus dem Fenster seines Hauses in Newport. Er spähte nach dem Abendzug aus, weil er seinen Sohn darauf vermutete. Er sah den Zug auf die Brücke hinausfahren, sah auch das Blinken der Lichter, sobald sie in die Mittelträger einfuhren. Aber wieder wurde der Blick im entscheidenden Moment unterbrochen, denn ein Nachbarhaus versperrte die Aussicht an ebendieser Stelle. Rechts davon konnte Robertson die Brücke weiterlaufen sehen und erwartete das Auftauchen des Zuges: Doch der kam nicht zum Vorschein. Während Robertson noch hinschaute, sah er deutlich zwei Säulen von Gischt hoch über die Brücke aufsteigen (vergl. auch Zeuge 10) und war sich sicher, die Navigationslichter der Brückenträger fallen und dabei aufleuchten zu sehen wie durch einen Dunstschleier. Dann lag das Bauwerk in völligem Dunkel.

on and he waited for the train to reappear. Yet it failed to come. While he was still watching he clearly saw two columns of white spray rise high above the structure (compare also eyewitness Nr. 10) and he was sure that he had seen the position lights of the girders tumble, still shining through the foamy spray. Then the structure was engulfed in total darkness.

Here, we must return once more to Thomas Barclay, the signalman in his southern signal box *(Fig. 28)*. On Sundays, he was not on continuous duty, but only at certain times when he had to let through the occasional train. It was not a bad job since his home was nearby. On this night, John Watt, the foreman surfaceman in charge of the section from Leuchars Junction to Tay Bridge – and himself **eyewitness Nr. 9** – was staying with Barclay at his house. When it was time to go to the box and to set the signal for the *Edinburgh*, Watt came along. By now the storm was at its peak – a continuous horrid howl underlined by bursts of incredible force. The men were relieved to find shelter in the box. The stove had burned low since the last train and Barclay started to rekindle the fire. Just then the *Edinburgh* was "offered" from St. Fort. (In British railway parlance this sounds somewhat more endearing than the German "gemeldet" ["reported"]; perhaps in those days, before the advent of automatic signals, one was inclined to deal with trains in almost human terms.) Barclay "accepted the train" and "offered it on" to his mate Sommerville at the northern box. Only then he pulled the signals, letting the train pass on to the bridge, and already he spotted its front light coming around the Peacehill bend (see *Fig. 27*). He hurried down the steps and held up the baton for stoker John Marshall to fetch as the engine was going by. The lit carriages rumbled by picking up speed, and when the brake van was swallowed up by darkness, the rattling of the train merged with the roar of the storm.

Barclay turned to his fire again, while John Watt stood by the window as many times before, his eyes trailing the train going up the bridge. Even though the signal box was shaking under the gales, neither Watt nor Barclay were much concerned about the bridge. By now the *Edinburgh* was about 200 yards (183 m) from the south curve and sparks began to streak off the wheels. Watt saw the train continue its fiery ride across the bridge, until it was about $^3/_4$ of a mile (1200 m) from the box. At first, the train's dark shape could still be made out, but gradually it was blurred in the dark – only the van's three red tail lights and the friction sparks indicated its position. Suddenly there was a single brilliant flash out near the girders and the red lights

Schließlich kommen wir noch einmal auf Thomas Barclay, den Signalmann in der Südkabine *(Abb. 28)* zurück: Er war sonntags nicht durchgehend im Dienst, sondern kam nur zu den bestimmten Zeiten, wenn die seltenen Züge durchzulassen waren – kein schlechter Arbeitsplatz, da sein Haus ganz in der Nähe lag. An diesem Abend war John Watt, Streckenmeister des Abschnittes von Leuchars bis zur Brücke – und **Augenzeuge Nr. 9** – bei Barclay. Als es Zeit wurde, zur Kabine zu gehen und dem «Edinburgh» das Signal zu geben, ging Watt mit. Der Sturm war jetzt auf seinem Höhepunkt – ein fortwährendes dumpfes Brüllen, durchsetzt mit Windstößen von unglaublicher Gewalt. Die Männer waren froh, den Schutz der Kabine zu erreichen. Das Feuer im Ofen war seit dem letzten Zug heruntergebrannt und Barclay ging daran, frisches nachzulegen. Da wurde auch schon der «Edinburgh» von St. Fort «angeboten» – was in der englischen Signalsprache um einiges fürsorglicher klingt als das deutsche «gemeldet». (Aber vielleicht pflegte man zu jener Zeit, vor der Automatisierung der Signalanlagen, in solch nachgerade menschlicher Weise mit Zügen umzugehen.) Barclay «nahm den Zug an» und «bot ihn weiter» an seinen Kollegen Sommerville auf der Nordkabine. Dann erst zog er die Signale, welche den Zug bis zur Brücke heranfahren ließen, und sah auch schon das Frontlicht um die Peacehill-Kurve kommen (vergl. *Abb. 27*). Er stieg die Stufen hinunter und hielt den Signalstab in die Höhe, damit ihn der Heizer Marshall im Vorbeifahren der Lokomotive greifen konnte. Die erleuchteten Wagen rumpelten vorbei, nahmen Fahrt auf, und als der Gepäckwagen von der Dunkelheit verschluckt wurde, ging auch das Poltern des Zuges im Sturmgetöse unter.

Barclay machte sich wieder ans Feuer, während John Watt wie so häufig am Fenster stand und zusah, wie der Zug auf die Brücke hinausspurte. Obwohl die Kabine im Wind vibrierte, war Watt ebensowenig wie Barclay beunruhigt über die Brücke. Inzwischen war der «Edinburgh» etwa 200 Yard (183 m) von der Südkabine entfernt und Funken begannen von den Wagenrädern zu fliegen. Watt beobachtete auch noch, wie der Zug seine feurige Fahrt über die Brücke fortsetzte, bis er etwa eine $^3/_4$ Meile (1200 m) von der Kabine entfernt war. Zunächst waren noch die dunklen Umrisse des Zuges auszumachen, aber mit der Zeit verschwamm seine Form mit der Finsternis – nur noch die drei roten Rücklichter des Gepäckwagens und die Reibungsblitze zeigten seine Position an. Plötzlich gab es einen einzigen grellen Blitz draußen bei den Hochträgern und die roten Lichter verschwanden: «Meine Augen wurden durch den Blitz von den roten Lichtern wie weggerissen und ich sah sie dann gar nicht mehr»,

vanished at once: "My eyes were taken off the red lights by the flash, and I never saw them again", Watt reported later. Barclay was still attending the fire when Watt turned around saying there seemed to be something wrong with the train. The signalman peered intently across the dark bridge, yet he did not share the fears of his friend. The lights would naturally have vanished beyond the hump at midspan as usual, and he should just wait for the train to swing out of the curve again at the far end. But no train appeared nor could the familiar ring of Sommerville's signal bell be heard, indicating safe arrival. Barclay tried the telephone first, then the telegraph, but there was no reply.

Only now did the men become alarmed and prepare to cross the bridge on foot. But only after a few yards, the wind forced them to their knees and to turn back. John Watt ran along the tracks to St. Fort. But stationmaster Morris was not there, so Watt set out for another four miles (6.4 km) to Leuchars Junction. Stationmaster Robertson had already lived through a difficult evening. The *Edinburgh* had just left, when the west wing of the station house where the children were sleeping, was being shaken so much by the storm that he had the family move to the east wing. Around 11 p.m. he locked up his station, sat down by the chimney and was just about to take off his boots, when there came a knock at the door. John Watt staggered across the threshold gasping, "Oh man, the bridge is down!" But the stationmaster did not believe him: So concerned was he about Watt's appearance and behaviour that he sent for the village doctor to look him over. The doctor listened to Watt's story and felt he might be telling the truth, whereupon Robertson grabbed the doctor's gig and raced down to St. Andrews. At the station there, it was known already that the bridge had collapsed. So Robertson turned around at once – well aware that he would be expected on duty at his own station.

Thus Robertson was at his post at 4 o'clock in the morning – the Monday after – when the special train from Edinburgh arrived at Leuchars Junction with a group of NBR officials. Harshly, the gentlemen had been awakened with the bad news. They were rushed by cab to Waverly Station, whence a train brought them down to Granton. From there followed the troublesome journey across the Forth on Bouch's famous *Leviathan;* for this trip it had been loaded with extra wagons for better stability on rough waters. Finally there was the travel through Fife. At Leuchars, several members of the engineering division disembarked, among them Sir Thomas Bouch: "He was in a pitiful state of mind", remarked Robertson who had recognised him in the group. The railway officials had a short consultation at the station office. Since Morris

sagte Watt später. Barclay war noch mit dem Feuer zu Gange, als Watt sich zu ihm umdrehte und sagte, mit dem Zug scheine etwas passiert zu sein. Der Signalmann spähte hinaus über die dunkle Brücke, aber er teilte nicht die Ängste seines Kollegen: Die Lichter wären über dem Buckel der Brückenmitte verschwunden wie jedesmal, und er solle nur abwarten, bis der Zug aus der Kurve am Nordende herausschwingen würde. Aber der Zug kam nicht wieder, noch ertönte das übliche Klingelsignal von Sommerville zum Zeichen der sicheren Ankunft. Barclay versuchte zuerst das Telefon und dann den Telegraphen, aber ohne Erfolg.

Jetzt erst waren die Männer alarmiert und machten sich auf den Weg über die Brücke. Aber schon nach wenigen Metern zwang sie der Wind in die Knie und zur Umkehr. John Watt rannte auf dem Schienenweg nach St. Fort. Aber Stationsvorsteher Morris war nicht da und Watt beschloß, die 4 Meilen (6,4 km) nach Leuchars zu Fuß weiterzulaufen. Der dortige Stationsvorsteher Robertson hatte bereits einen schweren Abend hinter sich. Kurz nach Abfahrt des «Edinburgh» wurde der Westflügel des Stationshäuschens, wo die Kinder schliefen, so vom Sturm gebeutelt, daß die Familie in den Ostflügel umquartiert werden mußte. Gegen 23.00 Uhr schloß der Vorsteher seinen Bahnhof ab, setzte sich ans Kaminfeuer und zog eben die Stiefel aus, als ein Pochen von der Haustür erscholl. John Watt stand auf der Schwelle: «O Mann», keuchte er, «die Brücke ist runter!» Doch Robertson glaubte ihm nicht: So besorgt war er über Watts Aussehen und Benehmen, daß er nach dem Dorfarzt schickte, ihn zu untersuchen. Der Doktor hörte sich die Geschichte des Streckenmeisters an und meinte, er könne wohl die Wahrheit sagen. Robertson lieh sich den offenen Wagen des Doktors und fuhr nach St. Andrews. Aber am Bahnhof dort wußte man schon, daß die Brücke eingestürzt war. Robertson machte kehrt – er wußte, man würde ihn im Dienst auf seinem eigenen Bahnhof erwarten.

Und so war Robertson vor Ort, als um 4.00 Uhr morgens – dem Montag danach – der Sonderzug mit den *NBR*-Gesellschaftern aus Edinburgh in Leuchars eintraf. Die Herren waren mit der bösen Nachricht aus ihren Betten geholt worden. Droschken hatten sie in aller Eile zur Waverly Station gebracht, ein Zug beförderte sie hinunter nach Granton. Dann kam die beschwerliche Fahrt über den Forth mit Bouch's berühmtem «Leviathan», der für diese Fahrt, zwecks besserer Stabilität auf rauher See, mit Extrawagen beschwert wurde. Schließlich folgte noch die Reise durch Fife. Mehrere Mitglieder der Ingenieur-Abteilung stiegen in Leuchars aus, darunter auch Sir

34
The Tay Bridge after the disaster, in an illustration of the year 1880, looking towards Dundee [19, p. 210].

34
Die Tay-Brücke nach dem Einsturz in einer Zeichnung aus dem Jahre 1880, mit Blick in Richtung Dundee [19, S. 210].

had already reported before from St. Fort that about 300 passengers had been on the train, the following – and first official – news-release was drawn up:

"From reports made to us here of the terrible calamity at the Tay Bridge, it appears that several of the large girders of the bridge, along with the last train from Edinburgh, were precipitated into the river about half past seven last night. There were, I deeply deplore to say, nearly three hundred passengers, besides company's servants in the train, all of whom are believed to have perished. The cause of the accident is not yet ascertained."

Thus the first report had gone out before anybody of the official party had actually been to the scene of the accident. The wording well illustrates the state of mind of those concerned – as it would have been quite impossible to transport anywhere near 300 passengers on the Edinburgh. Later it became evident that Morris in his excitement had counted all the tickets in the ticket box, including those collected from other trains. A later count proved that there had been no more than 75 people on this train. Only 46 bodies, however, would ever be recovered.

It is not without an element of irony that the NBR offi-

Thomas Bouch: «Er war in einem Mitleid-erregenden Zustand», sagte Robertson, der ihn in der Gruppe erkannte. Die Gesellschafter hielten eine kurze Besprechung im Bahnhofsbüro. Nachdem Morris schon von St. Fort gemeldet hatte, daß etwa 300 Passagiere auf dem Zug waren, wurde der folgende – und erste – Pressebericht verfaßt:

«Nach den uns hier überbrachten Berichten über das schreckliche Unglück auf der Tay-Brücke scheint es, das mehrere der großen Brückenträger, zusammen mit dem letzten Zug aus Edinburgh, in den Fluß gestürzt sind, ca. um 19.30 Uhr letzten Abend. Es waren, ich beklage zutiefst, es zu sagen, annähernd 300 Fahrgäste nebst Angestellten der Company im Zug, von denen allen man annehmen muß, daß sie umgekommen sind. Die Ursache des Unglücksfalles ist noch nicht ermittelt.»

Damit war diese erste Meldung herausgegangen, bevor noch irgendjemand von der Gruppe selbst am Unglücksort eingetroffen war. Der Wortlaut ist auch bezeichnend für die Gemütsverfassung der Betroffenen – denn es wäre ganz und gar unmöglich gewesen, auch nur annähernd 300 Passagiere in dem «Edinburgh» unterzubringen.

cials now had to follow the same cumbersome route, which already generations of travellers had to take before the erection of the great bridge. To get to Dundee, the special train could now only take the old route along the Tay – running from Wormit via Newport and on to Tayport. There the stationmaster led the honourable group to the pier where the steamer *James Cox* lay ready for the crossing to Dundee. But the sea was again too stormy for a passage, so coaches were hired back to Newport on the assumption of finding the *Dundee* there, but she was at the other shore. Returning to Tayport once more, the party went aboard the *James Cox* again, once the storm had eased. The small steamer fought its course upstream through waves littered with drifting debris from the bridge decking. When they approached Dundee, the moon came out as if to illuminate the horrid scene of the disaster *(Fig. 34)*.

Eyewitness Nr. 10: Finally, there was one further important eyewitness, the tenth: His testimony was never used in court, yet it could well have exonerated, at least partially, Sir Thomas Bouch. W. B. Thomson was an engineer and shipbuilder with the Caledon Shipyard in Dundee. On the day of the accident, he had left the Free Church at Broughty Ferry after the evening service at 7:06 p.m. that night. He went over to the beach to look at the river, but he was unable to turn the last corner at the riverfront as the storm was spewing sand and pebbles through the air. He reached only as far as the corner of James Place where he sought shelter for 15 minutes:

> "While there, I distinctly saw two luminous columns of mist or spray travelling across the river in the direction of the wind. Another one formed directly in front of me just in an instant. It appeared to rise from the center of the river. I never saw anything like this before, and looked round to see if anyone was near me, but seeing nobody thought it was better to take shelter. I went inside the railing in front of James Place and held on by one of them thinking that the column, which was advancing towards me, was to carry everything before it. It passed over me. It was spray from the river, not solid water. It struck the house behind me with a hissing noise. The height of each of these columns I should think was 250 to 300 ft [76–91 m]. On the following morning I found the windows of my house (about 170 ft [52 m] above the river) coated with salt. I never saw this before in a westerly wind, and can only account for this by one of these columns passing over the house. My

Später stellte sich heraus, daß Morris in seiner Aufregung alle Fahrkarten in der *ticket box* seiner Station gezählt hatte, worunter auch Tickets von anderen Zügen waren. Eine spätere Zählung ergab, daß sich höchstens 75 Menschen auf dem Zug befanden. – Allerdings hat man nur 46 Tote jemals geborgen.
Es ist nicht ohne Ironie, daß die Weiterfahrt der Gesellschafter auf demselben mühsamen Weg erfolgte, den schon Generationen vor dem Bau der großen Brücke nehmen mußten: Um jetzt nach Dundee zu gelangen, blieb dem Sonderzug nur die alte Strecke am Tay entlang – von Wormit über Newport bis nach Tayport. Dort führte der Stationsvorsteher die hohen Gäste zum Pier, wo der Dampfer «James Cox» zur Überfahrt nach Dundee bereitstand. Aber die See war wieder zu stürmisch für eine Passage geworden. So mietete man Kutschen zurück nach Newport in der Hoffnung, die «Dundee» dort anzutreffen: Die jedoch befand sich auf der anderen Seite. Bei der erneuten Rückkehr nach Tayport hatte der Wind nachgelassen und man begab sich an Bord der «James Cox». Der kleine Dampfer focht sich seinen Weg stromaufwärts durch Wogen bedeckt mit Treibholz der Brückenbohlung. Als er sich Dundee näherte, kam der Mond hervor und erhellte schaurig den Ort der Zerstörung *(Abb. 34)*.

Zeuge 10: Schließlich hatte es noch einen weiteren wichtigen Augenzeugen – den zehnten – gegeben: Seine Aussage wurde vor Gericht nie verwendet und hätte möglicherweise Sir Thomas Bouch zumindest teilweise entlastet. W. B. Thomson war Ingenieur und Schiffbauer im *Caledon Shipyard,* Dundee. Am Unglückstag verließ er nach dem Abendgottesdienst um 19.06 Uhr die Freie Kirche in Broughty Ferry. Er ging hinunter zum Strand, um nach dem Fluß zu sehen, fand es aber unmöglich, die letzte Ecke an der Uferfront zu nehmen, sosehr wirbelte der Sturm Sand und Steine durch die Luft. Er erreichte nur die Ecke von James Place und suchte Schutz für 15 Minuten:

> «Während ich dort stand, sah ich deutlich zwei leuchtende Säulen aus Dunst oder sprühender Gischt über den Fluß ziehen. Eine weitere formte sich ganz plötzlich genau vor mir: Sie schien sich direkt aus der Mitte des Flusses zu erheben. Ich hatte nie zuvor etwas Derartiges gesehen und schaute herum, ob noch jemand in meiner Nähe war; aber da ich niemand sah, schien es mir besser, in Deckung zu gehen. So schritt ich hinter das Geländer von James Place und hielt mich daran fest in der Erwartung, die Säule würde auf ihrem Weg zu mir alles mit sich reißen. Sie ging über mich hinweg – es war Gischt vom Fluß, kein dichtes Wasser – und traf

35 >
The scene of the accident at the Firth of Tay – recorded probably a few days after the disaster by an unknown photographer. Parts of the fallen girders resting on a sand bank appear above the waters during low tide. Inside they hold captive the train with 75 bodies [30, p. 492].

35 >
Der Unglücksort am Firth of Tay – vermutlich wenige Tage nach dem Einsturz von einem unbekannten Fotografen aufgenommen. Teile der abgestürzten Träger liegen auf einer Sandbank und erscheinen bei Ebbe über den Fluten. Im Inneren halten sie den Zug mit 75 Toten gefangen [30, S. 492].

theory is that the North end of the bridge gave way first, the failure being caused by one such column rising or passing under it. Such a thing would tend to lift the girders from their piers, or that the wind, which had the effect of lifting that column of spray, had also the effect of raising the girders from the piers, thereby overcoming part of the resistance offered by the lateral pressure of the wind. My theory put briefly is that there was a heavy upwards pressure as well as a lateral pressure in the wind that night. I never saw anything like it. My opinion is that the spray rising from the plunge of the falling girders and train could not rise to the height at which Provost Robertson saw his columns, and that Provost

das Haus hinter mir mit einem zischenden Geräusch. Die Höhe jeder der beiden Säulen schätzte ich auf 250–300 Fuß (76–91 m). Am folgenden Morgen fand ich die Fenster meines Hauses (etwa 170 Fuß [51,82 m] über dem Wasserspiegel gelegen) mit Salz bedeckt: So etwas hatte ich noch nie vorher bei einem Westwind beobachtet und kann mir das nur dadurch erklären, daß eine dieser Säulen über mein Haus gewandert sein muß. Meine Theorie ist, daß das Nordende der Brücke zuerst nachgegeben hat und der Einsturz durch das Aufsteigen bzw. Darunterdurchlaufen einer solchen Wassersäule verursacht worden ist. So etwas würde ja dazu neigen, die Träger von ihren Stützen hochzu-

Robertson's columns must have been of the same description as those I saw." [26, pp. 120, 121]

This almost sounds as if there could have been the very rare case of a *cyclone* racing along the Scottish Firth. It remains unexplained why the very thorough investigations of the court did not pay any attention to this unusual phenomenon – even though it had been observed clearly by at least two eyewitnesses.

The Day After

The disaster had come with such force and swiftness over Dundee – and indeed over the rest of the world – that the public was as if paralysed. It was hardly conceivable that in a single stroke 75 people, together with train and bridge, should have been destroyed. The scene the day after was quite unreal *(Fig. 35)*. The deadly silence after the storm, the smooth lead-like waters had covered everything up as if all had been a bad dream. Nothing was left to see but the buckled piers and the open ends of the girders high up, pointing into the empty void.
The whole night through until the early morning, scores of frightened people had waited at the banks of

heben – bzw. würde bewirken, daß der Wind mit seiner Kraft, eine solche Gischtsäule hochzuwirbeln, auch die Träger von den Pfeilern heben könnte, indem er dabei einen Teil des Widerstandes gegen den Seitendruck des Windes aufheben würde. Meine Theorie ist also, einfach gesagt, daß es einen starken Aufwärtsdruck und einen Seitendruck des Windes in dieser Nacht gab. Ich habe dergleichen noch nie gesehen. Meine Meinung ist, daß das Spritzwasser von den fallenden Brückenträgern und dem Zug niemals bis zu der Höhe hochschießen konnte, die Provost Robertson in seinen Wassersäulen gesehen hat – und daß Provost Robertsons Säulen von derselben Art gewesen sein mußten, wie ich sie sah.» [26, S. 120, 121]

Das klingt so, als könnte es sich hier in der Tat um den seltenen Fall einer *Windhose* auf dem schottischen Meeresarm gehandelt haben: Es bleibt unerfindlich, warum die sehr gründlichen Untersuchungen des Gerichtes auf dieses Phänomen – obwohl von mindestens zwei Zeugen deutlich beobachtet – in keiner Weise eingegangen sind.

36
The Tay Bridge after the disaster – seen from a similar viewpoint as in *Fig. 28* [2, p. 109].

36
Die Tay-Brücke nach dem Einsturz – von ähnlichem Blickwinkel gesehen wie *Abb. 28* [2, S. 109].

the Tay and in the crowded streets near Tay Bridge Station. The police had cordoned off the harbour so nobody would search for survivors on his own. Still, there were many boats on the water, their crews calling out into the night, in the desperate hope of finding some survivors who might be clinging to the shattered piers. Hardly anybody went to sleep that night, the windows stayed lit and the people kept staring out at the Firth, as if they hoped for a miracle whereby the bridge would perhaps re-erect itself and the train would steam safely into Dundee after all. With dawn approaching, the first dire news began to go out to the agencies – by the still functioning telegraph of the CR:

Terrific Hurricane
***
Appalling Catastrophe at Dundee
***
Tay Bridge down
***
Passenger Train hurled into River
***
Supposed loss of 200 lives

By now, all hope for survivors had to be given up. Queen Victoria had her private secretary wire to the Provost of Dundee:

> "The Queen is inexpressibly shocked and feels most deeply for those who have lost friends and relatives in this terrible accident."

For generations of families in Dundee, indeed for all of Scotland, the Tay Bridge Disaster has the familiarity of a household word, and serves as a historic turning point. It became the custom to recall family events as having occurred before or after the disaster. – Even today, some 110 years after the tragedy, the proverbial "man on the street" in Dundee is well aware of the significance of this event – "after all, it's part of our city's history" (so an automobile mechanic, 1989).

The contemporary diary of a certain John Inglis [12] from Edinburgh contains this entry for the day of *December 29, 1879* – following a note about the bad storm of the day before:

> "...read THE SCOTSMAN and learned that a train going towards Dundee hat been last night about 7 o'clock hurled over the Tay Bridge by the force of the wind, the whole of the over arching girdles(!) in the center of the bridge being carried bodily off so that not a vestige of the whole was visible. The number of passengers in the train, who of course all perished, was at first es-

Der Tag danach

Die Katastrophe war mit solcher Wucht und Schnelligkeit über Dundee – und über den Rest der Welt – hereingebrochen, daß die Zeitgenossen wie gelähmt waren: Daß mit einem Schlage 75 Menschen samt Zug und Brücke ausgelöscht sein sollten, war kaum faßbar, und die Szene am nächsten Tag hatte etwas Unwirkliches *(Abb. 35)*. Die Grabesruhe nach dem Sturm, das glatte bleigraue Wasser hatten alles zugedeckt wie einen bösen Traum. Nichts mehr war zu sehen außer den abgeknickten Pfeilerstümpfen und den offenen Enden der Brückenträger hoch oben, die ins leere Nichts hinausragten.

Die ganze Nacht bis in den frühen Morgen hatten Gruppen angstvoller Menschen an den Ufern des Tay und in den überfüllten Straßen nahe Tay Bridge Station ausgeharrt. Die Polizei bildete einen Ring um den Hafen, damit niemand auf eigene Faust mit Booten die Suche nach Überlebenden aufnehmen sollte. Dennoch trieben viele Boote im Wasser und die Insassen riefen in die Nacht hinaus in der verzweifelten Hoffnung, daß sich einige Überlebende an die Pfeilerreste geklammert haben könnten. Kaum jemand ging zu Bett, die Fenster blieben erleuchtet und die Leute starrten hinaus auf den Fluß, als ob sie ein Wunder erwarteten, durch das sich die Brücke vielleicht von selbst wiederaufrichten und der Zug wohlbehalten in Dundee einlaufen würde. Mit anbrechendem Morgen begannen die ersten dürren Nachrichten an die Zeitungen hinauszugehen – über den noch intakten Telegraphen der *CR:*

Schrecklicher Hurrikan
***
Entsetzliche Katastrophe bei Dundee
***
Tay-Brücke eingestürzt
***
Personenzug in Fluß geschleudert
***
Verlust von 200 Menschenleben befürchtet

Inzwischen war jede Hoffnung auf Überlebende geschwunden. Queen Victoria ließ ihren Privatsekretär an den Bürgermeister von Dundee kabeln:

> «Die Königin ist unaussprechlich schockiert und fühlt zutiefst mit denen, die Freunde und Verwandte in diesem schrecklichen Unglück verloren haben.»

timated at 300 but later in the day it was estimated at a much lower figure, viz. 70. The talk has of course been about little else all day, the accident being almost unprecedented in the annals of earthly disasters...."

On the day after the disaster, it was above all Sir Thomas Bouch who was a tragic figure. Whereas he had been the wealthy and highly respected engineer on Sunday, on Monday morning his work and his future lay in shambles. The initially sober, and balanced articles in most Scottish papers had even offered a consolation of sorts for him, such as that which appeared in *The Evening Times* in Glasgow:

"Sir Thomas Bouch had necessarily to use his own judgement in regard to points of novelty on which the experience of either engineers would give him but small assistance. His work, taking into account its surprising cheapness, has been universally pronounced a success, and there can be little doubt that it would in ordinary circumstances have become the parent of many similar structures. Encouraged by the Tay Bridge the engineers would have boldly essayed to answer the call of enterprising capitalists by offerring to throw arches over any estuary however wide. But it is inevitable that this accident should produce circumspection and caution."

And the *Standard* was eager to report right after the accident, that it was beyond doubt that the engine driver had gone against the explicit instructions of the railway administration in going a speed on the bridge that was bound to cause the disaster. The terrible accident would prove anew that there were limits which mankind could not exceed unpunished, but that the fame of British engineering would not be seriously affected by this deeply deplorable accident; the speedy reconstruction of the bridge being a matter of course, etc. – Only the public inquiries of the months to follow would reveal that the reputation of the highly praised British technology had indeed suffered.

This common tenor was followed also by the *Courier* in Dundee, in emphasising that the new bridge should now be built for a double line of railway:

"It is clear that if we are to have an erection on which the public will venture their lives it must be one of a broader base, of less height and one less top-heavy than that which was wrecked on Sunday evening."

By Monday noon, the news of the disaster had gotten to most of the international agencies. From then on, a growing stream of journalists and illustrators set out toward far-away Dundee, and in London the excited crowds were demanding further information. The

Für Generationen von Familien in Dundee, ja in ganz Schottland, war das *Tay Bridge Disaster* zu einem Haushaltswort, einer historischen Wendemarke geworden. Man pflegte Familienereignisse als vor oder nach der Katastrophe geschehen zu datieren. – Selbst heute noch, 111 Jahre danach, weiß der sprichwörtliche «Mann auf der Straße» durchaus Bescheid über die Bedeutung dieses Ereignisses – «after all, it's part of our city's history» (so ein Automechaniker unserer Tage, 1989).

Das zeitgenössische Tagebuch eines gewissen John Inglis [12] aus Edinburgh vermerkt – nach einer Notiz am Vortage über den schlimmen Sturm – für den *29. Dezember 1879* die folgende Eintragung:

«...Den *Scotsman* gelesen und erfahren, daß ein Zug auf dem Wege nach Dundee letzten Abend ca. 7 Uhr von der Gewalt des Sturmes über die Brücke geschleudert worden ist, wobei alle der übergewölbten Träger in der Mitte der Brücke zur Gänze fortgetragen wurden, so daß keine Spur vom Ganzen mehr zu sehen war. Die Zahl der Passagiere im Zug, die natürlich alle umgekommen sind, wurde zuerst auf 300 geschätzt, aber später am Tag auf eine viel niedrigere Zahl geschätzt, nahe 70. Natürlich wurde über kaum etwas anderes den ganzen Tag gesprochen, da ein solcher Unglücksfall nahezu ohne Beispiel in den Annalen irdischer Katastrophen sein muß...»

Am Tage nach dem Desaster war aber zuallererst Sir Thomas Bouch eine tragische Figur: War er am Sonntag noch der wohlhabende und hochangesehene Ingenieur gewesen, so lagen am Montagmorgen sein Werk und seine Zukunft in Trümmern. Die zunächst ruhig abwägenden Artikel im Großteil der schottischen Presse hatten sogar etwas wie Trost für ihn bereit – so die *«Evening Times»* in Glasgow:

«Sir Thomas Bouch mußte sich notgedrungen auf sein eigenes Urteil bezüglich jener Neuerungen verlassen, für welche die Erfahrungen anderer Ingenieure ihm kaum Hilfestellung boten. Sein Werk ist angesichts der überraschend niedrigen Kosten weltweit als ein Erfolg gewertet worden, und es kann kaum Zweifel geben, daß es unter normalen Umständen der Ahnvater vieler ähnlicher Bauwerke geworden wäre. Ermutigt durch die Tay-Brücke hätten die Ingenieure mit Wagemut versucht, dem Ruf unternehmender Kapitalisten mit weiteren Angeboten zu folgen, Brücken über Meeresbuchten gleich welcher Größe zu schlagen. Aber es ist unvermeidlich, daß dieser Unfall nach Umsicht und Vorsorge verlangt.»

37
The Tay Bridge after the disaster with "the terribly long gap", as seen from the heights above Wormit. Remarkable is the then still rather "airy" configuration of the entire bridge-head in the left foreground – the signal box being part of the ironwork itself, freely suspended in the fork branching off to Tayport [26, p. 137].

37
Die Tay-Brücke nach dem Einsturz mit der «schrecklich langen Lücke», von der Höhe über Wormit gesehen. Auffallend ist die damals noch sehr «luftige» Ausbildung des gesamten Brückenkopfes links vorne – die Signalkabine ist Teil des Gitterwerkes und hängt freischwebend in der Abzweigung nach Tayport [26, S. 137].

NBR announced that the destruction of the bridge was an act of God and that no human power was involved. The government instructed the Board of Trade to set up at once an inquiry into the causes of the accident. The President of the Board of Trade had chosen his investigators within hours; before the week was over, they were heading for Dundee.

The officials of the NBR had already decided on Monday afternoon on their return to Edinburgh, that the bridge should be rebuilt by Thomas Bouch. This was a brave but hasty decision indeed, as public opinion was already turning against that bridge. Although *The Glasgow Evening News* attacked those in danger of losing their heads over the disaster, *The Scotsman* wrote already two days after the disaster:[9]

> "Almost 3/4 of a mile of what had been just the day before the most magnificent bridge of the world, has collapsed.... The lesson of this accident is not as much the necessity for greater care on part of those responsible for railway traffic; rather it is the application of a larger factor of safety in constructing such bridges like the one across the Tay, both for loading under pressure

9 *The Scotsman,* Edinburgh, 30. December 1879.

Und der «*Standard*» wußte gleich nach dem Unglück zu berichten, es sei außer Zweifel, daß der verunglückte Lokomotivführer gegen die ausdrücklichen Bestimmungen der Eisenbahn-Verwaltung mit einer Geschwindigkeit über die Brücke gefahren sei, welche die Katastrophe herbeiführen mußte. Das entsetzliche Unglück weise aufs neue darauf hin, daß es Grenzen gäbe, die der Mensch nicht ungestraft überschreite, daß aber der Ruf der englischen Technik von diesem tief bedauerlichen Unfall nicht ernstlich berührt werde. Der schleunigste Wiederaufbau der Brücke sei eine selbstverständliche Sache. – Daß der Ruf der so hoch gepriesenen englischen Technik in der Tat angegriffen war, machten erst die Untersuchungen der folgenden Monate deutlich.

Dem allgemeinen Tenor schloß sich auch der «*Courier*» in Dundee an und betonte, daß die neue Brücke zweispurig gebaut werden solle:

> «Es ist klar, daß wenn wir ein Bauwerk haben sollen, dem die Öffentlichkeit ihr Leben anvertrauen wird, es eine breitere Basis haben muß, von geringerer Höhe und weniger kopflastig sein muß als das, welches am Sonntagabend zerstört worden ist.»

and trial runs, going far beyond of what could normally be expected."

Diving Manoeuvres

The various diving operations of the days to follow presented a macabre scene. Working conditions at the riverbed were poor, the water was muddy and the divers were forced most of the time just to feel their way around in total darkness. The diving journeys were not coordinated – divers out of work were coming from as far away as Sunderland when they got the news that work was available on the Tay. Others had been hired by the railway company, still others by the harbour office and also some by Bouch himself. Quite a few were moving about in circles, often in places that had already been searched by their mates before. On "the morning after", diver Cox hit upon a continuous row of girders resting on their side near the third pier

Am Montag mittag war die Nachricht über das Desaster bei den meisten internationalen Agenturen eingetroffen. Von da an setzte sich ein wachsender Strom von Journalisten und Illustratoren nach dem fernen Dundee in Bewegung, und in London verlangten aufgeregte Menschenmassen nach weiteren Informationen. Die *NBR* verkündete, daß der Einsturz ein Akt Gottes und keine menschliche Macht darin verwickelt sei. Die Regierung wies die IHK an, unverzüglich eine Untersuchung über die Ursachen des Einsturzes anzustellen. Innerhalb weniger Stunden hatte der Präsident der IHK seine Investigatoren gewählt; noch vor Ende der Woche waren sie unterwegs nach Dundee.
Die Gesellschafter der *NBR* hatten schon am Montagnachmittag, auf dem Rückweg nach Edinburgh beschlossen, daß die Brücke von Bouch wieder aufgebaut werden solle: Eine kühne Entscheidung, denn schon begann sich die öffentliche Meinung gegen die Brücke zu erheben. Auch wenn «*The Glasgow Evening*

38
One of the large central girders resting on its side after recovery. To the left, a carriage still enclosed is seen standing upright on the side-wall of the girder, just as it was found at the bottom of the river [26, p. 155].

38
Einer der großen Mittelträger in Seitenlage, nach der Hebung. Links steht ein noch darin eingeschlossener Waggon auf der Seitenwand des Brückenträgers, so wie man ihn am Flußbette vorgefunden hat [26, S. 155].

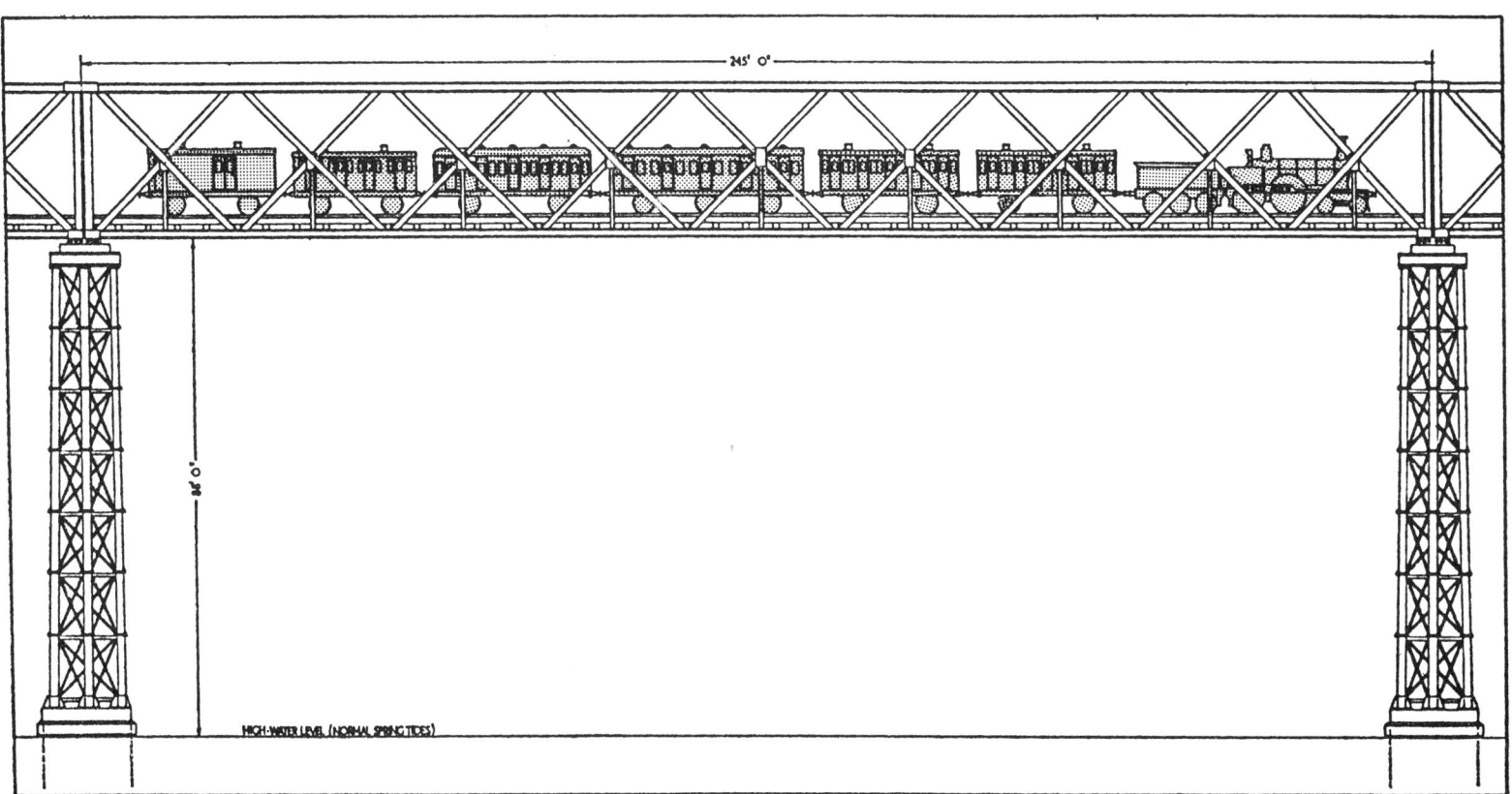

39
Assumed position of the train shortly before the fall of the Tay Bridge – in the fifth of the large central girders, between piers Nr. 32 and 33 [26, p. 89]. (The sequence of carriages shown in this drawing is not quite correct – the three-wheeled first-class carriage should be one position further in front.)

39
Rekonstruierte Position des Zuges kurz vor dem Einsturz der Tay-Brücke – im 5. der großen Mittelträger, zwischen Pfeiler Nr. 32 und 33 [26, S. 89]. (Die Reihenfolge der Waggons ist in dieser Zeichnung nicht ganz korrekt wiedergegeben – der 1. Klasse-Wagen [3-achsig] sollte um eine Position weiter vorne stehen.)

of the high girders; their underside, with the rails still attached, was pointing westward – but he did not find any train. On December 30, Cox went down again, this time near the fourth pier, following the girders for about 30 ft (9.20 m) to the north. There he came upon a carriage standing upright on what had been the east side of the girder, without roof and with broken windows and doors. *(Fig. 38)*. The interior was just an empty carcass void of compartments, except for broken wood, cushions and oil cloth. He came up clutching a shred of oil cloth and it was recognised as part of the first-class carriage. This was important evidence; it proved that the engine and one third-class carriage must be lying further on *(Fig. 39)*.

The next diving journeys centered on the stretch between the fourth and the fifth pier (Nrs. 32 and 33). On Wednesday, December 31, diver Simpson came upon a third-class carriage right away. It was leaning to the east, tight up against a girder, missing part of the side and almost all seats, the floor covered with gravel. On the second dive that day and near the sixth pier, Simpson finally found the locomotive Nr. 224 with tender lying on its side inside the girder, the smoke stack

News» jene attackierte, die in Gefahr waren, ihren Kopf über dem Desaster zu verlieren, schrieb «The Scotsman» schon 2 Tage nach dem Unglück:[9]

«Fast eine $^3/_4$ Meile von dem, was am Vortage noch die vortrefflichste Brücke der Welt war, ist eingestürzt... Die Lehre dieses Unglückes ist nicht sosehr die Notwendigkeit größerer Sorgfalt seitens derer, die für den Zugverkehr verantwortlich sind, sondern der Anwendung eines größeren Sicherheitsfaktors bei der Konstruktion solcher Brücken wie der über den Tay, und zwar für Druck- und Probebelastung, die weit über das hinausgeht, was man normalerweise erwarten dürfte.»

Tauchmanöver

Eine makabre Szene boten die mannigfachen Tauchversuche der folgenden Woche. Die Arbeitsbedingungen am Flußgrund waren schlecht, das Wasser trübe, und die Taucher mußten sich manchmal in völliger Dunkelheit herumtasten. Auch waren die Tauchgänge

9 The Scotsman, Edinburg, 30. Dezember 1879.

pointing to the east *(Fig. 40)*. The lower girder wall was covered by sand, the bracings broken and sand washed up to the boiler. Months later, the engine was recovered, as were the carriages. It was almost undamaged and still did service thereafter until the year 1924. The drivers used to call it affectionately *The Diver*.

Down at the bottom of the Tay, it had been proven now that the train was lying entirely enclosed within the tunnel-like central girders, about 30 ft (9.20 m) below water level at low tide, between the fourth and the fifth of the broken piers *(Fig. 41)*. This meant that the train could not in fact have fallen off the bridge *before* the girders had come down – such as by some damage caused perhaps to the girders by the storm: This possibility had been Sir Thomas Bouch's last ray of hope – indeed, it seemed as if he *wanted* the train still to be enclosed inside the girders. On the very night of the disaster he had declared to the *Daily Telegraph* that the train must have been derailed, either by pressure of the wind or by excessive speed, and must then have taken down the ironwork as well! If the matter had occurred in such or similar a way, then the designer of the bridge (whose world fame would lend plausibility to this explanation), would indeed be free of any blame.... For another four days the search for bodies went on, but only broken girders, pieces of railing or other fragments were found. Some of those relics may still be examined today in the section "Tay Bridge Disaster Display" at the Albert Institute and McManus Galleries in Dundee.

Public Inquiry

On Saturday, January 3, 1880, while the divers were still searching in the murky darkness of the Tay, a first public hearing was held at the Dundee Courthouse. The three members of the Board of Trade Inquiry were the Chairman and Wreck Commissioner[10], Henry Cadogan Rothery, a man of sharp reasoning and independent mind; Colonel William Yolland, engineer and Chief Inspector of Railways; and William Henry Barlow, President of the Institute of Civil Engineers and himself a well-known bridge builder. As would be proven in the course of the inquiry, very thorough investigations brought to light some hair-raising faults of material and workmanship, but also faults of design; in the end, Sir Thomas Bouch was a ruined man. The court of inquiry sat for 25 non-consecutive days and heard 120 witnesses, during the many months of these investigations.

10 A wreck commission usually was set up after the sinking of a major ship.

nicht koordiniert – arbeitslose Taucher kamen sogar aus Sunderland angereist, sobald sie hörten, daß es am Tay Arbeit gäbe. Andere waren von der Eisenbahngesellschaft angeheuert, wieder andere von der Hafenbehörde und auch einige von Bouch selbst. Manche bewegten sich im Kreis, oft an Stellen, die ihre Kollegen schon abgesucht hatten. Taucher Cox stieß am «Morgen danach», beim dritten Pfeiler der Hochträger, auf eine fortlaufende Reihe von Trägern in Seitenlage, die Unterseite samt Schienen nach Westen gerichtet – fand aber keine Spur vom Zuge. Am 30. Dezember tauchte Cox erneut, diesmal beim vierten Pfeiler und folgte den Trägern etwa 30 Fuß (9,14 m) nach Norden. Da traf er auf einen Wagen, aufrecht mit seinen Rädern auf dem stehend, was einmal die Ostseite des Fachwerkträgers war, ohne Dach und mit zerbrochenen Fenstern und Türen *(Abb. 38)*. Das Innere war nur noch eine leere Hülle ohne Abteile, bis auf zerbrochenes Holz, Kissen und Öltuch. Mit einem Stück Öltuch kam der Taucher wieder hoch und man erkannte es als Teil des 1. Klasse-Waggons. Das war ein wichtiges Beweisstück, denn damit war klar, daß die Lokomotive und ein 3. Klasse-Wagen weiter vorne liegen mußten *(Abb. 39)*.

So konzentrierten sich die weiteren Tauchgänge auf die Strecke zwischen dem 4. und 5. Pfeiler (Nr. 32 und 33). Am Mittwoch, den 31. Dezember stieß Taucher Simpson sogleich auf einen 3. Klasse-Wagen: Er lehnte schräg nach Osten, hart gegen einen Träger, ein Teil der Seitenwand und fast alle Sitze fehlten, der Boden mit Kies bedeckt. Beim zweiten Tauchgang des Tages entdeckte Simpson endlich beim 5. Pfeiler die Lokomotive Nr. 224 samt Tender, wie sie seitlich im Träger lag, den Schornstein nach Osten gereckt *(Abb. 40)*. Der untere Träger lag im Sand eingebettet, die Verstrebungen gebrochen, sanddurchschwemmt bis über den Boiler. Später wurde die Unglückslokomotive ebenso wie die Wagen geborgen: Sie war nahezu unversehrt und tat dann noch bis 1924 ihren Dienst. Die Fahrer gaben ihr fortan den liebevollen Namen «*The Diver*».

Unten am Grunde des Tay hatte sich erwiesen, daß der Zug zur Gänze innerhalb der tunnelartigen Trägermitte gefangen lag, bei Ebbe ca. 30 Fuß (9,14 m) unter Wasser, und zwar zwischen dem 4. und 5. der gebrochenen Pfeiler *(Abb. 41)*: Das bedeutete, daß der Zug nicht etwa schon vorher – z. B. durch eine vom Sturm verursachte Beschädigung der Trägerwand – von der Brücke gestürzt sein konnte: Daran klammerte sich Sir Thomas Bouch wie an einen letzten Rettungsanker – ja fast schien es, als *wollte* er, daß sich der Zug noch innerhalb der Träger befand. Denn er hatte noch in der Unglücksnacht dem «*Daily Telegraph*» erklärt, der Zug

40
The ill-fated engine Nr. 224 at Tayport after recovery. It was virtually undamaged and served for another 45 years under the name of *The Diver* [20, p. 207].

40
Die Unglückslokomotive Nr. 224 in Tayport nach ihrer Bergung. Sie war nahezu unversehrt und tat noch für die folgenden 45 Jahre unter dem Namen «*The Diver*» ihren Dienst [20, S. 207].

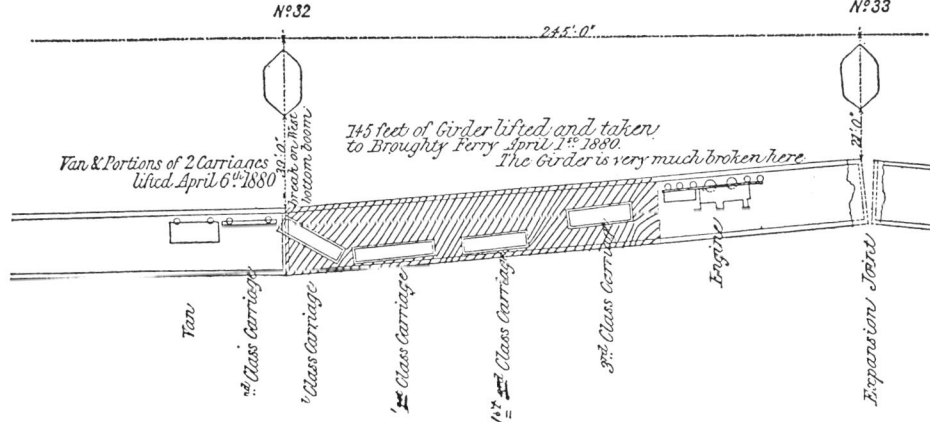

41
Official drawing showing the position of the fallen girders and the train at the river bed, to be presented at the inquiry [26, p. 171].

41
Offizielle Zeichnung zur Lage der Brückenträger und des Zuges am Flußgrund, zwecks Vorlage beim Untersuchungsgericht [26, S. 171].

müsse entweder durch den Winddruck oder zu hohe Geschwindigkeit zum Entgleisen gebracht worden sein und müsse dann das Gitterwerk mit in die Tiefe gerissen haben: Wenn die Sache sich so oder ähnlich abgespielt habe, so sei allerdings der Konstrukteur der Brücke, dessen weltberühmter Name dies ja voraussehen lasse, von jeder Schuld freizusprechen... Noch weitere vier Tage suchte man nach Leichen, fand aber nur geborstene Fachwerke, drei Hcoklampen des Gepäckwagens, Geländerteile und sonstige Bruchstücke. – Teile dieser Relikte sind noch heute in der Abteilung «*Tay Bridge Disaster Display*» des *Albert Institute and McManus Galeries* von Dundee zu besichtigen.

Öffentliche Untersuchung

Während die Taucher noch im schlammigen Dunkel des Tay herumsuchten, fand am Samstag, den 3. 1. 1880 eine erste Anhörung vor Gericht in Dundee statt. Die drei Mitglieder der IHK-Kommission waren der Vorsitzende und «*Wreck-Commissioner*»[10] Henry Cagodan Rothery, ein Mann von scharfem Verstand und unabhängigem Geiste; Colonel William Yolland

---

10 Eine «Wrack-Kommission» wurde für gewöhnlich nach dem Untergang großer Schiffe einberufen.

42 a
This rare photograph by George Washington Wilson shows the demolition of Bouch's Tay Bridge to the right, with the former trunk-line to Newport – and right next to it the new bridge by Barlow, built 1882–1887. The new, larger signal box is just under construction here and is still standing today (compare *Fig. 42 b*). The contract for the new bridge had been awarded to William Arrol who was already supervising the construction of the Forth Bridge at that time. The photograph was probably taken a few months before the official opening of the new bridge on June 13, 1887 [3, p. 18].

42 a
Diese seltene Aufnahme von G. W. Wilson zeigt den Abbau der Bouch'schen Brücke rechts, mit der alten Abzweigung nach Tayport – und parallel daneben die neue Brücke Barlows, 1882–1887 erbaut. Das neue größere Signalhaus ist gerade im Bau und steht bis heute (vergl. *Abb. 42 b*). Die Bauarbeiten für die neue Brücke waren William Arrol übertragen worden, der bereits den Bau der Forth-Brücke leitete. Die Aufnahme entstand vermutlich wenige Monate vor der offiziellen Eröffnung der neuen Brücke am 13. Juni 1887 [3, S. 18].

42 b
The same scene as in *Fig. 42 a*, as it appears today. To the right of the new bridge, only the bases of the piers can be seen tracing the course of the old bridge. In the background the city of Dundee – just a few chimneys of the once flowering jute industry still exist today.

42 b
Dieselbe Szene wie in *Abb. 42 a* in heutigem Zustand. Rechts neben der neuen Brücke markieren nur noch die Pfeilerstümpfe im Wasser den Verlauf der alten Brücke. Im Hintergrund die Stadt Dundee – nur noch wenige Kamine der einst blühenden Jute-Industrie sind zu erkennen.

43 a
View from the Tay Bridge shortly after reconstruction, towards Wormit. The new signal box has just been completed, (compare *Fig. 42 a*), replacing the two smaller boxes of the original bridge. "Wormit Junction" – as this photograph by G. W. Wilson is entitled – in those days still combined the trunk-line going off to the left towards Tayport, later abandoned. In the back, the stately homes of merchants and sea-captains are seen dotting the hillside above Wormit [3, p. 18].

43 a
Blick von der Tay-Brücke kurz nach dem Wiederaufbau in Richtung Wormit: Das neue Signalhaus ist gerade fertiggestellt (vergl. *Abb. 42 a*) und ersetzt die 2 kleineren Kabinen der ersten Brücke. «Wormit Junction» – so der Titel dieser Aufnahme von G. W. Wilson – verband damals noch die Abzweigung nach links mit Tayport, später aufgelassen. Im Hintergrund die stattlichen Kaufmanns- und Kapitänshäuser auf der Höhe über Wormit [3, S. 18].

43 b
Despite all the changes, many of the historic houses at Wormit are still standing today – such as this Victorian beauty with a magnificent view of the bridge. The same house may be seen, partially covered, in the upper center of *Fig. 43 a*. This "Villa Architecture" and the contemporary ironwork of Bouch's bridge follow the same vocabulary.

43 b
Trotz aller Veränderungen stehen heute noch viele der damaligen Häuser von Wormit – so dieses viktorianische Prachtstück, mit bestem Blick auf die Brücke. Dasselbe Haus ist halbverdeckt in der oberen Bildmitte von *Abb. 43 a* erkennbar: Die Villen-Architektur und das seinerzeitige Gitterwerk der Bouch'schen Brücke folgen derselben Formensprache!

The purpose of this first hearing was to collect those facts that could be ascertained on the spot. This included the accounts of the signalmen at the south end, but also those of the guards, stationmasters and passengers who had gotten off on time; further those of the captain of the *Mars,* of the old admiral at Scotscraig, the various eyewitnesses and finally the CR man who had actually seen the bridge go down. On Monday, the second day of the hearing, the first body was found. It was guard David Johnston of the NBR – the silver watch in his pocket had stopped at 7 : 16 p.m. [21, p. 131]. The body was placed in the refreshment room of Tay Bridge Station, the face uncovered. During the following days, this bright hall of polished wood and mirror glass was to become the bizarre mortuary where relatives had to identify their dead, or the meagre shreds of clothing which might give a name to those still missing. On the third day, the court had finished the preliminary inquest, after all local witnesses and those immediately involved in the events of December 28, 1879, had been heard.

als Ingenieur und Chefinspekteur der Eisenbahnen; und William Henry Barlow, selbst ein bekannter Brückenbauer, als Präsident des Ingenieurverbandes. Wie sich im Verlauf des Verfahrens zeigen sollte, brachten äußerst gründliche Untersuchungen haarsträubende Material- und Unterhaltsfehler, aber auch Konstruktionsfehler zutage, und am Ende war Sir Thomas Bouch ein ruinierter Mann. Das Untersuchungsgericht saß insgesamt 25 (nicht zusammenhängende) Tage und hörte 120 Zeugen während der vielen Wochen seiner Untersuchungen.
Der Sinn der ersten Anhörung war, zunächst jene Tatsachen zu ermitteln, die sich vor Ort feststellen ließen. Dazu hörte das Gericht die Aussagen vom Signalmann am Südende bis zu den Schaffnern, Stationsvorstehern und Reisenden, die noch rechtzeitig ausgestiegen waren, vom Kapitän der «Mars», dem alten Admiral auf Scotscraig bis zu den Augenzeugen und schließlich jenem *CR*-Mann, der die Brücke wirklich fallen sah. Am Montag, dem zweiten Anhörungstag, fand man den ersten Toten. Es war Schaffner David

By the time the court reconvened on February 26, 1880, in Dundee, 33 bodies had been recovered from the Firth. Later, more would be found, but 29 were never recovered. The NBR had offered 2 Pounds for each body found, and a Relief Fund was installed with a donation of 1,000 Pounds by the NBR; Bouch gave 250 Pounds, and once the sum had reached 3,000 Pounds, the ladies of Dundee organized refined music evenings to enlarge this amount. – The very last claim to this fund was brought forth in 1938 by the aged sister of a guard on the *Edinburgh.*

In every Scottish community at the time of Queen Victoria, there were three men respected and revered above anybody else: the minister, the doctor and the teacher. In times of low wages, without public welfare, the sudden death of a breadwinner meant nothing less than a catastrophe to the family. Thus the members of these three professions set out to write warm letters to the Relief Fund on behalf of their protégés – letters which still today are moving documents of the social conditions and the impact of such calamities upon the people affected [26, pp. 108–110].

The ongoing inquiry lasted until March 3, and now astonishing revelations about design, maintenance and operation of the bridge came to light. Starting with the last point, "operation" had carried the rule – ever since the opening of the bridge by General Hutchinson – that the maximum tolerated speed would be 25 m.p.h. The drivers therefore denied steadfastedly under questioning ever to have exceeded this limit; the regular slowing down to 2 m.p.h. (3.2 km/h) at each bridgehead – necessary for collecting the baton – made it impossible to reach 40 miles on the bridge. But the accounts of the passengers told a different story, and although the drivers kept insisting never to have raced the ferries across the Tay, it was quite evident that many citizens regarded this as a regular practice. If, for instance, the train and the ferry had started at the same time from Newport to Dundee, the train would have arrived earlier. But if the train were late, the driver could still draw even with the ferry or even overtake it. Quite a few passengers actually seemed to enjoy the high speeds during those races, although one railwayman recalled how he felt a sudden jerk whenever the speeding train entered the southern curve of the bridge, causing the carriage to lean over to the side, before it fell back into the vertical. In any event technical progress in locomotive design had hardly halted: by 1879; already 70 m.p.h. (112 km/h) could easily be obtained, thanks to newly developed engines.

Johnston von der *NBR* – die silberne Uhr in seiner Tasche war um 19.16 Uhr stehengeblieben [21, S. 131]. Die Leiche wurde im Erfrischungsraum von Tay Bridge Station aufgebahrt, das Gesicht unbedeckt. In den folgenden Tagen sollte dieser helle Saal aus poliertem Holz und Spiegelglas die bizarre Leichenhalle werden, wo Verwandte ihre Toten bzw. jene armseligen Kleidungsreste identifizieren konnten, welche den noch immer Vermißten einen Namen geben sollten. Am dritten Tag hatte das Gericht die vorläufige Anhörung beendet, nachdem alle örtlichen Zeugen zum Sturm und unmittelbaren Geschehen des 28. Dezembers 1879 vernommen waren.

Als das Gericht am 26. 2. 1880 erneut in Dundee zusammentrat, hatte man 33 Tote aus dem Fluß geholt. Später barg man noch weitere, aber 29 wurden nie gefunden. Die *NBR* hatte 2 Pfund für die Bergung jeder Leiche ausgesetzt, und es wurde ein Hilfsfond eingerichtet, zu dem die *NBR* 1000 Pfund gab: Bouch gab 250 Pfund, und als die Summe bei 3000 Pfund stand, veranstalteten die Damen von Dundee vornehme Musik-Abende, um den Betrag zu vergrößern. – Der allerletzte Anspruch auf diesen Fond wurde noch 1938 von der betagten Schwester eines Schaffners auf dem «Edinburgh» eingereicht.

In jeder schottischen Gemeinde zur Zeit von *Queen Victoria* gab es drei Männer, die man über alles respektierte und verehrte: den Pfarrer, den Doktor und den Lehrer. In jener Zeit niedriger Löhne, ohne öffentliche Versorgung, war der plötzliche Verlust des Brotverdieners eine Katastrophe. So begannen die Vertreter der 3 Berufe jene Bittbriefe an den Hilfsfond für ihre Schützlinge zu schreiben – Briefe, die noch heute eindringliche Dokumente der sozialen Verhältnisse und dafür sind, was solche Tragödien für die Menschen bedeuteten [26, S. 108 – 110].

Die laufende Anhörung dauerte noch bis zum 3. März, und jetzt trat Aufsehenerregendes über Konstruktion, Unterhalt und Benutzung der Brücke zutage. Um mit dem letzten Punkt anzufangen, war «Benutzung» dieser Brücke seit ihrer Freigabe durch Hutchinson gleichbedeutend mit einer «Geschwindigkeit» von maximal 25 Meilen/Stunde (40 km/h). Die befragten Zugführer weigerten sich hartnäckig, jede Überschreitung dieser Grenze zuzugeben. Denn die regelmäßigen Langsamfahrten von 2 Meilen (3,2 km/h) an den Brückenköpfen – erforderlich um dort den Stafettenstab zu greifen – machten es unmöglich, auf der Brücke dann 40 Meilen/Stunde (64 km/h) zu erreichen. Aber die Aussagen der Fahrgäste widerlegten das, und obgleich die Fahrer weiterhin verneinen, sich Wettläufe mit den Fähren über den Fluß geliefert zu haben, wurde doch klar, daß viele Leute das für eine ganz normale Praxis hiel-

Beaumont's Egg

Although the question of speed could not be definitely resolved there was no question that workmanship at the Wormit foundry was marked by horrifying and dangerous negligence. Indeed, the faulty casting of the piers was found to be an essential cause of the disaster. But the real scandal rested in the manner in which the many "soft spots" in the iron cast at the foundry had been camouflaged, thus deceiving the customers. In silent horror the court listened to the story of a mysterious substance called "Beaumont's Egg... which had no strength whatever"[11]; it was widely used to fill in blow-holes occurring as they may in cast iron. The surface would later to be painted over with oven silver. This egg-like substance consisted of bee's wax, fiddler's rosin, and very fine iron scraps – a mixture which was then melted together and enhanced by a little lamp black. For the melting process, a red-hot iron bar was used, allowing the mass to flow into the holes and harden "like metal". Once hardened, the spot was rubbed smooth with a stone, and after painting no difference could be made out. This mixture would not melt in the sun, but it could be picked out with the point of a pen knife if one wanted to. Only when the column took a sudden blow, did the stuff fall out.

Fergus Ferguson, foreman at the Wormit foundry, admitted the practice, since he himself had placed the miraculous mixture at the disposal of his men – "in a wee box that lay between the turning shop and the moulding shop on a brick wall", as a foundry worker recalled. And the men did as they were told. Another one admitted to have made the substance himself, out of material that he bought with money given to him by Ferguson. Others confirmed having seen blow-holes that were only $1/2$ inch (13 mm) deep on the surface, but may have been up to 2 inches (51 mm) and more inside the casting, and that, finally, castings had gone to the job site honey-combed as a Swiss cheese – and this was the material intended for the then largest bridge of the world! Mr. Ferguson, still a youngish but arrogant and rough man, had nothing but contempt for the uncallused upper class who knew nothing of his trade and were now questioning his ability and honesty. He denied ever having used that mixture but

11 "Beaumont's Egg", a corruption of "Beaumontage" which is any material used for filling cracks in metal or woodwork, acceptable in some cases but not when used as a bluff to hide bad craftsmanship. Said to be named after Elie de Beaumont, the French geologist.

ten. Wenn z. B. Zug und Fähre gleichzeitig von Newport nach Dundee abfuhren, kam der Zug 10 Minuten früher an; hatte er aber Verspätung, konnte der Fahrer noch mit der Fähre gleichziehen oder sie überholen. Nicht wenigen Fahrgästen schienen die hohen Geschwindigkeiten bei diesen Wettkämpfen Spaß zu machen. Aber ein Eisenbahner erzählte, daß er bei der Einfahrt der rasenden Züge in die Südkurve der Brücke jedesmal einen plötzlichen Schlag verspürte, der den Waggon seitlich überhängen ließ, bevor er wieder in die Senkrechte zurückkippte. Schließlich hatte auch der technische Fortschritt im Lokomotivbau nicht haltgemacht: 1879 konnte man dank neuentwickelter Maschinen ohne weiteres schon 70 Meilen/Stunde (112 km/h) erreichen.

«Beaumont's Egg»

Wenn auch die Sache mit der Geschwindigkeit nicht vollends geklärt werden konnte, so gab es kaum Zweifel, daß die Arbeit in der Wormit-Gießerei von erschreckender und gefährlicher Nachlässigkeit war. In der Tat erwies sich der mangelhafte Eisenguß der Pfeiler als eine wesentliche Unglücksursache. Das eigentlich Skandalöse daran war aber die Art, wie man in der Gießerei «weiche Stellen» kaschiert und so die Abnehmer bewußt getäuscht hatte: In starrem Staunen hörte das Gericht die Geschichte von einer mysteriösen Substanz – genannt *«Beaumont's Egg... which had no strength whatever»*[11], mit der man damals Luftlöcher und Dellen, wie sie im Gußeisen vorkommen können, ausfüllte und außen mit Ofensilber überpinselte. Diese Ei-ähnliche Masse bestand aus Bienenwachs, Geigenharz *(fiddler's rosin)* und sehr feinen Eisenspänen – eine Mixtur, die man dann einschmolz und mit etwas Lampenruß versetzte. Zum Schmelzen diente ein rotglühender Eisenstab, damit die Masse in die Löcher fließen und sich wieder «wie Metall» verfestigen konnte. Hartgeworden, rieb man die Stelle mit einem Stein glatt und konnte nach dem Anstreichen keinen Unterschied mehr erkennen. Auch schmolz diese Mixtur nicht an der Sonne, jedoch konnte man sie mit der Spitze eines Taschenmessers herauspicken, wenn man wollte. Nur wenn die Säule einen plötzlichen Schlag erhielt, fiel das Zeug heraus.

11 «Beaumont's Egg»: Eine Verballhornung von «Beaumontage»: Gemeint ist jedes für die Füllung von Rissen in Metall oder Holz geeignete Material – in manchen Fällen durchaus akzeptabel, jedoch nicht, wenn es schlechte Werkarbeit vertuschen soll. Angeblich nach Elie de Beaumont, einem französischen Geologen, so benannt.

admitted to have known of course of its existence. He admitted – unaware of the ramifications of what he was saying – to have casted much iron according to his own judgement of what was right; and that there was a great waste of iron at the foundry, due to bad casting – information which however he managed to present as proof of his conscientiousness. Questioned harder, he admitted that the original number of 30 to 40 columns that had to be scrapped in fact lay closer to 60 to 70; under further questioning, this figure finally came to "about 200, for one defect or another". How many columns were not sorted out but sent to the site – stuffed full with the magic egg – could only be guessed at in the end!

There were many stories going around the public about the working conditions at the foundry. It was common knowledge that workmanship there was sloppy, that there was no real supervision and the workmen did as they liked. Quite a few citizens thought the bridge was bound to fall down some day. Also, the *Daily Chronicle* had surmised soon after the accident that the total destruction of the piers, of which only small humps jutting above the water remained, would lead to the conclusion that the *piers* were the essentially weak spot of the whole structure, unable to withstand the extraordinary gales of that night. It should be hoped that the inexcusable negligence which had not only claimed so many lives, but had also dealt such a grave blow to the reputation and the honour of British engineering, would be proven and brought to public attention without mercy *(Fig. 44 and 45)*.

By this stage of the inquiry, it had become quite evident that Bouch had left his supervision to the firm of Hopkins, Gilkes & Co, which in turn had relied on Mr. Groethe, who in turn – as far as the casting of iron was concerned – had entrusted his responsibility to foreman Ferguson, according to Hammond. According to Eyth, however, he had repeatedly sent faulty columns back to the workshop:

"A considerable number of columns has to be returned to the foundry, which Lavalette [de Bergue] is accepting with commendable grace. No column shall pass which is not faultless to me, as far as material and workmanship goes. Unfortunately, I can't keep my eyes everywhere. As long as Lavalette [de Bergue] was in complete health, I could rest assured. Even though he is the contractor, he is a man upon whom one can build like on our rocks."

Rejection of faulty casting [4, p. 436]
("Lavalette" = de Bergue)

Fergus Ferguson, Werkmeister der Wormit-Gießerei, gab den Sachverhalt zu, hatte er doch selbst seinen Arbeitern das Wundermittel bereitgestellt – «*in a wee box that lay between the turning shop and the moulding shop on a brick wall»,* wie ein Gießer sagte. Und die Arbeiter taten, wie ihnen geheißen. Ein anderer gab an, die Substanz selbst aus dem Material hergestellt zu haben, das er mit Geld von Ferguson gekauft hatte. Weitere Arbeiter bestätigten, daß man oft Luftlöcher gefunden habe, die an der Oberfläche nur $1/2$ Inch (13 mm) ausmachten, im Inneren des Gußstückes aber bis zu 2 Inch (51 mm) und mehr betragen konnten; und daß schließlich Gußrohre zur Baustelle gegangen waren – so hundertfach durchlöchert wie ein Schweizer Käse – und das für die damals größte Brücke der Welt! Mr. Ferguson, ein noch relativ junger, aber arroganter und grober Mann hegte nichts als Verachtung für weißhändige Nobelmänner, die nichts von seinem Handwerk verstanden und jetzt seine Könnerschaft und Redlichkeit in Frage stellten. Er verneinte, die Substanz je verwendet zu haben, hätte aber natürlich von ihrer Existenz gewußt. Er gab zu – ohne zu wissen, was er damit sagte – daß er viel Eisen nach seinem eigenen Urteil, was richtig sei, gegossen habe; auch, daß es auf der Gießerei eine große Verschwendung von Eisen wegen mangelhaftem Guß gegeben hatte – was er aber wie einen Beweis für seine Gewissenhaftigkeit anzubringen wußte. Unter hartem Nachfragen gab er zu, daß die Zahl von ursprünglich ca. 30 bis 40 Säulen, die er als unbrauchbar verschrotten mußte, in Wahrheit näher bei 60 oder 70 lag; nach weiterem Insistieren waren es schließlich «wegen des einen oder anderen Fehlers ungefähr 200» geworden. Wie viele nicht ausgesondert, sondern zur Baustelle geschickt worden waren – vollgestopft mit dem Wunder-Ei – konnte man nur noch erraten!

In der Bevölkerung erzählte man sich ohnehin so allerhand über die Zustände in der Gießerei. Es war allgemein bekannt, daß dort schlampig gearbeitet wurde, es keine richtige Aufsicht gab und die Arbeiter machten, was sie wollten. Mancher Bürger dachte, daß die Brücke eines Tages so oder so herunterfallen würde. Auch hatte der «*Daily Chronicle*» schon frühzeitig vermutet, daß die vollständige Zerstörung der Pfeiler, von denen nur noch kleine Reste aus dem Wasser ragten, darauf schließen lasse, daß die *Pfeiler* der eigentliche Schwachpunkt der ganzen Konstruktion waren und den außerordentlichen Stürmen jener Nacht nicht standzuhalten vermochten. Es sei zu hoffen, daß die sträfliche Leichtfertigkeit, welcher nicht nur soviel Menschen zum Opfer gefallen seien, sondern die namentlich auch dem Ansehen und der Ehre des englischen Ingenieurwesens einen schweren

44

"It is clear... that (the structure) must be one of a broader base, of less height and one less top-heavy..." The tall central piers of Bouch's old Tay Bridge (to the right) were still standing side-by-side Barlow's new piers for some time, until they were demolished down to the masonry base [19, p. 215].

44

«Es ist klar, ... daß (das Bauwerk) eine breitere Basis haben muß, von geringerer Höhe und weniger kopflastig sein muß...» Die hohen Mittelpfeiler von Bouchs alter Brücke (rechts) standen noch für einige Zeit Seite an Seite mit den neuen Pfeilern Barlows und wurden später, bis auf die gemauerten Pfeilersockel, abgetragen [19, S. 215].

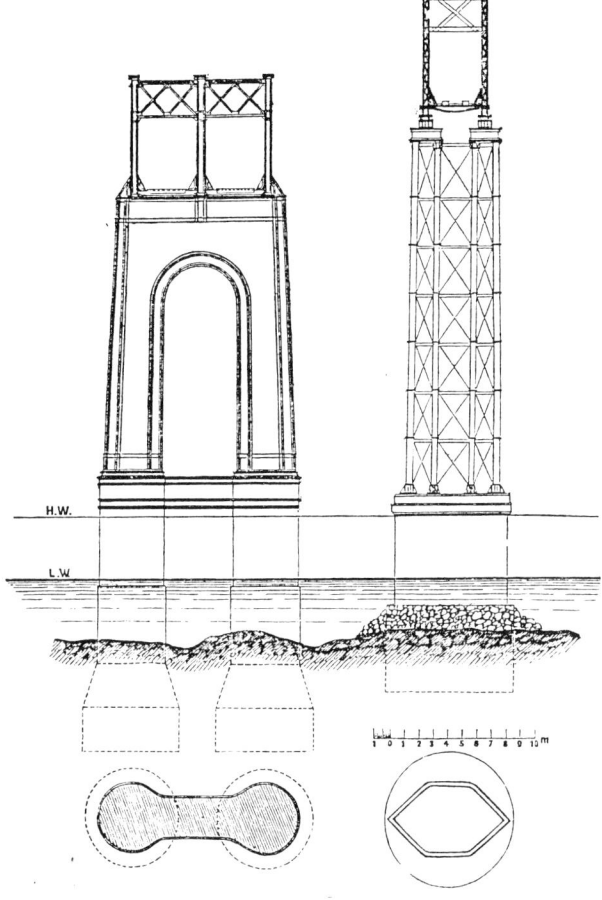

Furthermore, it had even been left to Mr. Ferguson to decide for himself on the thickness of the walls of the columns – and he would then have given rather a bit more iron, as he said, than too little. With such supervision – or as the Rothery Report stated – with the absence of all supervision, the further course of events was no longer surprising.

Henry Noble's Role

When the court reconvened on April 19, 1880, for the third stage of the inquiry, this time in London, the solicitors of the contractors and of Bouch were present. During the next 16 days, a thorough and exhaustive investigation of design, construction and maintenance of the bridge took place. The first witness was perhaps, next to Bouch, the saddest figure of the whole inquiry – Henry Abel Noble, formerly Bouch's assistant and then resident inspector of the bridge during its short lifespan. Now his pitiful qualifications for

Schlag versetzt habe, nachgewiesen und in rücksichtsloser Weise an den Pranger gestellt werde *(Abb. 44 und 45).*

Spätestens in dieser Phase der Untersuchungen war klargeworden, daß Bouch seine Aufsichtspflicht der Baufirma Hopkins, Gilkes & Co überlassen hatte, diese wiederum verließ sich auf Mr. Groethe, welcher seinerseits, soweit es den Eisenguß betraf, dem Werkmeister Ferguson vertraute, zumindest nach Hammond, – jedoch nach Eyth immer wieder schadhafte Säulen in die Werkstatt zurückgehen ließ:

> Eine ziemliche Anzahl der Säulen muß in die Gießerei zurückwandern, was sich Lavalette mit löblicher Ergebung gefallen läßt. Es soll mir wenigstens keine passieren, die nicht nach Material und Ausführung tadellos ist. Leider kann man die Augen nicht überall haben. Solange Lavalette vollständig gesund war, konnte ich ruhiger sein. Er ist zwar der Unternehmer, aber ein Mann, auf den man bauen kann wie auf unsre Felsen.

Reklamation schlechter Gußarbeit [4, S. 436]
(«Lavalette» = de Bergue)

Im übrigen war es Mr. Ferguson sogar überlassen geblieben, die Wandstärke der Säulen selbst zu bestimmen – und er habe dann lieber, wie er versicherte, etwas mehr Eisen hinzugegeben als zuwenig. Mit solcher Aufsicht – oder, wie der *Rothery-Report* sagt, mit solchem Fehlen jeglicher Aufsicht, ist der weitere Lauf der Dinge nicht verwunderlich.

Henry Noble's Rolle

Als das Gericht am 19. April 1880 zur dritten Untersuchungsphase zusammentrat, diesmal in London, waren die Anwälte der Unternehmer und Bouch zugegen. Während der folgenden 16 Tage kam es zu einer gründlichen und entscheidenden Überprüfung von Entwurf, Bau und Unterhalt der Brücke. Der erste Zeuge war vielleicht neben Bouch die traurigste Figur der ganzen Untersuchung – Henry Abel Noble, einst Bouch's Assistent und dann ortsansässiger Inspekteur der Brücke während ihrer kurzen Lebenszeit. Seine Redlichkeit stand nie in Frage, lediglich seine kümmerlichen Qualifikationen für diese Aufgabe. Sein stärkstes Motiv schien der bewundernswerte Wunsch, seinem Arbeitgeber Geld zu sparen: Einmal hatte er selbst einige Tauchmanöver unternommen, anstatt die Gesellschaft um Vergütung für einen Berufstaucher anzugehen. Als ausgebildeter Maurer hatte er keinerlei Erfahrung im Eisenbau; aber es war

45
The subsequent shifting of the undamaged girders, from Bouch's old Tay Bridge unto Barlow's new piers, 1885/86. To the extreme right, the configuration of cast-iron piers in the old bridge can be seen [23, plate 49].

45
Das spätere Versetzen der unbeschädigten Fachwerkträger von Bouchs alter Tay-Brücke auf die neuen Pfeiler Barlows, 1885/86. Ganz rechts erkennt man die Ausbildung der gußeisernen Pfeiler der alten Brücke [23, Tafel 49].

46
The south curve of the new Tay Bridge as it appears today – seen from the same vantage point as in the historic photograph by G. W. Wilson, summer of 1879 (compare *Fig. 28*). The maximum speed on the bridge has been raised, over the last 100 years, from 25 to 35 miles/hour (40 to 56 km/h); this seems rather modest progress, considering the early dreams in those days to build a tunnel under the British Channel.

46
Die Südkurve der neuen Tay-Brücke in heutigem Zustand – von gleichem Standpunkt wie in der historischen Aufnahme von G. W. Wilson aus dem Sommer 1879 gesehen (vergl. *Abb. 28*). Die zulässige Geschwindigkeit auf der Brücke ist, wie man sieht, in 100 Jahren immerhin von 25 auf 35 Meilen/Stunde (40 auf 56 km/h) angehoben worden. Angesichts der Träume der damaligen Erbauer von einem baldigen Tunnel durch den Ärmelkanal ist das ein recht bescheidener Fortschritt.

this task were on trial, never his integrity. Indeed, his strongest motive seemed to have been an extraordinary desire to save money for his employers. Once he had even done some diving himself instead of asking his company to pay for a professional diver. As a trained mason and inspector of brickwork, he had no knowledge whatsoever of iron structures; during all his duty as bridge inspector from May 1878 to December 1879, Noble had been left totally to his own and without guidance from anyone with engineering skill. Indeed, he had not even been instructed by the NBR or the bridge company to actually report any defects to those responsible for their remedy.

Conscientious as he was, Noble had taken it upon himself to check the cast-iron columns more closely; and of course he had discovered faults – they were all too evident. He found alarming cracks in the columns of four piers below the high girders – just wide enough to insert a sheet of paper. Some of those cracks were from 4 to 7 ft long (1.22–2.13 m). He tore a page from his notebook, wetted it with saliva and pasted it across the crack; then he waited for the next train, and when the paper failed to tear he concluded that the crack had not widened. Then he pushed a thin wire (from the cap of a ginger-beer bottle) into the crack and was relieved that it did not get through to the pier's concrete core. One must admit that this man was resourceful! In any case, he assumed his bridge to be safe. And Bouch's confidence in him was unshakable – on several oc-

auch während seiner ganzen Tätigkeit von Mai 1878 bis Dezember 1879 nie ein Fachmann erschienen, um ihn etwa zu unterstützen oder gar selbst die Eisenkonstruktion zu überprüfen. Ja, Noble war völlig auf sich selbst gestellt und ohne Weisung der NBR bzw. der Brückengesellschaft geblieben, etwaige Schäden auch weiterzumelden.

Gewissenhaft, wie er war, hatte Noble von sich aus damit begonnen, die gußeisernen Säulen näher zu untersuchen, und natürlich auch Schäden gefunden – sie waren ja überdeutlich. So entdeckte er alarmierende Risse in den Säulen von vier Pfeilern der Mittelträger – gerade weit genug, um ein Blatt Papier hindurchzuschieben. Manche dieser Risse waren zwischen 4 bis 7 Fuß lang (1,22 – 2,13 m). Also riß er eine Seite aus seinem Notizbuch, befeuchtete sie mit Speichel und klebte sie über den Riß; dann wartete er, bis oben der nächste Zug durchfuhr, und als das Papier nicht einriß, folgerte er, daß der Spalt sich nicht erweiterte. Dann stieß er einen dünnen Draht (vom Verschluß einer Ginger-Bierflasche) in den Spalt und war erleichtert, daß dieser nicht bis zum Betonkern der Brücke durchdrang: Man muß einräumen, daß sich der Mann zu helfen wußte! Jedenfalls glaubte er, daß die Brücke sicher war. Und Bouch's Vertrauen auf ihn war nicht zu erschüttern – hatte doch Noble wiederholt aus eigener Tasche Metallstücke besorgt, um lose Teile an den Trägern zu sichern.

Auf Noble's Bericht über die Risse ließ Bouch sie, ebenso wie Mauerwerksrisse an 12 Pfeilersockeln, mit schmiedeeisernen Ringen und Faßbändern überbrücken. Noch eine Woche vor dem Unglück hatte er auf einem seiner seltenen Besuche diese Reparaturen begutachtet, sich aber nicht übermäßig besorgt über die Risse gezeigt. Seine angegriffene Gesundheit erlaubte ihm auch nicht mehr so häufige Besuche, wie er wollte; und nach Eyth «wird der alte Herr täglich behaglicher und eingebildeter. Die Brücke war zuviel für sein moralisches Gleichgewicht» [4, S. 466]. Zu guter Letzt entdeckte man, daß der Gezeitenstrom das Flußbett an einigen Pfeilerfüßen ausgewaschen hatte, worauf diese Hohlstellen mit losen Steinen aufgefüllt wurden. – All diese Remeduren ergaben aber doch in ihrer Summe für die Kommission ein Bild kaum vorstellbarer Sorglosigkeit beim Unterhalt eines so großen Bauwerkes – oder, wie wir heute sagen würden, von «Pfusch am Bau».

Abgesehen von den offensichtlichen Fehlern beim Eisenguß war der Durchmesser der verstrebten Rohre von einem Ende zum anderen sehr unterschiedlich ausgefallen, und die Löcher für die Bolzen waren oft zu weit und nicht zylindrisch gebohrt, so daß diese locker saßen. Schon ein Jahr zuvor waren schreckliche schla-

47
Sir Thomas Bouch (1822–1880) in his later life [21, p. 118]. This portrait-drawing communicates something about the character of this Victorian. According to Beckett [1], he had been shouldered with an unfair measure of blame for the collapse of the Tay Bridge. Blessed with a considerable breadth of imagination, Bouch remains – despite all his personal tragedy – one of the key figures behind the origin of both bridges, at the Forth and the Tay.

47
Sir Thomas Bouch in späteren Jahren (1822–1880) [21, S. 118]. Die Portrait-Zeichnung vermittelt etwas vom Charakter dieses Viktorianers. Nach Beckett [1] war ihm ein unfaires Maß an Schuld für den Brückeneinsturz aufgeladen worden. Bei aller persönlicher Tragik bleibt Bouch eine Haupt- und Schlüsselfigur für die Entstehungsgeschichte beider Brücken – am Forth und am Tay.

casions, Noble hat dug into his own pocket to pay for pieces of metal in order to secure loose parts on the girders.

Upon receiving Noble's reports about the cracks, Bouch had them strapped with rings and hoops of wrought iron. Similarly, the masonry cracks appearing in 12 bases of the piers had been fixed. Only one week before the accident, Bouch had examined those repairs on one of his rare visits, but he did not seem overly concerned about the cracks. His ill health would not allow him as many visits as he would have liked; yet according to Eyth, "the old gentleman is turning more comfortable and complacent by the day. The bridge was too much for his moral equilibrium." [4, p. 466]. Finally, Noble had discovered that the current of the tide had scoured the riverbed at the footings of some piers, whereupon the hollows were filled with loose stones. – Taken together, all of these repairs created for the commission an image of near-incon-

gende Geräusche aufgetreten, wenn ein Zug über die Brücke fuhr [13, S. 160]. Und Mr. Noble bekannte, daß nach jedem Zug ein paar Eisenteile aus dem Gestänge gefallen waren [25, S. 232]. Die kreuzförmigen Windverbände und Zugstangen zwischen den Pfeilersäulen (es gab 7 solche Verbände in jedem der hohen Pfeilertürme) hatten sich immer mehr gelöst, so daß die ganze Brücke zitterte und schwankte, wenn ein Zug zu rasch darüberging. Signalmann Barclay hatte es kommen sehen, seit Monaten, wie er später aussagte. Erst eine Woche vor dem Unglück sei wieder ein Dutzend Bolzen herausgefallen [4, S. 485].

Bouch's Zeugnis

Schließlich, am 30. April 1880 wurde Sir Thomas Bouch vorgeladen, und trotz der Belastung der letzten Monate und des wachsenden Risikos für seine Reputation oder Selbstachtung, schlug er sich tapfer, diesmal

48
The former residence of Sir Thomas Bouch, today: Nr. 6 Oxford Terrace in Edinburgh's West End.

48
Der ehemalige Wohnsitz von Sir Thomas Bouch heute: Nr. 6 Oxford Terrace in Edinburgh's West End.

ceivable negligence in the maintenance of so large a structure.

Aside from the obvious faults in casting, the diameter of the braced columns had turned out quite uneven from one end to the other, and the lugs for the bolts had often been drilled too wide and not cylindrically, so the bolts were sitting loose. Already the year before, terrible banging noises had occurred whenever a train was going across the bridge [13, p. 160]. And Mr. Noble admitted that with every train passing, a few iron pieces had fallen off the girders [25, p. 232]. The bracings and wind-ties between the columns (there were seven such bracings in each of the tall piers) had loosened more and more, so the bridge trembled and shook whenever a train went overhead too fast. Signalman Barclay had seen it coming, months ago, as he recalled later. Just a week before the accident, yet another dozen of the bolts had fallen out [4, p. 485].

Bouch's Testimony

At last, on April 30, 1880, Sir Thomas Bouch was summoned to the court. Despite the stress of the preceding months and the growing risk to his professional reputation and self-esteem, he carried himself gallantly, even with a touch of justifiable pride this time. To the question as to how many bridges he had built, he declared "I do not suppose anybody has built more." He defended his assistants as loyally as he upheld this type of bridge, it being the result of twenty years' experience. The collapse was not due to faulty design, he asserted and during the days in Dundee, shortly after the accident, he had formed his own assessment of how it might have happened:

> "Well, I have thought about it very anxiously, and my opinion is fixed now; that it was caused by the capsizing of one of the last, or two of the last carriages – that is to say, the second-class carriage and the van; that they canted over against the girder.... Practically the first blow would be the momentum of the whole train until the couplings broke. If you take the body of the train going at that rate, it would destroy anything." [21, p. 135]

Thus, according to Bouch, *first* part of the train would have been lifted or toppled over from the rails by a sudden gale, and only *then,* the train would have wrecked the trusses of the eastern girder, bringing down the whole bridge in the end. This version was Bouch's "derailment theory", which he saw supported by the fact that the last carriages were indeed much more severely destroyed than the ones in front, still coupled to the engine after the fall. His theory meant, in fact, that

49
The sketch by John Waddell showing the grinding marks upon the girder fragments recovered [26, p. 165].

49
Die Skizze von John Waddell zeigt die Schleifspuren an den geborgenen Träger-Fragmenten [26, S. 165].

auch mit einem Anflug berechtigten Stolzes. Auf die Frage, wie viele Brücken er entworfen habe, erklärte er: «Ich glaube nicht, daß irgend jemand mehr gebaut hat.» Er verteidigte seine Mitarbeiter ebenso loyal wie diesen Brückentyp – das Ergebnis 20jähriger Erfahrung. Der Einsturz war nicht auf den Entwurf zurückzuführen, und während der Tage in Dundee, gleich nach dem Unglück hatte er sich seine eigene Meinung darüber gebildet:

> «Nun, ich habe darüber sehr angestrengt nachgedacht und meine Meinung steht jetzt fest: Daß es verursacht worden ist durch das Umkippen eines der letzten, oder zwei der letzten Waggons – nämlich des 2. Klasse-Wagens und des Gepäckwagens; daß diese sich seitlich gegen die Träger geworfen hatten... Praktisch der erste Stoß würde das Gesamtmoment des Zuges ausmachen, bis die Kupplungen brechen.

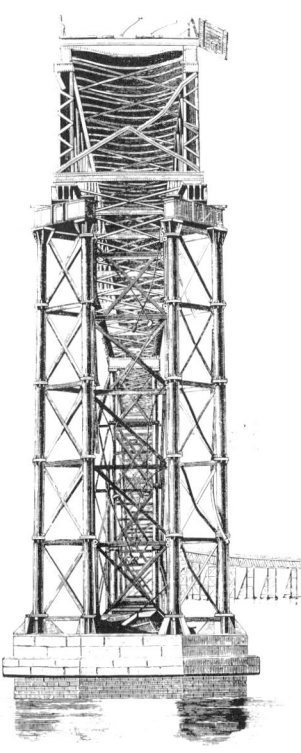

50
The end of the northern half of the Tay Bridge, intact after the collapse [19, p. 212].

50
Das Ende der beim Einsturz stehengebliebenen Nord-Hälfte der Tay-Brücke [19, S. 212].

the derailing train must have brought down the bridge, not vice-versa. This was the reason Bouch had been so adamant, albeit in vain, during the lifting manoeuvre in Dundee, not to have the engine brought up first, but to have it raised *together* with the large segment of the girder wherein it was lying. Still on April 19, things did not look too bad at all for this theory. Bouch had received the following telegram from Mr. Armit, the salvage man:

> "Have lifted portions of the east lattice number four girder with strong evidence of train having been running against it. Photograph now being taken."

Sketches of the fragments of girders brought up *(Fig. 49)* showed horizontal scratch marks which, according to Bouch, could only have been caused by the train running against the girder. These sketches had been prepared in Dundee by John Waddell, a railway contractor from Edinburgh, and were sent to London during the hearings. Bouch's assistants and solicitors were already regaining new hope – not the least of their hopes was based on the cautious manner by which Rothery was questioning the designer of the bridge that failed.

But it was to no avail – instead, the proceedings seemed to have reached a state now where the fate of an individual no longer mattered; instead other interests gained priority. The national disgrace that was felt over the numerous faults in supervision and maintenance, was perceived as a shameful calamity to the engineering profession, indeed to the nation as a whole. The indisputably thorough investigations had made public, point by point, an unimaginable sequence of blunders and downright errors, which some people would rather have left in the dark. A careful assessment of the actual extent of Bouch's guilt, partial guilt or innocence now had to recede behind the national task "to clean up". For this task Chairman Rothery was the right man.

The court of inquiry finally turned to the decisive question of the wind pressure. This issue had raised tremendous excitement ever since the disaster, not just among engineering circles, and it had been widely discussed by the papers. Bouch's frank answers buried any hope he may still have held to get out of this trial without blame:

> Question: Sir Thomas, did you in designing this bridge, make any allowance at all for wind pressure?
> Answer: Not specially.
> Question: You made *no* allowance?
> Answer: Not specially.
> Question: Was there not a particular pressure

Wenn man die Masse des Zuges mit solcher Geschwindigkeit annimmt, würde sie alles zerstören.» [21, S. 135]

Damit wäre also *zuerst* ein Teil des Zuges durch eine Sturmböe aus den Gleisen gehoben bzw. umgekippt worden, hätte dann seitlich das Fachwerk des östlichen Hauptträgers zertrümmert und so letztlich die gesamte Brücke zum Einsturz gebracht. Für diese «Entgleisungs-Theorie» Bouch's sprach auch, daß die letzten Waggons tatsächlich viel stärker zerstört waren als die vorderen, welche nach dem Sturz noch an die Lokomotive gekuppelt waren. Seine Theorie bedeutete, daß der entgleisende Zug die Brücke zum Einsturz gebracht haben mußte, nicht umgekehrt. Deshalb war auch Bouch bei den Hebungen vor Dundee besonders darauf erpicht gewesen, allerdings vergeblich, daß nicht zuerst die Lokomotive, sondern diese *zusammen* mit dem großen Trägersegment, in dem sie lag, geborgen werden sollte. Und noch am 19. April standen die Dinge gar nicht schlecht für seine Theorie; er hatte folgendes Telegramm von Mr. Armit, dem Bergungsmann erhalten:

> «Habe Teile des Ostfachwerkes Nr. 4 Träger gehoben mit deutlichem Beweis, daß Zug dagegen gelaufen ist. Foto wird gerade gemacht.»

Abb. 49 zeigt Skizzen der gehobenen Trägerfragmente mit horizontalen Schleifspuren, die laut Bouch nur vom Zuge stammen konnten. Die Skizzen waren in Dundee von John Waddell, einem Eisenbahnunternehmer aus Edinburgh, angefertigt und noch während der Anhörungen nach London geschickt worden, Bouch's Mitarbeiter und Anwälte begannen schon, neuen Mut zu schöpfen – nicht zuletzt auch wegen der eher vorsichtigen Art und Weise, mit der Rothery den Schöpfer der eingestürzten Brücke befragte.

Aber es half alles nichts – vielmehr schien das Verfahren inzwischen einen Punkt erreicht zu haben, wo nicht mehr das Schicksal eines einzelnen, sondern weitergehende Interessen in den Vordergrund traten: Die nationale Schande, welche man über die vielen Versäumnisse bei Bauaufsicht und Unterhalt empfand, wurden als Schmach für den Berufsstand der Ingenieure – ja für die britische Nation angesehen. Die ohne Zweifel gründlichen Untersuchungen hatten, Zug um Zug, eine bislang kaum für möglich gehaltene Kette von Schlampereien und handfesten Fehlern ans Licht der Öffentlichkeit gebracht, die so mancher lieber im Dunkel gelassen hätte. Eine sorgfältige Abwägung über das tatsächliche Ausmaß von Schuld, Mitschuld oder Unschuld im Falle Bouch's trat in den Hintergrund vor der nationalen Aufgabe, «aufzuräumen» – und dafür war der Vorsitzende Rothery ein guter Bürge.

had in view by you at the time you made the design?
Answer: I had the report of the Forth Bridge.

Wind Pressure

It should be noted that during the decade from 1870 to 1880, British engineers held only very vague conceptions about the actual wind pressures on structures; in France and North America, progress in this field had been more rapid, where certain methods for calculation had been developed already. This lagging behind in Britain had already become evident during the questioning of Albert Groethe. He had been summoned from Spain where he was working at the time. He admitted that in his opinion the bridge had been destroyed solely by force of the wind. On the question about the allowance customary for wind pressure, he admitted that his notions regarding this point had been very erroneous at that time; they had been corrected substantially by what he had learned since the fall of the Tay Bridge.

Bouch himself, in considering the wind pressure, had followed the over 100-year-old figures of the Royal Astronomer John Smeaton of 1759 – according to the motto, "in case of doubt, ask Greenwich": These figures called for 6 lbs/sq ft (29.26 kg/m^2) for "high winds", 9 lbs/sq ft (43.88 kg/m^2) for "very high winds" and 12 lbs/sq ft (58.51 kg/m^2) for "storm or tempest" – while his contemporary French and American engineers were already working with 50 to 55 lbs/sq ft (approx. 256 kg/m^2)!

Still, there was a letter addressed to Bouch by the then Royal Astronomer, Sir George Airy, at the Greenwich Observatory of April 1873, when the Tay Bridge had long been started. Apparently, the letter was preceded by an inquiry from Bouch regarding the wind pressure for the Forth Bridge, then in the planning stage; but Bouch seemed to have considered this answer also applicable to the Tay Bridge. The letter states that according to the knowledge of the Royal Astronomer, the wind pressure acting upon very limited surfaces and for very limited times does amount sometimes to 40 lbs/sq ft (195 kg/m^2), and in Scotland probably more. Thus it could be concluded that, when calculated over the entire length of the bridge, a plane surface would be exposed to a *maximum wind pressure of 10 lbs/sq ft (48,75 kg/m^2)*; he would not think that a wind pressure of 40 or 50 lbs acting on only a limited portion of a span could pose a danger, as such pressures could be due to "irregular swirlings of the air" [8, p. 32]. This proved once again how little was actually known about the real effects of wind

Das Untersuchungsgericht wandte sich zuletzt auch der entscheidenden Frage nach dem Winddruck zu. Das brisante Thema hatte seit dem Einsturz gewaltiges Aufsehen erregt, nicht nur bei der Fachwelt, und war in den Zeitungen weidlich ausgeschlachtet worden. Bouch's offene Antworten dazu beendeten jede Hoffnung, die er noch gehabt haben mag, aus dieser Sache ohne Schuld herauszukommen:

> Frage: Sir Thomas, haben Sie beim Entwurf dieser Brücke irgend eine Sicherheits-Toleranz für Winddruck berücksichtigt?
> Antwort: Nicht speziell.
> Frage: Sie haben *keine* Toleranz vorgesehen?
> Antwort: Nicht speziell.
> Frage: Gab es da nicht eine bestimmte Druckbelastung, die Sie im Blick hatten, als Sie den Entwurf anfertigten?
> Antwort: Ich hatte den Bericht von der Forth-Brücke.

Windbelastung

Man muß sich klarmachen, daß damals, d.h. im Jahrzehnt von 1870 – 1880, bei britischen Ingenieuren erst sehr vage Vorstellungen über die tatsächlichen Windkräfte auf Bauwerke herrschten; in Frankreich und Nordamerika war man auf diesem Gebiete schon weiter fortgeschritten, hatte auch schon bestimmte Berechnungsfaktoren dafür entwickelt. Dieser Rückstand war bereits bei der Vernehmung von Albert Groethe deutlich geworden: Nachdem man ihn aus Spanien, wo er inzwischen arbeitete, vorgeladen hatte, bekannte er, daß seiner Meinung nach die Brücke einzig vom Wind zerstört worden war. Auf die Frage nach der üblichen Sicherheitstoleranz für Winddruck gab er zu, daß seine Vorstellungen zu dem Punkt damals sehr irrig gewesen seien; sie wären wesentlich korrigiert worden durch das, was er seit dem Einsturz der Tay-Brücke erfahren habe.

Bouch selbst hatte sich bei der Berechnung der Windlast nach dem Motto «*in case of doubt, ask Greenwich*» an die über 100 Jahre alten Zahlen des königlichen Astronomen John Smeaton vom Jahre 1759 gehalten: 6 Pfund/Quadratfuß (29,26 kg/m^2) für «starke Winde», 9 Pfund/Quadratfuß (43,88 kg/m^2) für «sehr starke Winde» und 12 Pfund/Quadratfuß (58,51 kg/m^2) für «Sturm oder Orkan» – während zeitgenössische französische und amerikanische Ingenieure bereits mit 50 – 55 Pfund (ca. 256 kg/m^2) Windlast arbeiteten! Jedoch gab es einen Brief an Bouch von dem damaligen königlichen Astronomen, Sir George Airy am Greenwich Observatorium vom April 1873, als die Tay-Brücke längst im Bau war. Wie es scheint, war dem

pressure on large spans. Once an allowance with the factor of 2 is applied, the actual wind pressure of 10 lbs/sq ft which was then held to be valid – at least in Britain – would have meant a *permissible wind pressure of 20 lbs/sq ft* (97.52 kg/m^2) for calculation.

Upon further questioning, Sir Airy thought the greatest known wind pressures of recent years in Britain may have been around 50 lbs/sq ft (243.80 kg/m^2). Possibly this could have been proven by his instruments during a particularely fierce storm in December of 1872, – but "unfortunately our recording pencil broke at a particular point". Therefore he would be unable to say what wind pressure in the night of the "Great Storm" might have been along the Tay, as there were no measuring instruments in Dundee. At his residence in Greenwich, the highest wind pressure had been only 10 lbs/sq ft [8, p. 32].

Bouch admitted under further questioning not to have taken any special provisions or safety measures against exceptional wind loads. Indeed, there was not

Bouch's wind theory
"Sir William Bruce" = Sir Thomas Bouch [4, pp. 437, 473]

"The worst was not the simple load-bearing capacity. The lattice girders are quite in order, I think. Even later on, when the tall central piers had to be spaced further apart, this part of the task was dealt with in such a way that we didn't need to worry. But we were totally in the dark about how to calculate the air pressure against the whole structure. Bruce didn't want to hear anything about this. 'Wind, wind!' he called whenever I brought up the subject; 'Anything that supports six heavy locomotives hovering up in mid-air, no wind will topple over!' This was his theory, and there is something to be said for that. During my weak moments, I was literally clinging on to it. Still, one knew, and still knows today, abysmally little about the air pressure of a storm. We assume 20 lbs/sq ft [97.52 kg/m^2]. With that my piers as I had designed them originally, must be standing like rocks. Subsequently, when the bridge had already been completed halfway across the bay, I learned that the government engineers in France were using 40 lbs/sq ft [195 kg/m^2]. And just a year ago, an acquaintance from America wrote to me that they were using 50 lbs/sq ft [243.80 kg/m^2] over there, and the American engineers are not overly cautious, as the whole world knows. – But the question only came up in earnest much later, once everything was already well under construction. Nobody, not even myself, paid much attention to it in the beginning. We believed in Bruce [Bouch], and Sir William [Sir Thomas] believed in himself and his intuition. During the last days, when the calculations on which the whole bridge project is based, came to an end, I still had a spirited battle with myself – which safety coefficient might I trust?
In Greenwich, at the Royal Observatory, they also know nothing, not even how fast a good Scottish gale is blowing. During the best experiments they can remember, the measuring instrument broke down regularly at the critical moment, and until they had it fixed up again, the storm was over."

Brief eine Anfrage Bouchs bezüglich des Winddruckes für die neu zu planende Firth-of-Forth-Brücke vorausgegangen – worauf Bouch aber dann die Antwort auch auf die Tay-Brücke bezogen hat. Jedenfalls heißt es darin, daß nach dem Wissen der Astronomen der Winddruck auf sehr begrenzte Oberflächen und sehr kurze Intervalle 40 Pfund/Quadratfuß (195 kg/m^2) betragen kann und in Schottland wahrscheinlich noch mehr. Woraus folge, daß auf die ganze Brückenlänge gerechnet, eine ebene Oberfläche einem *größten Winddruck von 10 Pfund/Quadratfuß* (18,75 kg/m^2) ausgesetzt sei; er glaube nicht, daß ein Winddruck von 40 oder 50 Pfund auf nur einen begrenzten Teil eines Brückenfeldes gefährlich wäre, zumal solche Belastungen von unregelmäßigen Luftwirbeln – «*irregular swirlings of the air*» – herrühren könnten [8, S. 32]. Was wieder einmal zeigte, wie wenig man im Grunde über die tatsächliche Wirkung von Windkräften auf weitgespannte Konstruktionen wußte. Legt man einen Sicherheitsfaktor von 2 zugrunde, so hätte der da-

Bouch's Wind-Theorie
«Sir William Bruce» = Sir Thomas Bouch [4, S. 437, 473]

„Das Schlimmste war nicht die einfache Tragfähigkeit. Mit den Gitterbalken ist, glaube ich, alles in Ordnung. Auch später, als die hohen Mittelpfeiler weiter gestellt werden mußten, wurde dieser Teil der Aufgabe so behandelt, daß wir keine Sorge zu haben brauchten. Aber in völligem Dunkel war man mit der Berechnung des Luftdrucks gegen die ganze Struktur. Bruce wollte hiervon überhaupt nichts wissen. „Wind! Wind!" rief er, wenn ich auf das Kapitel zu sprechen kam; „was sechs schwere Lokomotiven freischwebend trägt, wirft kein Wind um!" Das war seine Theorie, und sie läßt sich anhören. In schwachen Augenblicken habe ich mich selbst förmlich daran geklammert. Dabei wußte man und weiß noch heute blutwenig über den Luftdruck eines Sturms. Wir nehmen zwanzig Pfund auf den Quadratfuß an. Dabei müssen meine Pfeiler, wie ich sie ursprünglich projektiert hatte, wie Felsen stehen. Später, als die Brücke schon über die halbe Bucht fertig war, erfuhr ich, daß die Staatsingenieure in Frankreich vierzig Pfund annehmen. Vor einem Jahr erst schrieb mir ein Bekannter aus Amerika, daß sie dort auf fünfzig rechnen, und die amerikanischen Ingenieure sind nicht übermäßig vorsichtig, wie alle Welt weiß. – Doch tauchte die Frage erst später ernstlich auf, als schon alles in flottem Bau war. Niemand, auch ich nicht, kümmerte sich anfänglich darum. Wir glaubten an Bruce, und Sir William glaubte an sich und sein Gefühl. In den letzten Tagen, in denen die Berechnungen zum Abschluß kamen, auf denen das ganze Brückenprojekt aufgebaut ist, hatte ich noch einen lebhaften Kampf mit mir selber. Welchem Sicherheitskoeffizienten darf ich trauen?

In Greenwich, am Königlichen Observatorium, wissen sie auch nichts, nicht einmal, wie schnell ein gut schottischer Sturmwind läuft. Bei den besten Exempeln, an die sie sich erinnern, ist regelmäßig im kritischen Augenblick ihr Meßapparat zusammengebrochen, und bis er wieder im Gang war, war der Sturm vorbei.

even a continuous lateral wind bracing under the bridge deck. Nevertheless, experts at the inquiry were able to prove that *at least a wind pressure of around 40 lbs/sq ft* (195 kg/m²) would have been required to overturn the bridge with its piers. But such a force would correspond to a wind speed of 115 m.p.h. (185 km/h) – and it is unlikely that even in the night of the storm such an extreme figure had been reached. Still, this late calculation of 40 lbs wind pressure being necessary for the collapse, proved that Bouch's 20 lbs/sq ft calculation was only half of this amount – or reversely, that Bouch's bridge still possessed twice the stability it "should actually have had", so to speak, according to his own calculations. In sum, the under-estimated wind pressure *by itself* could not have been sufficient cause for the collapse.

These public investigations now drew for the first time attention to the great significance of wind pressure on so large a structure. Until then, one had largely been at the mercy of guesswork, which in turn could range in fine grading from "educated guesses" up to "complete guesswork" – or in other words, to a kind of "practical intuition"! Even Sir George Airy's confidence in such habits seemed shaken after the disaster, to the extent that in a paper on "Winds and Bridges", he suggested that from now "all calculations for the strength of a proposed structure should be based on the assumption of a pressure of 120 lbs/sq ft" (585 kg/m²), thus "establishing a modulus of safety".

This would have rendered any further bridge construction hopelessly uneconomical, and according to Hammond [8, p. 32], would, instead, have presented itself as a "factor of ignorance". Sir Airy's revelations upon further questioning by the commission were indeed disarming: To the question as to whether it were not true that until quite recently, the knowledge of engineers and scientists had not required any provisions against wind pressure, he replied, he could not say how it was – but new ideas would not always come into one's mind!

In the course of the investigations, there was an unusual teaming up between meteorology and astronomy, as revealed in the evidence given by George Stokes, Professor of Mathematics from Cambridge: He described storms of 90 m.p.h. (145 km/h) and wind loads bearing down with over 50 lbs/sq ft and across wide fronts. Also, strong gales should not just be considered as momentary whims of the wind; sometimes they would last for two or three minutes and with great force. And Henry Scott, Secretary of the Meteorological Council confirmed that there most certainly could occur great wind pressures along the Tay, across a front of perhaps

mals – zumindest in England – als gültig angenommene Winddruck von 10 Pfund eine *zulässige Windbelastung von 20 Pfund/Quadratfuß* (97,52 kg/m²) für die Berechnung bedeutet.

Immerhin meinte Sir Airy auf weiteres Nachfragen vor Gericht, daß die größten bekannten Windlasten der letzten Jahre in Großbritannien bei ungefähr 50 Pfund/Quadratfuß (243,80 kg/m²) gelegen hätten. Möglicherweise wäre das auch mit seinen Instrumenten schon bei einem besonders heftigen Sturm im Dezember 1872 bewiesen worden – «wäre da nicht unglücklicherweise der Bleistift des Windschreibers im entscheidenden Moment abgebrochen». Deshalb sei er auch außerstande, zu sagen, welcher Winddruck in der Nacht des «Großen Sturmes» am Tay geherrscht haben könne, denn es gab keine Meßinstrumente in Dundee. Dagegen hätte an seinem ständigen Wohnsitz in Greenwich der höchste Winddruck nur 10 Pfund betragen [8, S. 32].

Bouch räumte in der weiteren Untersuchung ein, keine besonderen Vorkehrungen oder Sicherheiten gegen außergewöhnliche Windbelastung vorgesehen zu haben. Ja, es gab nicht einmal eine durchgehende laterale Windaussteifung unter dem Brückendeck. Dennoch konnten Experten bei der Untersuchung nachweisen, daß *immerhin ein Winddruck von ca. 40 Pfund/Quadratfuß (195 kg/m²)* notwendig gewesen wäre, um die Brücke samt Stützen umzuwerfen: Diese Kraft würde aber einer Windgeschwindigkeit von 115 Meilen/Stunde (185 km/h) entsprechen – und es ist unwahrscheinlich, daß selbst in jener Sturmnacht ein derart hoher Wert erreicht worden ist. Immerhin war mit dieser Nachberechnung von 40 Pfund notwendiger Windlast für einen Einsturz klar, daß Bouchs rechnerische Windlast von 20 Pfund (97,52 kg/m²) um die Hälfte unter diesem Wert lag – oder andersherum interpretiert, daß Bouchs Brücke immer noch die doppelte Stabilität besaß, als sie nach seinen eigenen Berechnungen eigentlich «hätte haben dürfen»: Kurz: Der unterschätzte Winddruck *allein konnte* nicht die Ursache für den Einsturz gewesen sein.

Aber diese öffentlichen Untersuchungen machten jetzt zum erstenmal auf die große Bedeutung der Windlast bei solchen Großbauwerken aufmerksam. Bis dahin war man weithin auf «*guesswork*» angewiesen, das seinerseits in feinen Abstufungen von «*educated guesses*» bis zu «*complete guesswork*» – also bis zu einer Art von «praktischer Intuition» reichen konnte! Selbst Sir George Airy's Vertrauen in diese Gepflogenheiten war nach der Katastrophe so erschüttert, daß er in einem Beitrag über «Winde und Brücken» forderte, von nun an «alle Berechnungen zur Stabilität eines geplanten Bauwerkes auf eine Lastan-

250 ft. (76 m). This pressure could even exceed the above mentioned 50 lbs/sq ft and the wind speed could be above 90 m.p.h. – Such a wind velocity has actually been measured in modern times at the Forth road bridge.

Wind Pressure Upon Rolling Loads?

In the end, the inquiry had proven also, that one had simply "forgotten" to consider the *lateral wind pressure acting upon a moving train,* when calculating the wind load. As the train was enclosed inside the central girders, it offered to the wind an enlarged and additional surface of leverage, instead of the open web of the girders. Also, the driver may indeed have gone with excessive speed that night, as some minutes had to be made good which were lost due to the storm. *Did, in the end, the wind pressure acting upon the rolling train, cause the fall of the bridge?* The question can no longer be answered today, just as many other details remain in the dark – and thus there is much room for speculation. If indeed the train had not entered the High Girders at the moment of the heaviest gale, but had still been rolling atop the shorter spans, the train may quite possibly have been blown right off the bridge without a trace – doing no significant damage to the structure itself. This is at least a conceivable theory – and Bouch's desperate attempts to cling to his "derailment theory" no longer appear quite that bizarre. The impression remains that although the commission had worked thoroughly, it had not taken into consideration some other conceivable conclusions.

The Verdict

If one looks at all the faults of construction, material and maintenance in their totality, which had been known for months, it seems to be a miracle that this bridge – after the first test runs of September 22, 1877 – still managed somehow to withstand all stresses, albeit tenuously, for another 27 months.
On July 5, 1880, the court of inquiry submitted its findings to both Houses of Parliament. Rothery felt obliged to present his results and conclusions separately. Even though all three members did agree in principle, Yolland and Barlow shied away from harsh judgements – which is quite understandable in the case of Yolland, as we shall see later. But Rothery held the view "that we ought not to shrink from the duty, however painful it might be, of saying with whom the responsibility for this casualty rests."

nahme von 120 Pfund/Quadratfuß auszulegen – um so einen Sicherheitsmodul festzusetzen» (also 585 kg/m^2): Das hätte jeden weiteren Brückenbau hoffnungslos unwirtschaftlich gemacht, und wäre nach Hammond [8, S. 32] eher einem *«factor of ignorance»* gleichgekommen! Sir Airy's Aussagen auf weiteres Insistieren der Kommission sind in der Tat entwaffnend: Zur Frage, ob es denn nicht wahr sei, daß bis in allerjüngste Zeit der Kenntnisstand von Ingenieuren und Wissenschaftlern nach keinerlei Vorkehrungen gegen Winddruck verlangt habe, meinte er, er könne nicht sagen, wie es sich damit verhalte – aber neue Ideen kämen einem eben nicht immer in den Sinn!
Wie sonderbar schon damals die Ansiedlung der Meteorologie im Bereich der Astronomie war, machte George Stokes, ein Mathematikprofessor aus Cambridge deutlich: Er erläuterte Stürme von 90 Meilen/Stunde (145 km/h) sowie Windlasten, die mit über 50 Pfund/Quadratfuß (243,80 kg/m^2) und auf breiten Fronten daherkämen. Auch seien starke Böen keineswegs bloß als momentane Launen des Windes anzusehen, sondern würden manchmal für zwei oder drei Minuten und mit großer Kraft andauern. Und Henry Scott, Sekretär des *Meteorological Council* bestätigte, daß es mit Sicherheit großen Winddruck entlang des Tay geben könne, und zwar auf einer Breite von vielleicht 250 Fuß (76 m). Dieser Druck könne die obigen 50 Pfund noch übersteigen und die Windgeschwindigkeit sogar über 90 Meilen/Stunde betragen. – Eine solche Windgeschwindigkeit ist auch neuerdings an der Forth-Straßenbrücke tatsächlich gemessen worden.

Winddruck auf rollende Lasten?

Schließlich war mit der Untersuchung klargeworden, daß man letztlich «vergessen» hatte, bei der Berechnung der Windlast auch den *Seitendruck auf den fahrenden Zug* zu berücksichtigen: Denn dieser war ja im Mittelträger eingefangen und bot so dem Wind anstelle eines offenen Fachwerkes eine vergrößerte, zusätzliche Angriffsfläche. Zudem war der Zugführer möglicherweise wirklich mit überhöhter Geschwindigkeit gefahren, denn einige Minuten Verspätung infolge des Sturmes mußten aufgeholt werden: *Hatte also letztlich der Winddruck auf den fahrenden Zug die Brücke zum Einsturz gebracht?* Die Frage ist nicht mehr zu beantworten, ebensowenig wie manch andere Einzelheiten – und so bleibt viel Raum für alle möglichen Spekulationen: Hätte sich nämlich der Zug im Augenblick der stärksten Orkanböe noch nicht im Hochträger, sondern auf der Oberseite eines der kürzeren Brückenträger befunden, so wäre er möglicherweise vom Wind einfach über die Brücke gefegt wor-

Yet even Wreck Commissioner Rothery had to concede in the end that neither he himself nor his colleagues could ascertain conclusively what really had happened in that fateful night. Although there could be no doubt about some of the causes of the collapse, its precise sequence was never fully explained. Yolland and Barlow could only venture some restrained conjectures. Less cautious was Rothery:

> "What probably occurred was this: The bridge had probably been strained, partly by previous gales, partly by the great speed at which trains going north were permitted to run through the high girders. The result would be that, owing to the defects to which we have called attention, the wind ties would be loosened; so that when the gale of the 28th of December came on, a rocking motion would be set up between the two triangular groups into which the six columns forming each pier were divided. This would bring a great additional strain upon the wind ties between the 15-inch columns which connected the two groups of columns together, and which would receive comparatively little support from the ties between the outer 18-inch and the two nearest inner columns, owing to the angle which they made with the line of pressure." [8, p. 34]

The conclusion of the report of 1880 [22] arrives with logical consistency at the following judgement, and it is devastating to Sir Thomas Bouch:

> "We find that the bridge was badly designed, badly constructed and badly maintained and that its downfall was due to inherent defects in the structure which must sooner or later have brought it down. For these defects both in the design, the construction, and the maintenance, Sir Thomas Bouch is, in our opinion, mainly to blame. For the faults of design he is entirely responsible. For those of construction he is principally to blame in not having exercised that supervision over the work, which would have enabled him to detect and apply a remedy to them. And for the faults of maintenance he is also principally, if not entirely, to blame in having neglected to maintain such an inspection over the structure, as its character imperatively demanded."

The contractors and the NBR were being blamed as well, but the primary responsibility was put unto Bouch. His hopes, which had been furthered by the support of some colleagues at the inquiry, to now go on building the new Forth Bridge were shattered: He was told flatly that his services were no longer required. Thus the professional career of one of the

den, ohne diese selbst nennenswert zu beschädigen. Zumindest bleibt das eine denkbare Theorie – und Bouch's verzweifelte Versuche, seine «Entgleisungstheorie» zu halten, erscheinen gar nicht so abwegig. Es bleibt der Eindruck, daß die Kommission zwar gründlich gearbeitet, aber einige denkbare Schlußfolgerungen gar nicht in Erwägung gezogen hat.

Das Verdikt

Sieht man aber all die seit Monaten bekannten Bau-, Material- und Unterhaltsfehler im Zusammenhang, so ist es eigentlich ein Wunder, daß diese Brücke – nach der ersten Probefahrt am 22. September 1877 – noch 27 Monate lang allen Belastungen, wenn auch «mit Mühe» hatte standhalten können.

Am 5. Juli 1880 übergab das Untersuchungsgericht seine Ergebnisse den beiden Häusern des Parlamentes. Rothery fühlte sich verpflichtet, seine Ergebnisse und Schlußfolgerungen gesondert vorzulegen. Auch wenn alle drei Mitglieder im Prinzip übereinstimmen, so scheuten doch Yolland und Barlow vor harten Verurteilungen zurück – was im Falle Yollands nur allzugut verständlich ist: Siehe dazu den Abschnitt «NACHSPIEL». Aber Rothery war der Ansicht, «daß wir uns nicht scheuen sollten, so schmerzlich es auch sein mag, zu sagen, wer die Verantwortung für diesen Unglücksfall trägt».

Aber selbst *«Wreck Commissioner»* Rothery muß schließlich einräumen, daß weder er selbst noch seine Kollegen mit letzter Sicherheit sagen könnten, was wirklich in jener verhängnisvollen Nacht passiert war. Wenn es auch keinen Zweifel über die Ursachen des Einsturzes geben konnte, sein genauer Ablauf wurde niemals völlig geklärt und konnte für Yolland und Barlow nur in Vermutungen rekonstruiert werden. Weniger vorsichtig äußerte sich Rothery:

> «Was wahrscheinlich geschah, ist folgendes: Die Brücke war wahrscheinlich bereits überbeansprucht worden, teils durch frühere Stürme, teils durch die große Geschwindigkeit, mit der man die nordwärts laufenden Züge durch die Hochträger hatte fahren lassen. Das Ergebnis wäre dann, daß wegen der von uns aufgeführten Schäden sich die Windverbände gelockert hätten; so daß mit dem Aufzug des Orkans vom 28. Dezember eine sich aufschaukelnde Bewegung in Gang gesetzt wurde, und zwar zwischen den beiden Dreiecksgruppen, in die man die 6 Säulen jeder Pfeilerkombination aufgeteilt hatte. Das würde eine große zusätzliche Belastung auf die zwischen den 15-Inch-Säulen eingebauten Windverbände ausüben, welche

51 >
The long line of piers' stumps of Bouch's old Tay Bridge, running next to the massive piers of Barlow's new bridge, up to this day. At the top left, Bouch's re-used girders can be seen. His old footings today serve as breakwater for the tides.

51 >
Die lange Reihe der Pfeilerstümpfe von Bouchs alter Tay-Brücke (rechts) begleitet noch heute die massiven Pfeiler der neuen Brücke Barlows. Oben links die wiederverwendeten Fachwerkträger Bouchs. Seine alten Pfeilerstümpfe dienen heute als Wellenbrecher.

die beiden Säulengruppen zusammenhielten – und welche ihrerseits vergleichsweise wenig Unterstützung von den Zugankern zwischen den äußeren 18-Inch-Säulen und den beiden nächsten inneren Säulen erhalten würden, wegen ihres (ungünstigen) Winkels zur Angriffslinie der Belastung.» [8, S. 34]

Der Schlußteil des Untersuchungsberichtes von 1880 [22] kommt mit lapidarer Folgerichtigkeit zu folgendem Urteil und ist vernichtend für Sir Thomas Bouch:

«Wir müssen feststellen, daß die Brücke schlecht geplant, schlecht gebaut und schlecht unterhalten worden war, und daß ihr Einsturz wegen der dem Bauwerk innewohnenden Schäden erfolgt war, welche es früher oder später zum Einsturz gebracht hätten. Für diese Schäden, sowohl in Planung, Bau und Unterhalt, muß nach unserer Ansicht Sir Thomas Bouch die Hauptschuld tragen. Für die Fehler in der Planung ist er zur Gänze verantwortlich, für die des Baues ist er prinzipiell verantwortlich, da er nicht jene Aufsicht über die Arbeiten ausgeübt hat, die es ihm ermöglicht hätte, die Schäden zu entdecken und zu beheben. Und für die Fehler des Unterhaltes ist er auch prinzipiell, wenn nicht gänzlich verantwortlich, da er es versäumt hat, eine solche Überwachung des Bauwerkes aufrechtzuerhalten, wie es dessen Bauart dringend erforderte.»

Die Unternehmer und die *NBR* wurden ebenfalls verurteilt, aber der Hauptschuldspruch war auf Bouch gefallen. Seine Hoffnungen, genährt durch die Unterstützung einiger Kollegen bei der Untersuchung, jetzt die neue Forth-Brücke weiterbauen zu können, waren zunichte: Man teilte ihm mit, daß seine Dienste nicht länger benötigt würden. Damit war die berufliche Karriere eines der erfolgreichsten und gefeiertsten Brückenbauers seiner Zeit am Ende.

Nach wie vor liefen zwar Bouch's Eisenbahnfähren mit regem Zuspruch über den Forth; und seine 300 Meilen (480 km) an Eisenbahnstrecken, Viadukten und Brücken, die er in Schottland und England über die Jahre gebaut hatte, dienten weiterhin dem aufblühenden Eisenbahnverkehr: Aber all diese Erfolge zählten nichts mehr nach der Veröffentlichung des Untersuchungsberichtes. Schon während der viermonatigen Verhandlungen war Bouch's Haar weiß geworden und nun war er auch als Person das Ziel beißender Kritik der Öffentlichkeit. Zuletzt als Ingenieur und Mensch vollkommen diskreditiert, zog er sich auf sein Landhaus in Moffat (Dumfriesshire) zurück. Dort starb er frühmorgens am 1. November 1880 an einer Erkäl-

most celebrated and successful bridge builders of his time had come to an end.

Bouch's railway ferries were still a most popular way of crossing the Forth; and the 300 miles (480 km) of railway, viaducts and bridges he had built in Scotland and England over the years, still served the prospering railway traffic. But all of these successes did not seem to count for anything now, after the Rothery Report had come out. Already during the court hearings, Bouch's hair had turned white, and now he had become the target even of bitter personal abuse and criticism by the public. Utterly discredited as an engineer and as a human being, he withdrew to his country house in Moffat, Dumfriesshire. There he died in the early morning of November 1st, 1880, from a cold which he had neither the strength nor the will to resist; this was four months after publication of the report, and ten months after the fall of his bridge. According to Murray [15, p. 30], he died from "acute melancholia", at the age of 58, and he was buried in Edinburgh's Dean Cemetery.

In one respect, Sir Thomas Bouch had lived up to the Victorian ideal of self-reliance and success: this son of an ordinary sea-captain died a rich man, leaving behind an inheritance of over 200,000 Pounds Sterling. – As to the dead, this era had its own language. The words of a correspondent of the Dundee press – written soon after the fatal night on the Tay – could also have stood for Bouch: "Life is not lost which is given or sacrificed in the magnificent enterprises of useful industry."

To the 20 dead claimed by the construction of the Tay Bridge over the course of 6 years, the 75 dead in the disaster had to be added, plus the death of the first contractor de Bergue, and finally, the death of its designer – adding up to 97 casualties altogether.

Up to this day, the piers' stumps of the old Tay Bridge can still be seen marching across the Tay in a long line, like stepping stones *(Fig. 51)*, – although some citizens of the time preferred to speak of grave stones instead. A more fitting memorial to Bouch may be seen in the girders still serving the new Tay Bridge today.

tung, der zu widerstehen er weder die Kraft noch den Willen hatte, – vier Monate nach Veröffentlichung des Berichtes, bzw. zehn Monate nach dem Einsturz seiner Brücke: Er war nach Murray [15, S. 30] an «akuter Melancholie» gestorben, im Alter von 58 Jahren, und liegt in Edinburghs Dean Cemetery begraben.

Doch war Sir Thomas Bouch in einem Punkt dem viktorianischen Ideal von Sebsthilfe und Erfolg treu geblieben: Dieser Sohn eines einfachen Kapitäns starb als reicher Mann und hinterließ ein Vermögen von über 200 000 Pfund Sterling. – Was die Toten angeht, hatte jene Epoche ihre eigene Sprache. Die Worte eines Korrespondenten der Presse von Dundee – geschrieben nach der Unglücksnacht am Tay – hätten auch für Bouch stehen können: «Das Leben ist nicht verloren, welches in den großartigen Unternehmungen nützlicher Industrie gegeben oder geopfert wird.» Zu den 20 Toten, die der Bau der Brücke in 6 Jahren gefordert hatte, waren die 75 Todesopfer des Zuges gekommen, dazu der Tod des ersten Unternehmers de Bergue, und nun noch das Ende ihres Erbauers – 97 Tote insgesamt.

Noch heute schreiten die Pfeilerstümpfe der alten Tay-Brücke in langer Linie über den Tay wie Trittsteine *(Abb. 51)* – wenn auch mancher Zeitgenosse eher von Grabsteinen sprach. Ein besseres Denkmal für Bouch kann man aber in den Trägern sehen, die noch heute in der neuen Tay-Brücke ihren Dienst tun.

Chapter II

The Interim Between the Bridges

Sequel

Upon publication of the report of inquiry, Major General Hutchinson of the Board of Trade was asked what he thought of the findings. After all, he had cleared the bridge as safe for traffic in February of 1878 – a bridge that had now been proven so disastrously faulty. Even before the report had been made public, St John Vincent Day, an independent engineer from Glasgow, had already given a much publisised lecture to the Institution of Engineers and Shipbuilders in Scotland; at this occasion (by the end of January 1880) and also shortly after the accident, he had declared:

> "I fail, in view of what I have seen for myself of the character of the structure, to comprehend how Major General Hutchinson could report that 'the iron work has been well put together *in the columns* and the girders!' Did General Hutchinson really closely inspect the columns? Did he ascertain the depth and variation in depth of the holding-down bolts? If so his report is, to say the least, incorrect, for it has been abundantly shown how unfit was the structure of the columns for the functions imposed on them."

Mr. Day was harshly attacked by the colleagues gathered there for his accusations – long before the court had arrived at similar conclusions. Mr. Day also criticised Hutchinson for having neglected the wind pressure during his safety inspection: "The bridge was opened without any knowledge or observation on his part of the effect of wind on the structure." Under cross-examination in court, Hutchinson grew less convinced about the bridge's safety than he had been in his report. To the direct question whether he considered the design of the bridge to be satisfactory, he gave this answer: "I would rather put it this way; that the design was not unsatisfactory". He admitted to having conducted an only superficial inspection, that he had based his assessment of the bridge only on its exterior appearance and that he had not tested it for wind pressure. But he emphasised that he had spent "anxious care and thought" on the bridge and had not found any flaws during his three-day inspection. He made it clear, too, that the bridge would have been safe, if only maintenance and workmanship had been in order. He

Kapitel II

Zwischen den Brücken

Nachspiel

Mit der Veröffentlichung des Untersuchungsberichtes wurde General Hutchinson von der IHK befragt, wie er zu diesem Bericht stünde. Denn die Brücke, welche sich als so fehlerhaft erwiesen hatte, war ja von ihm selbst im Februar 1878 für verkehrssicher erklärt worden. Und St. John Vincent Day, ein unabhängiger Ingenieur aus Glasgow, hatte schon Ende Januar 1880, kurz nach dem Einsturz, in einem aufsehenerregenden Vortrag beim Verband der Ingenieure und Schiffsbauer Schottlands in Glasgow, folgendes erklärt:

> «Ich vermag angesichts dessen, was ich selbst über die Art des Bauwerkes festgestellt habe, nicht zu verstehen, wie Generalmajor Hutchinson berichten konnte, ‹daß die Eisenarbeiten *in den Säulen* und Trägern gut ausgeführt worden sind!› Hat General Hutchinson die Säulen wirklich genau inspiziert? Hat er die Tiefe und unterschiedliche Tiefe der Ankerschrauben geprüft? Wenn ja, so ist sein Prüfbericht, gelinde ausgedrückt, unkorrekt, denn es hat sich ja überdeutlich gezeigt, wie untauglich die Konstruktion der Säulen für die Funktion war, der sie dienen sollten.»

Mr. Day war damals von den versammelten Kollegen wegen dieser Vorwürfe heftig angegriffen worden – lange bevor das Gericht zu denselben Ergebnissen gekommen war. Auch kritisierte Day das Versäumnis von Hutchinson, bei der Sicherheitsprüfung den Winddruck nicht berücksichtigt zu haben: «Die Brücke war von ihm ohne jede Kenntnis oder Begutachtung über die Wirkung von Wind auf das Bauwerk freigegeben worden.» – Im Kreuzverhör vor Gericht war Hutchinson dann weniger zuversichtlich über die Sicherheit der Brücke als in seinem Bericht. Auf die direkte Frage, ob er den Entwurf der Brücke für zufriedenstellend halte, meinte er: «Ich würde ihn eher als nicht unbefriedigend bezeichnen.» Er gab zu, daß seine Inspektion nur oberflächlich war, er seine Einschätzung der Brücke nur von der äußeren Erscheinung gebildet und sie auch nicht gegen Winddruck geprüft habe. Aber er betonte, daß er der Brücke «angestrengte Sorge und Überlegung» gewidmet und während seiner dreitägigen Inspektion keinen Fehler entdeckt habe. Er stellte

presented it as a good deed to have imposed a speed limit of 25 m.p.h. (40 km/h) for safety's sake, and he emphasized that he could hardly be made responsible for defects which were caused by failure to observe this restriction. Regarding wind pressure, he declared:

> "I was anxious to see how the lateral stiffness of the piers might be affected by the action of a high wind upon the side of a train in motion over the bridge. This I had intended to get if possible an opportunity of doing before the traffic commenced running, but I was laid aside by serious illness shortly after the inspection and before my recovery the bridge had been opened for traffic."

Naturally, there was a great deal of anxiety among professional circles and the public that a major work of engineering had failed so early after its official safety check by a government body. While a storm was brewing over this, the President of the Board of Trade, Joseph Chamberlain, sent an amendment to Hutchinson's report to both Houses of Parliament. In this he confirmed his complete confidence in Hutchinson and stated that his inspection of the Tay Bridge "had not been such as to forfeit their [the Houses of Parliament's] confidence". He went on to say that the Board of Trade had no power to safeguard the maintenance of the bridge after it had been passed for use, and Hutchinson could not be blamed for defects that were not evident at the time of inspection, nor for faults of material and workmanship which were covered by paint after completion of the works – and most certainly not for defects which only appeared after his inspection. Where the design was concerned, it may be said that Hutchinson should have seen how weak

auch klar, die Brücke wäre sicher gewesen, wären nur Unterhalt und Ausführung zufriedenstellend gewesen. Er stellte es als eine gute Tat hin, sicherheitshalber eine Höchstgeschwindigkeit von 25 Meilen/Stunde (40,2 km/h) verhängt zu haben und betonte, daß er nicht zur Rechenschaft gezogen werden könne für Schäden infolge Nichtbeachtung dieser Einschränkung. Zur Sache mit dem Winddruck erklärte er:

> «Ich war gespannt zu sehen, wie die Quersteifigkeit der Pfeiler durch die Wirkung eines starken Seitenwindes auf einen Zug beansprucht werden würde, der gerade die Brücke überquert. Das wollte ich bei einer sich bietenden Gelegenheit tun, bevor der Verkehr aufgenommen würde, wurde aber durch eine ernste Erkrankung kurz nach der Inspektion daran gehindert; und vor meiner Genesung war die Brücke bereits freigegeben.»

Es gab erhebliche Unruhe in der Öffentlichkeit und Fachwelt, daß ein wichtiges Ingenieurbauwerk so bald nach seiner offiziellen Genehmigung versagt hatte. Als sich ein Sturm zusammenbraute, schickte der Präsident der IHK, Joseph Chamberlain, ein Protokoll zu Hutchinsons Bericht an beide Häuser des Parlaments. Darin bekräftigte er sein volles Vertrauen auf Hutchinson und bestätigte, daß dessen Inspektion der Tay-Brücke «keineswegs dazu angetan war, sein (des Parlaments) Vertrauen zu verwirken». Die IHK habe keine Macht, nach der Freigabe den Unterhalt der Brücke zu kontrollieren und Hutchinson könne nicht für Schäden verantwortlich gemacht werden, die bei der Inspektion nicht feststellbar waren, bzw. für Arbeits- und Materialfehler, die nach Abschluß der Arbeiten unter dem Anstrich verborgen waren, – und schon gar nicht für Schäden, die erst nach seiner Inspektion auftraten.

52
Panorama photograph of the Tay Bridge as it appears today.

52
Panoramafoto der Tay-Brücke in heutigem Zustand.

the bridge essentially was, and that he should have stated as much; however, obvious flaws and its inherent weakness were still in doubt. The task of an inspector in judging a design, would be to check that the design did not violate those rules and precautions which practice and experience had proven to be imperative for safety. But if he were to make himself responsible for every novel design and try to introduce new rules and practices not accepted by the profession, he would be removing a responsibility from the civil engineer and taking it on himself, a power which Parliament had never entrusted with him.

If in the end one were to entrust just any public office with the power and duty to correct and guarantee all designs of responsible railway engineers, this would in effect transfer control away from the very enterprise which had done so much for the country – and one would thus substitute the real responsibility of the engineer for the unreal and elusive responsibility of a public agency, etc. – We may ask ourselves, what *else* the President of an institution under his own command should have said. Chamberlain completed his nimble defence (not without shedding a few crocodile tears for Bouch) with these words:

> "At the present moment there is no one who is more deserving of pity than the civil engineer who designed and constructed the Tay Bridge and who, as the law now stands, is held responsible for its defects. With this case in view it is in the highest degree improbable that any civil engineer entrusted with a similar task in future will commit similar errors."

Despite his many words, Chamberlain remained silent on the decisive issue of Hutchinson's omission to consider the wind pressure. Another point speaking

Was den Entwurf angeht, so könne man zwar sagen, daß Hutchinson gesehen haben müßte, wie schwach die Brücke im Grunde war, und das auch hätte berichten müssen, aber das sei eben, zumindest, zweifelhaft. Die Aufgabe eines Inspekteurs sei es, beim Entwurf darauf zu achten, daß die Konstruktion nicht die Regeln und Vorsichtsmaßnahmen verletzt, welche Praxis und Erfahrung als sicherheitsnotwendig bewiesen haben. Würde er sich selbst für jeden neuartigen Entwurf verantwortlich machen und selbst versuchen, neue Regeln und nicht von der Fachwelt anerkannte Praktiken einzuführen, würde er dem Bauingenieur eine Verantwortung abnehmen und auf sich selbst laden, die das Parlament ihm gar nicht auferlegt hat. Würde man schließlich irgendein öffentliches Amt mit der Macht und Pflicht beauftragen, all die Entwürfe der verantwortlichen Eisenbahn-Ingenieure zu korrigieren und zu garantieren, so hieße das, genau jenes Unternehmertum zu kontrollieren, welches so viel für unser Land getan hat – und man würde aus der wirklichen Verantwortung des Ingenieurs die unwirkliche und nicht faßbare Verantwortung einer öffentlichen Stelle machen, usw. – Man mag sich fragen, was denn der Präsident einer ihm unterstellten Behörde anderes hätte sagen sollen. Seine nicht ungeschickte Rechtfertigung schließt er, nicht ohne ein paar Krokodilstränen für Bouch, mit den Sätzen:

> «Im Augenblick gibt es niemand, der Mitleid mehr verdient als der Bauingenieur, der die Tay-Brücke entworfen und gebaut hat, und der nach gültigem Recht für ihre Schäden die Verantwortung trägt. Angesichts dieses Falles ist es höchst unwahrscheinlich, daß irgendein Bauingenieur bei einer ähnlichen Aufgabe in der Zukunft ähnliche Fehler begehen wird.»

against Hutchinson was the fact that he was not the only railway inspector of the Board of Trade; this important task could have been delegated. Finally, there was no reason not to conduct the wind tests after the official opening of the bridge. Only in January of 1879, Hutchinson had been to Dundee investigating a minor complaint by a passenger. If he had conducted his checks then and examined the bridge even only superficially, he would have been forced to notice the serious faults which had become so obvious to any layman – and the course of events might perhaps have been altered.

The public, moreover, was "not amused" by Chamberlain's self-complacent conclusion that this disaster would probably teach a lesson to engineers; after all, the Board of Trade was considered the watchdog of the nation and the railway travellers rightfully regarded as totally safe any structure which had been passed. The words of the President now pointed out quite clearly that such a guarantee had never been implied. W. T. Arrol, the widely respected contractor of Bouch's Forth Bridge and also of the new one by Fowler and Baker, expressed what many were thinking when he said to the members of the Institution of Engineers and Shipbuilders in Scotland that the Board had skirted its responsibility in blaming the collapse on poor workmanship.

Finally, the following should be noted in judging Bouch's role: In his "neglect" to take into account the wind pressure, he had been simply following the instructions issued by the very Board which was now sitting in judgement over him: the officer who had told him that he need not regard the wind load, was no one else than Colonel Yolland himself, one of Bouch's judges today. Bouch being anything but easily satisfied, had taken it upon himself to get accurate figures about the wind pressure and a proper expertise on his design. Already on October 5, 1869, ten full years before the disaster, he had already written to Yolland.

> "In calculating the strains of malleable iron girders will you kindly tell me what you take the live load as per running foot for spans over a hundred feet and is it necessary to take *the pressure of wind* into account for spans not exceeding 200 feet span, the girders being open lattice work? My own opinion is that one and a quarter tons per foot run for live loads is sufficient for spans over 100 feet, and that it is not necessary to take the force of the wind into account where open work girders are used and spans less than 200 feet. I merely ask this information that I may act in accordance with the views of the Board of Trade."

Trotz der vielen Worte schwieg sich Chamberlain aus über den entscheidenden Punkt, daß Hutchinson die Windlast nicht berücksichtigt hatte. Auch war Hutchinson nicht der einzige Inspekteur der IHK; man hätte die wichtige Aufgabe delegieren können. Schließlich gab es auch keinen Grund, die Windversuche nicht nach der Übergabe durchzuführen. Im Januar 1879 war Hutchinson wegen der nichtigen Beschwerde eines Fahrgastes in Dundee gewesen. Hätte er seine Tests dann angestellt, und die Brücke auch nur oberflächlich untersucht, so hätte er die krassen Schäden sehen müssen, die jedem Laien deutlich geworden waren – und der Gang der Dinge wäre vielleicht anders verlaufen.

Auch die selbstgefällige Schlußfolgerung, das Desaster würde den Ingenieuren eine Lehre sein, kam nicht gut an; denn die IHK wurde als der Wachhund der Nation angesehen und die Bahnbenutzer betrachteten jedes genehmigte Bauwerk als absolut sicher. Die Angaben des Präsidenten machten nun recht deutlich, daß ein solches Versprechen nie damit verbunden war. W.T. Arrol, der angesehene Bauunternehmer von Bouch's Forth-Brücke, sowie derjenigen von Fowler und Baker, drückte aus, was viele dachten, als er den Mitgliedern des Ingenieur- und Schiffsbauer-Verbandes von Schottland erzählte, die IHK habe sich ihrer Verantwortung entzogen, indem sie den Brückeneinsturz auf schlechte Bauausführung zurückführte.

Schließlich ist noch folgendes zur Rolle Bouch's festzuhalten: Er hatte bei seinem «Versäumnis», den Winddruck zu berücksichtigen, dies deshalb unterlassen, weil er von genau derselben Kammer, die jetzt über ihn zu Gericht saß, dazu angehalten worden war: Der Beamte, der ihm gesagt hatte, er brauche den Winddruck nicht zu berücksichtigen, war niemand anders als Oberst Yolland selbst gewesen, einer seiner heutigen Richter. Bouch, alles andere als selbstzufrieden, hatte eigene Schritte zu exakten Daten über den Winddruck und einer fachmännischen Beurteilung seines Entwurfes unternommen. Schon am 5. Oktober 1869, also 10 Jahre vor dem Einsturz, hatte er an Yolland geschrieben:

> «Würden Sie mir für die Festigkeitsberechnung von schmiedeeisernen Trägern freundlicherweise mitteilen, was Sie als Nutzlast für den laufenden Fuß bei Spannweiten über 100 Fuß ansehen, und ist es notwendig, den *Druck des Windes* bei Spannweiten bis zu 200 Fuß zu berücksichtigen, wobei es sich um offene Gitter-Träger handelt? Meine eigene Meinung ist, daß $1^{1}/_{4}$ Tonne/lfd. Fuß für Nutzlasten ausreicht bei Spannweiten über 100 Fuß, und daß es nicht notwendig ist, die Stärke des Windes bei Ver-

Three days later, Colonel Yolland's answer arrived; it did not seem to have caused him much of a headache:

> "A ton and a quarter foot run will be sufficient for spans over 100 feet, and *we do not take the force of the wind into account* when open lattice girders are used for spans not exceeding 200 feet."

To this we have to add the information given to Bouch by the Royal Astronomer mentioned earlier, regarding a maximum wind pressure on plane surfaces like those of a bridge, of 10 lbs/sq ft – which would equal a wind load of 48.76 kg per square meter.

Lastly, Bouch's plans had also been examined and approved by two outstanding Westminster engineers, T. E. Harrison and J. M. Hepper. Harrison declared that "the bridge was sufficiently strong in all its parts", and Hepper summed up his totally positive assessment by stating:

> "I can therefore have no hesitation in stating my complete conviction of the efficacy of the design in every particular, whether in regard to the foundations or the superstructure."

So much for the state of professional expertise at the time, and for the judgement about Bouch's design by his fellow engineers – *before the disaster!*

Furthermore, it had been a grave error by Hutchinson, according to Hammond [8, p. 34], to have imposed a speed limit of 25 m.p.h. at all (40.2 km/h) when he passed the bridge for traffic. He must have known that in those days Daniel Gooch's famous locomotives of the Great Western Railway were already doing twice or three times that speed. Hammond thus hints at the possibility that these fast engines must have posed a great challenge to the drivers, but also to the railway companies, not to pay much attention to this restriction. According to Hammond, it was inexcusable under the circumstances for Hutchinson to have passed the structure for traffic, and his responsibility for the collapse would have to be regarded, in all fairness, to have been at least as high as that of Bouch.

And Beckett [1, p. 112] states that Bouch had probably been charged unfairly with such a major portion of blame – for a disaster which had mainly been caused by poor workmanship and maintenance. – Perhaps we are only today in a position, after the distance of some 100 years, to judge more fairly the role of Thomas Bouch.

wendung von offenen Gitterträgern und Spannweiten bis zu 200 Fuß zu berücksichtigen. Ich stelle diese Anfrage nur, um in Übereinstimmung mit den Ansichten der IHK vorzugehen.»

Drei Tage später kam Oberst Yolland's Antwort, die ihm kaum viel Kopfzerbrechen bereitet zu haben schien:

> «$1^1/_4$ Tonne/lfd. Fuß wird für Spannweiten über 100 Fuß ausreichen, und *wir berücksichtigen die Stärke des Windes nicht* bei Verwendung offener Gitterträger und Spannweiten unter 200 Fuß.»

Dazu kommt die früher erwähnte Information des königlichen Astronomen über den maximalen Winddruck auf ebene Flächen, wie die der Brücke, von 10 Pfund/Quadratfuß – was einer Windlast von 48,76 kg auf den Quadratmeter entsprechen würde.

Schließlich war Bouch's Entwurf auch noch von zwei hervorragenden Westminster-Ingenieuren, T. E. Harrison und J. M. Hepper, geprüft und gutgeheißen worden. Harrison erklärte, daß «die Brücke in all ihren Teilen von genügender Stärke» war, und Hepper schloß seinen zur Gänze positiven Bericht mit der Feststellung:

> «Ich kann daher keine Bedenken haben, meine vollständige Überzeugung über die Leistungsfähigkeit des Entwurfes in all seinen Teilen festzustellen, sowohl bezüglich der Fundierung als auch des Oberbaus.»

Soviel zum Stand des Fachwissens der Zeit und zum Urteil der Berufskollegen über Bouch's Entwurf – *vor dem Einsturz!*

Darüber hinaus war es nach Hammond [8, S. 34] ein schwerer Fehler von Hutchinson gewesen, bei der Freigabe der Brücke jene Höchstgeschwindigkeit von 25 Meilen/Stunde (40,2 km/h) überhaupt verhängt zu haben – denn er mußte wissen, daß zu jener Zeit Daniel Gooch's neue Lokomotiven auf der *Great Western Railway* schon das Zwei- bis Dreifache fuhren. Hammond deutet damit an, daß aller Voraussicht nach diese schnellen Maschinen für die Fahrer, aber auch für die Bahngesellschaft, eine Herausforderung sein würden, diese Grenze nicht zu beachten. Nach Hammond war es unter diesen Umständen unentschuldbar für Hutchinson, das Bauwerk dennoch für den Verkehr freigegeben zu haben, und seine Verantwortung für den Einsturz sei gerechterweise als mindestens gleichgroß anzusehen wie diejenige von Bouch.

Und Beckett [1, S. 112] stellt fest, daß man Bouch wahrscheinlich in unfairer Weise einen so großen Schuldanteil angelastet habe – für ein Desaster, welches in der Hauptsache auf schlechte Bauausführung und mangelnden Unterhalt zurückgegangen war. Vielleicht sind wir erst heute, mit dem Abstand von über

Reconstruction

The subsequent reconstruction of the Tay Bridge (1882–1887) following the plans of William H. Barlow was put into the hands of the very able William Arrol of Glasgow. We will hear more of him in the chapter on the Forth Bridge. Examination of Bouch's Tay Bridge had shown that the old girders of the shorter spans – all of them had remained intact after the fall of the central portion – could well be re-used. They only had to be shifted sideways onto the new piers, which were erected just 60 ft (18.29 m) to the west of the old ones. *(Fig. 45)* This meant using the same spans with an identical spacing of piers, as the new footings had to rest on axis with the old ones. Thus the old stumps of Bouch's bridge were now acting as breakwater for the new piers!

The new bridge was built to accomodate now two tracks of railway; thus Bouch's girders were re-used for the two exterior sides, while a new set of identical girders had to be constructed for the inner sides. The new piers were considerably more massive than the old ones, with a gate-like configuration and fully enclosed structure. The centre of the river was spanned by 13 new large girders of wrought iron and of the same length, but with arched upper beams, in place of the former parallel box girders. Thus the great bridge was completed in 1887 for the second time – almost 10 years after completion of the first one.

During reconstruction, also the overall height was reduced by 11 ft (3.36 m), resulting in a more stable and mightier but also a more earth-bound appearance of the new Tay Bridge. (Compare panorama photograph, *Fig. 52.*) The exciting novelty and slender elegance of the original structure is certainly gone now; actually, the new bridge seems even more ordinary than the old one had been – considering its modest spans. Basically, it resembles just a safe and very long viaduct. Being double-tracked and double-footed, it is much broader to look at than the light and tall-legged spider-web dreamed up by Bouch.

Previous Bridge Failures

At this point it seems fitting to recall the experience gained from previous bridge failures, elsewhere. The significance of wind pressure – especially when combined with the oscillation of a structure – had become evident with the Tay Bridge Disaster, although not for the first time. This phenomenon had actually been known before, although it mostly affected the then revolutionary suspension bridges. A number of spectacular bridge failures during the first half of the 19th

53
Sir Samuel Brown's Brighton Chain Pier in 1836 after its collapse during a storm [18, p. 95].

53
Sir Samuel Browns Brighton Chain Pier 1836 nach dem Einsturz in einem Sturm [18, S. 95].

century had all occurred in similar fashion. Each time, extreme vibrations caused by dynamic loads of wind pressure had immediately preceded the collapse itself.

In 1817, the first Dryburgh Abbey Bridge was struck in this manner; in 1831, Sir Samuel Brown's Broughton Bridge went down and in 1836 also his Brighton Chain Pier *(Fig. 53)*. And Telford's famous Menai Suspension Bridge had been hit three times: first in 1825, when it was partially demolished during construction, then again in 1836 and 1839, when one third of the suspension rods broke and the bridge had to be closed for some time to all traffic between London and Holyhead [18, p. 169].

But then in those days, only part of the problem was recognised with regard to the behaviour of bridges

schwingung eines Bauwerkes – war zwar mit dem «Tay Bridge Disaster» deutlich geworden, jedoch nicht zum erstenmal. Man kannte das Phänomen eigentlich schon vorher, wenngleich es meist den damals revolutionären Hängebrücken galt: Eine ganze Reihe spektakulärere Einstürze in der ersten Hälfte des 19. Jahrhunderts war nach ähnlichem Muster abgelaufen. Jedesmal waren extreme Schwingungen durch dynamische Nutz- oder Windlast dem Einsturz unmittelbar vorausgegangen.

So traf es 1817 die erste *Dryburgh Abbey Bridge,* 1831 Sir Samuel Browns *Broughton Bridge* und 1836 auch noch sein *Brighton Chain Pier (Abb. 53).* Und Telford's berühmte *Menai-Brücke* traf es gleich dreimal: Zuerst 1825 als Teilzerstörung noch während der Bauzeit, und dann wieder 1836 und 1839, als ein Drittel der Auf-

during vibrations: live loads and wind pressures were still calculated as if they presented *static* loads, without consideration for the oscillation or autovibration of a structure, caused by one or more loads *of impact*. That is to say, the calculations were done without a dynamic "impact coefficient". Indeed, it was to take another 100 years, until the construction of the George Washington Bridge across the Hudson River (1923–1931), a span of over 1,000 metres, that the lessons from a simple principle of physics were learned: dynamic loads do not need to be excessive to cause collapse; the problem arises whenever the loads act *rhythmically* on the structure, coinciding with a solid body's own frequency of vibrations. Which means that the effect of a small force acting over an extended period of time may be equal to the effect of a large force acting over a short time span. Actually, everybody was familiar with this "rocking principle", but nobody had thought of applying it to bridge design up to that date.

The Angers Disaster

The sequence of bridge failures during the first half of the 19th century found its tragic peak in the collapse of the Basse-Chaine-Suspension Bridge in Angers, France, on April 16, 1850, with 226 casualties *(Fig. 54)*. Following the crossing by a squadron of husars, the bridge was still in trembling motion, when a battalion of infantry was already approaching in close formation – but did not follow orders to open up in sections, because of heavy rain. Under the marching steps of the soldiers, the bridge experienced such vibrations during a stormy westerly wind that it suddenly crashed into the Maine River, with the total load of 478 troops.
Yet this very bridge had been constructed just 11 years before (1836–1839) with special care by Chaley and Bordillon, two of the best suspension-bridge builders of their time. Because the cast-iron piers at the right shore were totally destroyed, a fault of casting was assumed at first. Later it was established that one of the two primary suspension cables of the bridge spanning 102 meters had broken – deep down below, at the point of fluctuating water level. The anchoring cables had simply rusted through, as the lime mortar had never fully protected the cables running in anchoring tubes, from the water. This meant not just an error in construction for this particular bridge, but far worse – it meant a grave *fault of the system* itself, which was then being used for all similar suspension structures: since 1831 bridges had been constructed along the "System Vicat" with anchoring cables running inside a mortar bed.

hänger riß und die Brücke längere Zeit für den Verkehr zwischen London und Holyhead geschlossen werden mußte [18, S. 169].
Allerdings hatte man damals beim Schwingungsverhalten von Brücken erst einen Teil des Problems erkannt. Denn immer noch berechnete man Nutz- und Windlasten so, als ginge es dabei um *statische* Belastungen, ohne Rücksicht auf die Oszillation oder Eigenschwingung einer Konstruktion, als Folge einer oder mehrerer *stoßweiser* Belastungen; d.h. man arbeitete noch ohne den dynamischen «Stoßzuschlag». Es sollte noch weitere 100 Jahre, bis zum Bau der *George Washington-Brücke* über den Hudson River (1923 – 1931) mit über 1000 m Spannweite dauern, bis man die Lehren aus einem an sich einfachen statischen Prinzip zog: Dynamische Belastungen müssen keineswegs groß sein, um Einstürze herbeizuführen, sobald die Lasten *rhythmisch* auf die Konstruktion einwirken und mit der eigenen Schwingungsfrequenz dieses Körpers übereinstimmen. D.h., der Effekt einer kleinen Kraft über einen längeren Zeitraum hinweg kann genauso groß wie die Wirkung einer großen Kraft über eine kurze Zeitspanne sein. Jeder kannte an sich dieses «Schaukel-Prinzip», aber keiner hatte es bis dahin auf Brücken angewandt.

Das Angers-Desaster

Die Kette von Brückeneinstürzen in der ersten Hälfte des 19. Jahrhunderts fand ihren tragischen Höhepunkt mit dem Einsturz der *Basse-Chaine-Hängebrücke in Angers* in Frankreich am 16. April 1850, mit 226 Toten *(Abb. 54)*. Nach dem Durchritt einer Husarenschwadron befand sich die Brücke noch in zitternder Bewegung, als bereits ein Infanteriebataillon in geschlossener Kolonne anrückte, – dabei aber dem Befehl, sich in Sektionen aufzulösen, wegen eines heftigen Regens nicht folgte; unter dem Marschtritt der Soldaten geriet die Brücke bei stürmischem Westwind in solche Schwingung, daß sie plötzlich mit dem ganzen Kontingent von 478 Mann in den Maine-Fluß stürzte.
Dabei war gerade diese Brücke erst 11 Jahre vorher (1836 – 1839) mit besonderer Sorgfalt von Chaley und Bordillon, den zwei besten Hängebrückenbauern ihrer Zeit, errichtet worden. Da die gußeisernen Pfeilertürme am rechten Ufer vollständig zerstört waren, vermutete man zunächst einen Gußfehler. Doch dann zeigte sich, daß eines der beiden Hängekabel dieser 102 m langen Brücke gerissen war – und zwar tief unten an der Verankerung – im Bereich wechselnden Wasserstandes: Die Ankerkabel waren schlichtweg durchgerostet, denn der Kalkmörtel hatte die in Anker-

54
Failure of the Basse-Chaine Suspension Bridge across the Maine in Angers, France, on April 16, 1850, with the loss of 226 lives [9, p. 24].

54
Einsturz der Basse-Chaine-Hängebrücke über die Maine in Angers, Frankreich, am 16. 4 1850 mit 226 Todesopfern [9, S. 24].

The result hit home like a bombshell, considering the several hundred cable suspension bridges which by then had been built in France and Switzerland. For all practical purposes, it meant the end for the first generation of suspension bridges and indeed for their designers – leading to a twenty-year moratorium on all suspension bridge construction in France. Only a government decree of 1870 ended the official moratorium, under the following conditions: the anchors for the suspension cables were to be constructed from now on only with solid iron rods, and under no circumstances with cables. Furthermore, all parts of the suspension system had to be accessible and fully visible for inspection at all times [18, p. 171]. This proves once again how key progress in the development

röhren laufenden Kabelstränge nie richtig vom Wasser abschirmen können. Das bedeutete nicht etwa nur einen Baufehler bei dieser einen Brücke, sondern weit schlimmer – einen gravierenden *Systemfehler* für alle gleichartigen Hängekonstruktionen: Sie waren nämlich seit 1831 nach diesem «*System Vicat*» mit den im Mörtelbett eingelassenen Ankerkabeln ausgeführt worden.

Das Ergebnis schlug wie eine Bombe ein – angesichts mehrerer hundert Kabelhängebrücken, die es damals bereits in Frankreich und der Schweiz gab. Praktisch bedeutete es das Ende für die erste Generation von Hängebrücken, sowie für ihre Erbauer – und führte zu einer zwanzigjährigen Pause jeglichen Hängebrückenbaues in Frankreich. Erst ein Regierungsdekret von

of bridge construction was to result from a bridge failure.

Practicioners versus Theoreticians – National Differences

With this experience, certain national differences in the European history of technology – such as that between Great Britain and France – were confirmed. The fascinating aspect of "the French way" for instance, lies not so much in this country's centralistic documentation of all new scientific knowledge as such – as if decreed by some sovereign act of central government in Paris; rather it stems from the strict belief in a *single* reason, a *single* truth derived from rational reasoning as being the only valid motive for technological advance. This is considered so "matter of fact" and generally compelling, that it is sanctioned and pontificated by the semi-military *"Corps royale des ponts et chaussees"* in Paris.
Compared to this, the "rugged individualists" representing the "British way" must indeed appear somewhat more dilettantish, with their manifold and uncoordinated experiments, with their charming and often futile sequence of educated guesses – even though these traits are coupled with admirable individual daring and perseverance! The German engineer Max Maria von Weber describes this pointedly in 1867 [19, p. 143]:

> "In Britain, there were no schools through which to learn engineering science; knowledge in mathematics and physics had to be gathered in a pitiful fashion. The masters of engineering science did create themselves empirically through their practice, most of them rising from its lowest ranks, or being thrown accidentally into technology, out of a different sphere of life with its appropriate education. Almost all of the great engineers of Britain, the originators of the entire railway craftsmanship, were type-cast in this way."

Also, a distinct class-consciousness was not exactly furthering the transmission of knowledge across such social boundaries. The leading figures in industrial enterprise were for many years people who may have gotten a good education, but no thorough technical training. In industry, there was a common notion – which still exists today – that a man possessing a good general education would easily be able to run an industrial enterprise. Those relatively few who had indeed studied engineering science, were highly respected by their workmen; but next to lawyers and accountants, they mattered less. A recent article in the

1870 beendete das offizielle Moratorium, mit folgenden Auflagen: Die Verankerungen der Hängekabel durften jetzt nur noch mit massiven Eisenstangen, keinesfalls aber mit Kabeln ausgeführt werden. Ferner mußten sämtliche Teile des Hängesystems zur jederzeitigen Inspektion offen sichtbar und frei zugänglich angeordnet werden [18, S. 171]. Womit wieder ein wesentlicher Entwicklungsschub im Brückenbau das Ergebnis eines Einsturzes sein sollte.

Empiriker gegen Theoretiker – Nationale Unterschiede

Mit dieser Erfahrung waren gewisse nationale Unterschiede in der europäischen Technikgeschichte – namentlich zwischen England und Frankreich – erneut bestätigt worden: Das eigentlich Faszinierende am «französischen Weg» ist ja nicht so sehr die zentralistische Erfassung neuer Erkenntnisse an sich – sozusagen als ein hoheitlicher Akt der Zentralverwaltung in Paris; vielmehr ist es der unbedingte Glaube an *eine* Ratio, *eine* vernunftbestimmte Wahrheit als die allein verbindliche Maxime für technisches Handeln. Sie gilt als so allgemeingültig und richtungweisend, daß sie in der halb-militärischen Elite des angesehenen *«Corps royale des ponts et chaussees»* behördlich sanktioniert und behütet wird.
Verglichen damit muß der englische Weg der «*rugged individualists*» mit ihren vielfältigen unkoordinierten Experimenten, mit ihrer charmanten Kette von «*educated guesses*» in der Tat fast dilettantisch erscheinen – wenn auch mit großem persönlichen Einsatz und Wagemut verbunden! Recht treffend schildert das Max Maria von Weber, 1867 [19, S. 143]:

> «In England gab es keine Schulen für die Erwerbung der Ingenieur-Wissenschaft; kümmerlich mußten mathematische und physikalische Kenntnisse zusammengesucht werden. Die Meister der Ingenieurkunst erzeugten sich empirisch aus der Praxis, meist von den untersten Schichten derselben emporsteigend, oder zufällig aus einer anderen Lebenssphäre mit zu dieser gehörigen Bildung in die Technik verschlagen. Fast alle großen Ingenieure Englands, die Schöpfer der gesamten Eisenbahnkunst hatten diese Provenienz.»

Zudem war ein ausgeprägtes Klassenbewußtsein nicht eben förderlich für die Weitergabe von Wissen über solche gesellschaftlichen Risse hinweg. Die leitenden Männer in Industriebetrieben waren viele Jahre lang Leute, die vielleicht eine gute Schulbildung, aber keine solide technische Ausbildung genossen hatten. In der Industrie galt – und gilt heute

German newspaper *Frankfurter Allgemeine Zeitung* [24], by drawing a line between technical trades and other professions, illustrates how this attitude prevails in British society today:

> "By industry, people working in the City are called *'paper pushers'*, while people working for instance in heavy industry are being put down as *'metal bashers'* .... But the cradle of British industry stood not in London or Canterbury, but in Manchester, Glasgow, Birmingham, Sheffield, Leeds and West Bromwich."

The differing approaches in these countries to technological training and education, but also of the engineer's social standing, deepened the polarisation between the Anglo-Saxon "practicioner" and the French "theoretician". This even affected the way in which the lessons derived from such bridge failures were dealt with – although such disasters were equally frequent in both countries.

The Lessons of the "Tay Bridge Disaster"

The Rothery Report of 1880 had proven sufficiently that notwithstanding the errors and omissions by workers and engineers, in the end the exceptionally strong vibrations of one stormy night had caused the collapse of the Tay Bridge – aided by great dynamic stress under rolling load. In other words: a bridge that had already been weakened but was still standing, was finally destroyed by the coincidence of an unusual combination of forces. Just like 29 years before at the Angers Bridge, also here two extreme stresses of differing type – both vertical and horizontal – had coincided in a fateful manner. The similarity of events is obvious and quite telling by itself for the history of modern bridge construction.

After this grave blow to the national prestige, nobody was inclined to let the matter rest for long. The damaged professional honour of British engineering had to be re-established. The first step was a drastic tightening-up of requirements by the Board of Trade regarding the permissible wind pressures for bridge construction, which then set these standards:

1) for railway bridges and viaducts, a maximum wind pressure of 56 lbs/sq ft (275 kg/m^2) is mandatory;
2) the structures must be designed in such a way as to withstand four times the pressure thus calculated, except that in cases where the force of the wind is compensated solely by the weight of the structure, a safety factor of two would suffice.

noch – vielfach die Vorstellung, daß ein Mann mit guter Allgemeinbildung allemal imstande sei, ein Unternehmen zu leiten. Die relativ wenigen, die Ingenieurtechnik studiert hatten, genossen zwar bei ihren Arbeitern hohes Ansehen, aber neben den Juristen und «*Accountants*» zählten sie weniger. Ein Zeitungsbericht aus unseren Tagen [24] beschreibt eine Grundhaltung der britischen Gesellschaft, wenn sie zwischen technischen und anderen Berufen unterscheidet:

> «Die Industrie bezeichnet die Menschen in der City als ‹*paper pushers*›, etwa: ‹Schreiberlinge›, wofür umgekehrt z. B. die in der Metallindustrie als ‹*metal bashers*›, Metallhämmerer, heruntergemacht werden... Aber die Wiege der britischen Industrie stand nicht in London oder Canterbury, sondern in Manchester, Glasgow, Birmingham, Sheffield, Leeds und West Bromwich.»

Die unterschiedlichen Wege bei der technischen Aus- und Weiterbildung, aber auch in der gesellschaftlichen Stellung des Ingenieurs, vertieften die Polarisierung zwischen dem angelsächsischen «Empiriker» und dem französischen «Theoretiker». Das wirkte sich auch in der Aufarbeitung solcher Brückeneinstürze aus – die wohl in beiden Ländern mit gleicher Häufigkeit vorkamen.

Die Lehren aus dem «Tay Bridge Disaster»

Mit dem *Rothery-Report* von 1880 war hinlänglich klargeworden, daß ungeachtet der vorangegangenen Bauschäden und Versäumnisse, letztlich die ungewöhnlich starken Schwingungen einer Sturmnacht, zusammen mit großer dynamischer Beanspruchung durch rollende Last, den Einsturz der Tay-Brücke herbeigeführt hatten. Oder anders ausgedrückt: Eine bereits geschwächte Brücke, die aber noch stand, war erst durch das auslösende Moment einer ungewöhnlichen Kräfte-Konstellation vernichtet worden. Auch hier waren, wie schon 29 Jahre zuvor bei der Angers-Brücke, zwei extreme Belastungen unterschiedlicher – vertikaler und horizontaler – Art auf schicksalhafte Weise zusammengetroffen: Die Ähnlichkeit des Ablaufes ist nicht zu übersehen und kennzeichnet die Geschichte des modernen Brückenbaus.

Nach diesem schweren Schlag für das nationale Prestige war man keineswegs gewillt, die Sache lange auf sich beruhen zu lassen; es galt, die beschädigte Berufsehre des britischen Ingenieurwesens wiederherzustellen. Das erste Ergebnis war eine drastische Verschärfung der *Board-of-Trade*-Bestimmungen über den zulässigen Winddruck bei Brückenbauwerken:

It is easy to unterstand that such safety requirements meant the end for the airy elegance of railway bridges built up to this date. The year of 1880 marks the beginning of a new era for major engineering structures. Today, with hindsight one century later, it may seem easy to derive the appropriate lessons from such a disaster. But in those days, the lessons entailed a major re-thinking – not just of the required calculations and safety factors, but even of the basic conception of such bridges. It is beyond contention that this collapse was in the end "useful" to the further development of modern bridge construction – in all compassion for the victims. Without the experiences and consequences derived, the major breakthroughs of the coming decade – such as the erection of the Forth Bridge – would have been unthinkable. The most important lessons drawn from the failure of the Tay Bridge implied that in the future:

    1) wind pressure would have to be taken into consideration to a far greater degree than hitherto customary;

    2) one would have to gain insight into the auto-vibrations (oscillation) of large structures;

    3) the use of exact scientific methods of calculus, instead of free experimentation by using mostly smaller structures built previously would become mandatory, as would:

    4) the exact supervision during all phases of construction, especially in prefabrication by division of labour, coupled with a strict allocation of the respective responsibilities; and finally,

    5) the decisive role of responsible maintenance *after* the official clearance for traffic, with regular inspections, would have to be acknowledged.

In the years to follow the engineering profession was adamant in paying keen attention to these five points – as would be seen during construction of the Forth Bridge.

1) Für Eisenbahnbrücken und Viadukte ist ein maximaler Winddruck von 56 Pfund/Quadratfuß (275 kg/m^2) anzunehmen.

2) Die Bauwerke sind so zu dimensionieren, daß sie dem Vierfachen des danach berechneten Druckes widerstehen können, daß jedoch für die Fälle, in denen die Angriffskraft des Windes allein durch das Gewicht der Konstruktion aufgehoben wird, eine zweifache Sicherheit genügen soll.

Es versteht sich, daß mit solchen Sicherheitsbestimmungen die leichte Eleganz der bis dahin gebauten Eisenbahnbrücken ihr Ende gefunden hatte. Das Jahr 1880 markiert den Beginn einer neuen Zeit bei Großbauwerken. Dabei erscheint es uns heute einfach, mit dem Abstand von über 100 Jahren die entsprechenden Lehren aus einer solchen Katastrophe zu ziehen: Aber zu jener Zeit bedeuteten sie ein erhebliches Umdenken – nicht nur in den notwendigen Berechnungen und Sicherheits-Faktoren, sondern in der gesamten Grundkonzeption derartiger Brücken. Daß dieser Einsturz letztlich «nützlich» für die weitere Entwicklung des modernen Brückenbaus war, ist unbestritten – bei allem Schrecken über die Opfer. Denn ohne die gewonnenen Erfahrungen und Folgerungen wären die entscheidenden Neuerungen der kommenden Dekade – und damit der Bau der Forth-Brücke – gar nicht denkbar gewesen. – Die wichtigsten Lehren aus dem Einsturz der Tay-Brücke sind folgende:

    1) Die Berücksichtigung der Windlast bei weitgespannten Brücken, in viel höherem Maße als bislang üblich.

    2.) Die immerhin beginnende Einsicht in das Schwingungsverhalten (Oszillation) von Großbauwerken.

    3) Die Notwendigkeit exakter wissenschaftlicher Berechnungsmethoden, anstelle des freien Experimentierens anhand früherer, meist kleinerer Bauwerke.

    4) Die Unerläßlichkeit einer genauen Bauaufsicht in allen Arbeitsphasen, gerade bei arbeitsteiliger Vormontage, mit fester Zuweisung der jeweiligen Verantwortungen.

    5) Die entscheidende Rolle eines verantwortlichen Bauunterhaltes *nach* der offiziellen Freigabe, mit regelmäßigen Inspektionen.

In den folgenden Jahren setzte man alles daran, beim Bau der Forth-Brücke diese fünf Punkte mit größter Sorgfalt zu beachten.

Chapter III

The Railway Bridge Across the Firth of Forth

Background

Many of the reports and articles about the Forth Rail Bridge are indebted, in one way or other, to the work of Wilhelm Westhofen[12] [29]. As an engineer he had himself been working on the bridge, being responsible for the central of three piers, on Inchgarvie Island. His lucid account in 1890 of the entire sequence of construction was the first comprehensive description of this major bridge, of its early background and its designers. His account is still held today to be *the* authoritative documentation; over the course of 100 years, it has lost none of its dry charm:

> "It has at all times been a subject for controversy and a matter of difficulty to fix the precise boundary line between the river and the sea, that is to say exactly where the sea ends and the river commences. With regard to the Forth and its estuary, the same discussion has been carried on in Parliament and elsewhere with considerable warmth, but does not appear at the present moment to have got any nearer to settlement than in 1882."

With this contemplation, Westhofen begins his geographic description of the bridge in 1890.
The soul-searching question as to where beginning and end of the Firth of Forth lie – as a Firth is neither river nor sea but an extended bay, not unlike its linguistic

---

12 Wilhelm Westhofen (1842–1925) was educated in Mainz and Karlsruhe, before becoming chief draughtsman to a mechanical engineering firm in Mannheim. In 1867 he moved to London to join the firm of T. R. Crampton, where his varied work included experimental investigations into the proposed Channel Tunnel. In 1882 he became Assistant Engineer at the Forth Bridge, responsible for foundations, piers and the Inchgarvie superstructure. On completion of the bridge and publication of his account in 1890, he moved to South Africa to superintend a large bridge designed by Baker, later to become Head of the Engineering Branch of the Public Works Department in Cape Town.

Kapitel III

Die Eisenbahnbrücke über den Firth of Forth

Vorgeschichte

Viele der Berichte und Artikel über die Forth-Brücke gehen, in der einen oder anderen Weise, auf das Werk von Wilhelm Westhofen[12] [29] zurück. Er war als Ingenieur selbst am Bau der Brücke beteiligt und für den mittleren der drei Pfeilertürme (Inchgarvie) verantwortlich. Seine genaue Chronik des Bauablaufes war die erste zusammenhängende Darstellung dieser Großbrücke, ihrer Vorgeschichte und ihrer Erbauer. Sie gilt bis heute als *die* authentische Dokumentation – und hat auch im Laufe von 100 Jahren noch nichts von ihrem spröden Charme verloren:

> «Es war zu allen Zeiten ein Gegenstand von Kontroverse und Schwierigkeit gewesen, die genaue Grenzlinie zwischen Meer und Fluß zu bestimmen. ... Im Hinblick auf den Firth of Forth und seine Mündung ist dieselbe Diskussion im Parlament und sonstwo mit beträchtlichem Eifer geführt worden; sie scheint aber zum gegenwärtigen Zeitpunkt in keiner Weise einer Lösung näher gekommen zu sein als 1882.»

Mit dieser Betrachtung beginnt Westhofen seine geografische Lagebestimmung der Brücke im Jahre 1890.
Die obige Gewissensfrage nach Anfang und Ende des Firth of Forth – ein «Firth» ist weder Fluß noch Meer, sondern ein langgestreckter Meeresarm, ähnlich dem sprachverwandten «Fjord», aber ohne dessen steile

---

12 Wilhelm Westhofen (1842–1925) studierte in Mainz und Karlsruhe, und wurde dann Chef-Zeichner bei einem Maschinenbau-Unternehmen in Mannheim. 1867 ging er nach London zur Firma von T.R. Crampton, wo seine vielfältigen Arbeiten auch experimentelle Untersuchungen für den geplanten Ärmelkanal-Tunnel umfaßten. 1882 wurde er Hilfs-Ingenieur auf der Forth-Brücke, verantwortlich für die Gründungen, Pfeiler und den Inchgarvie-Oberbau. Nach Vollendung der Brücke und Veröffentlichung seines Werkberichtes von 1890, zog er nach Südafrika, um den Bau einer von Baker geplanten Großbrücke zu beaufsichtigen, und wurde danach Leiter der Ingenieurs-Abteilung des öffentlichen Bauamtes in Kapstadt.

relative "Fjord" though lacking its steep flanks – is resolved if an imaginary line is drawn from Anstruther in the north to Dunbar in the south, passing near May Island and Bass Rock, and if everything east of this is called sea, and west of it is called Firth. From there it is 32 miles (51.5 km) to the west to Queensferry, the location of the Forth Bridge *(Fig. 55)*.

On both sides of this great waterway were situated at that time – and still are today – hundreds of square miles of the most fertile land of the "Three Kingdoms" (England, Scotland and Ireland), as well as rich mineral resources. The only connection across this immense waterway was afforded by sailing vessels and steam ferries later on. There were three crossings: the eastern one, 5 miles (8 km) long, from Granton to Burntisland; the centre one between South and North Queensferry; and the western one, 15 miles (24 km) upstream near Kincardine, where the Firth narrows down to the River Forth. The next railway bridge was situated only at Alloa, 20 miles (32 km) to the west of Queensferry.

The long Granton-Burntisland passage was often unnavigable during storms, and even at the best of times, the disembarkation from train to boat and reverse was "... a source of considerable discomfort to passengers, and what is worse, a great waste of time" [29, p. 1]. Therefore there was nothing left but to accept the wide detour from Granton via Alloa or Stirling to Burntisland, with about 150 miles (241 km) of rail travel, even though the distance as the crow flies, is less than 18 miles (29 km).

When and with whom the idea of crossing the Forth first originated has to remain speculation – but there are plenty of stories around including the ones about the frequent pilgrimages of St. Margaret between Edinburgh, Linlithgow and Dunfermline at the time of the Norman Conquest; or about her son Alexander the First of Scotland, who in 1123 on business of State was driven by a storm to the island of Inchcolm (near Aberdour, *Fig. 55*), while attempting a crossing at Queensferry; there he founded a monastery in gratitude for the help of a hospitable hermite (thus providing for the occasion that such a mishap should strike him again). Or there is the story of that "spectral" party of unlucky passengers who were driven down the Hawes Brae with such speed by the storm, that horses, carriage and passengers went right off the pier into the water and none of them came out alive. Somewhat further back into history goes the saga of King Hungus who in returning from victorious warring to his native Kingdom of the Picts (today's Fife) crossed the narrows near Inchgarvie and impaled the head of his foe Athelstane upon a stake on the

Flanken – erledigt sich, wenn man eine imaginäre Linie von Anstruther im Norden nach Dunbar im Süden zieht, quer über May Island und Bass Rock, und alles östlich davon als Meer, und westlich davon als Firth bezeichnet. Von dort sind es noch 32 Meilen (51,5 km) westwärts nach Queensferry, und damit zur Lage der Forth-Brücke *(Abb. 55)*.

Auf beiden Seiten dieses großen Wasserweges lagen damals – und liegen heute noch – Hunderte von Quadratmeilen des fruchtbarsten Ackerlandes der «drei Königreiche» (England, Schottland und Irland), sowie große Bodenschätze. Die einzige Verkehrsmöglichkeit über dieses gewaltige Hindernis boten Segelschiffe und später Dampfer-Fähren. Davon gab es drei: die östliche, 5 Meilen (8 km) lang, von Granton nach Burntisland; die mittlere zwischen South- und North-Queensferry; und die westliche, 15 Meilen (24 km) stromaufwärts bei Kincardine, wo sich der Firth zum Flusse Forth verengt. Die nächste Eisenbahnbrücke lag erst bei Alloa, 20 Meilen (32 km) westlich von Queensferry.

Die weite Granton-Burntisland-Passage war bei Sturm oft unpassierbar, und auch zu den besten Zeiten war das Umsteigen von Bahn auf Boot und umgekehrt *«...a source of considerable discomfort to passengers and what is worse, a great waste of time»* [29, S. 1]. Also blieb nur der große Umweg von Granton über Alloa bzw. Stirling nach Burntisland mit etwa 150 Meilen (241 km) Bahnfahrt, während die Luftlinie – *«as the crow flies»* – kaum 18 Meilen (29 km) beträgt.

Wann und bei wem die Idee, den Forth zu überbrücken, erstmalig aufkam, muß Spekulation bleiben – doch Geschichten gibt es viele: So von häufigen Pilgerfahrten der heiligen Margaret zwischen Edinburgh, Linlithgow und Dunfermline zur Zeit der Normannenkriege; oder von ihrem Sohn Alexander I. von Schottland, der 1123 *«on business of state»* beim Versuch der Überfahrt nahe Queensferry vom Sturm auf die Insel Inchcolm (vor Aberdour, siehe *Abb. 55*) verschlagen wurde, und dort aus Dankbarkeit für die Hilfe eines Eremiten – und für den Fall, daß ihm solch Mißgeschick noch mal widerfahren sollte – ein gastfreundliches Kloster gründete; oder die Geschichte von jener gespensterhaften Gruppe Unglückseliger, die vom Sturm in ihrem Gefährt mit solchem Tempo den *Hawes Brae* hinuntergetrieben wurden, daß Pferde, Wagen und Insassen geradewegs über die Ufermauer ins Wasser rasten – *«and none of them came out alive»*. Um einiges weiter zurück geht die Geschichte von König Hungus, der nach erfolgreicher Kriegsfahrt auf der Rückkehr in sein heimatliches Königreich der Pikten (auf dem heutigen Fife) die Meerenge bei Inch Garvie überquerte und auf der Insel den Kopf seines

55
The railway line from Edinburgh to Dundee, including the Firth of Forth and Firth of Tay [16].

55
Die Strecke Edinburgh-Dundee mit Firth of Forth und Firth of Tay [16].

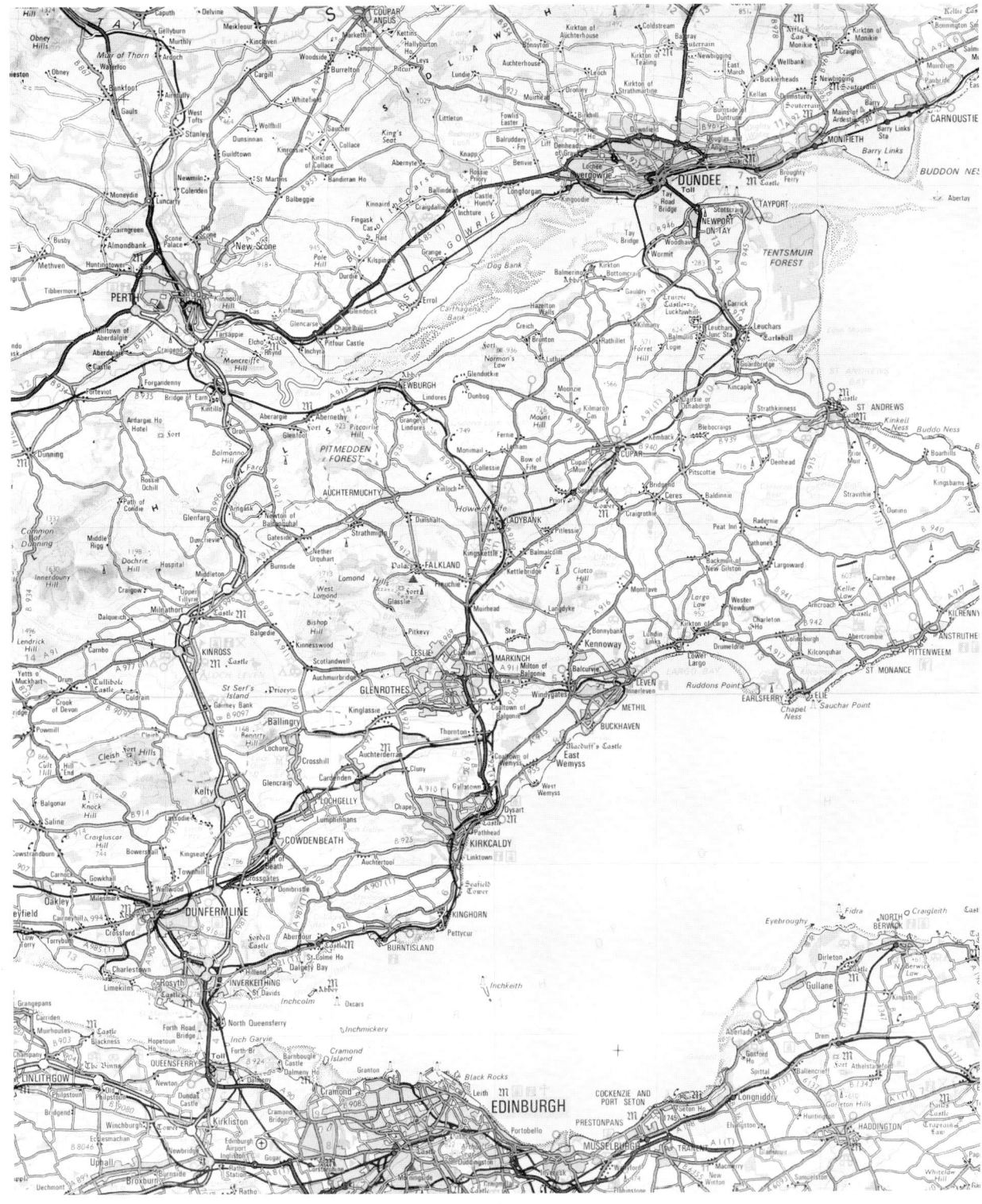

island; this dire warning served for many years to scare off anybody venturing further north across the Forth.

Despite all this, it can safely be assumed that since ancient times, some kind of ferry service did exist across these narrows, roughly following the course of today's bridge. During medieval times, the ferry rights were awarded as Royal feus and jealously guarded by the affected communities. The ferrymen were regarded an obstinate and rude lot, their treatment of passengers and equipment leaving a lot to be desired. Since they kept the ferry boats near their village in North Queensferry, the passengers from the south often had to wait for hours, in bad weather even for days – so they took refuge in the hospitable Hawes Inn at the southern shore, where it is still holding its own even to this day. How necessary a direct link was between the two shores is proven by the fact that in 1805 – before there were steamers or even trains – these sought-after ferry rights were awarded for a yearly rent of 2,000 Pounds Sterling, while "unlicensed outsiders" offering ferry services to goods and passengers in small boats, were collecting a yearly income estimated at 5,000 Pounds per annum.

Age of Wild Projects

By the end of the 18th century, the idea of a *tunnel* under the Firth of Forth seems to have come up in earnest for the first time. Some progressive people in Edinburgh, then illuminated by the spirit of enlightenment – were familiar with the successful experiments of tunnelling the Thames River in London. In 1805 three Edinburgh engineers proposed to drive a double tunnel underneath the riverbed, with two tubes of 15 ft (4.57 m) diameter each – "one for comers, one for goers" – a little to the west of Queensferry. According to these plans, each tube was to have a separate pedestrian footpath next to a carriageway. Evidently this project was seriously pursued, and in 1806 a "prospectus" was issued "by a number of noblemen and gentlemen of the first respectability and scientific character" – inviting the public to subscribe shares at 100 Pounds each. The tunnel was to have cost 160,000 Pounds, requiring four years to build. In 1807 a 120-page pamphlet came out, entitled "Observations on the Advantages and Practicability of making Tunnels under Navigable Rivers - applicable to the proposed Tunnel under the Forth", and illustrated with a map and cross-section. Still, nothing seems to have come of this undertaking.

But now the time had arrived when new roads and

56

The early design by James Anderson for a stay-cable bridge and, alternatively, a chain suspension bridge across the Firth of Forth, 1818 [29, p. 3].

56

Der frühe Entwurf von James Anderson für eine Schrägseil- bzw. Ketten-Hängebrücke über den Firth of Forth, 1818 [29, S. 3].

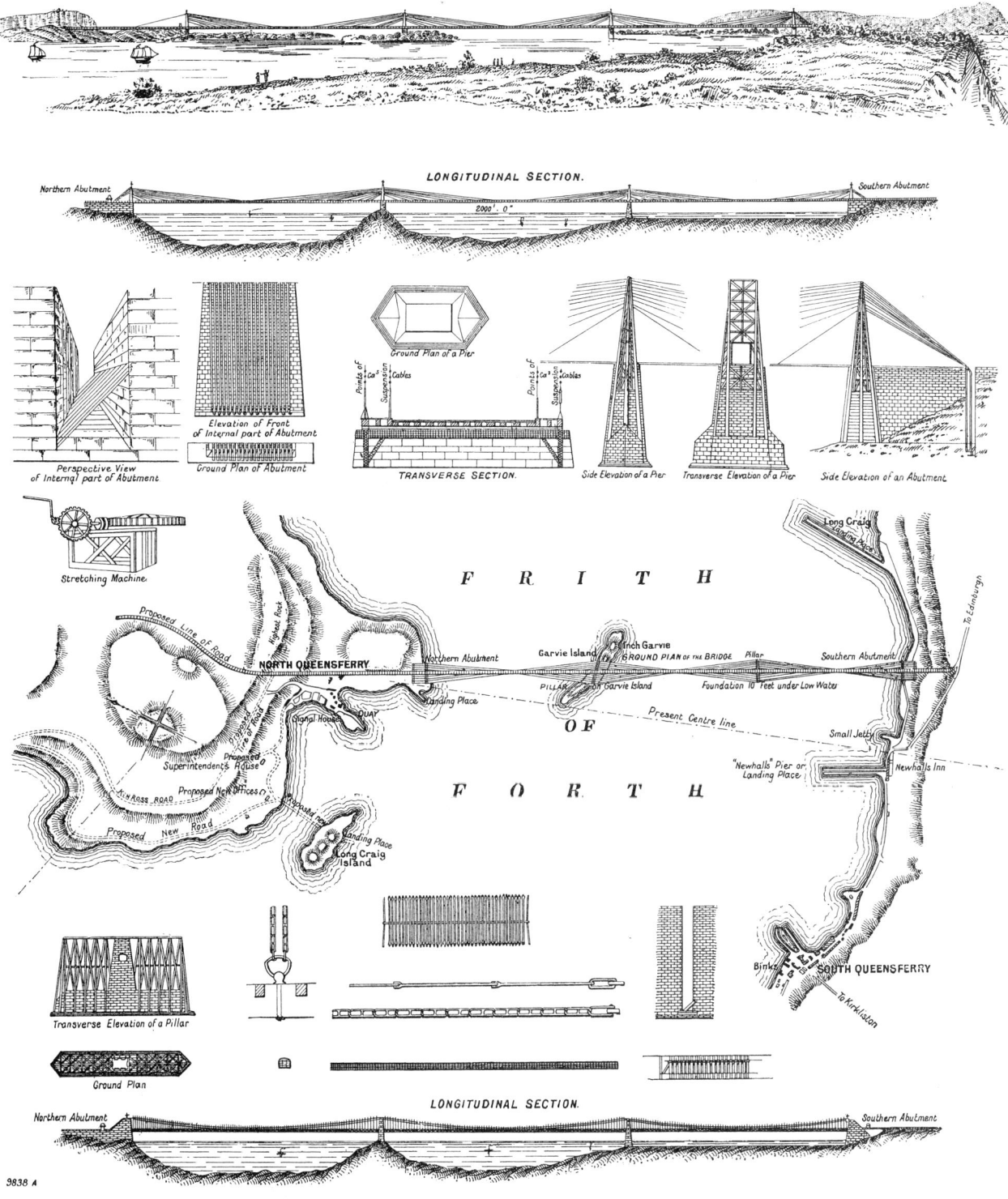

bridges began to open up areas all over Scotland, which had been considered simply inaccessible to transport on wheels. It was a time that not only allowed the engineer to remove natural obstacles; it actually turned this into his national duty!

The Role of James Anderson

Eleven years after talk of the tunnel project, there was a new and remarkable attempt to link the shores, the significance of which has long been underestimated. We speak of nothing less than a precursor design – even an early prototype – for a major bridge across the Forth. A pamphlet entitled "Report relative to the Design for a Chain Bridge thrown over the Firth of Forth at Queensferry" came out in 1818, penned by James Anderson, surveyor and civil engineer from Edinburgh. His vision for the old narrows conjured a modern miracle, a meeting of science and art:

> "The appearance and situation are altogether so favourable and so inviting for some work of art that it has often occurred to the reporter when he considered them attentively that a bridge of some description ought to be attempted." [15, p. 17]

Anderson's design proposed two early alternatives for a suspension bridge. One version may be considered a precursor of a modern stay-cable suspension bridge – albeit showing very low piers with extremely flat stay-cables; the other version proposed was a chain suspension bridge of a similarly extended appearance (Fig. 56). In both cases, the site as planned corresponded closely to the one of today, resulting in three almost equal spans of about 2,000 ft (609.60 m), even the main piers appeared in a similar position as they do today. The bridge started out nearly from the same point in North Queensferry, though it turned slightly southeast along its axis; thus one pier stood right in the middle of Inchgarvie, not at its northwesterly corner like today – while the southern end met the shore some 600 ft (183 m) further to the east, than is the case today.

In Anderson's design, however, the shallow southern portion of the bay (about one third of the bridge's total length) was crossed over by a third span – not just by a viaduct like today. This made for an altogether elegant and daring solution, without a doubt. But in view of the very low profile of the structure, it may be questioned whether his design could really have been executed in this low-slung and extended form. Even today's stay-cable and suspension bridges require about three times the height of the piers, when using the same spans. In his design, the clearance for ship-

ping underneath was to be 90 or 110 ft (27.43 or 33.53 m), depending on the alternative proposal chosen, in order to accomodate the tallest sailing masts. The main spans would have been 1,500 or 2,000 ft (457.20 or 609.60 m). In comparison, the Forth Bridge, as it was eventually built, had a clearance of 150 ft (45.72 m) and spans of 1,710 ft (521.71 m). Despite these differences it remains a tribute to the visionary foresight of James Anderson that his design anticipated the essential features of the final bridge by more than 60 years:

> 1) the almost identical site of the bridge, as required by
> 2) the utilization of Inchgarvie Island as the most sensible location for the main pier, situated right in the middle between the northern and the southern shipping channel; and
> 3) despite all the differences in construction, a similar overall appearance, with a profile rising and falling in three-fold rhythm.

It is essential to understand and evaluate this design in light of the state of knowledge of those times, not of our times.

Today, the Scottish civil engineer James Anderson, F.R.S. (Fellow Royal Society), 1792–1861, may rightly be called the ancestor and spiritual father of the Forth Bridge that we know now[13]. He had offered his completed and fully calculated plans already 62 years earlier to the government which, however, considered these proposals as too daring. Only in 1880, when the final bridge was to be started, did the government want to obtain these plans and calculations for a considerable amount of money from Anderson's widow (née Anna Watt, daughter of the Governor Robert Watt of Jamaica). She, however, had sold all his records as waste paper only shortly before!

In this context, a few biographic notes might be illuminating about this breed of British engineer and autodidact, a product of the early age of industrialisation. James Anderson had abandoned his first career choice of becoming a lieutenant in order to become a civil engineer; as he was very wealthy, he was able to set his own tasks and goals. He sold many military coastal defence inventions to the government. His

---

[13] A family chronicle by Anderson's grandson, Carl-James Bühring, building councillor in Leipzig, relates that British papers at the opening of the bridge in 1890 recalled James Anderson as its spiritual ancestor and the most eminent engineer of his time. – For this and some further references, as well as for the use of the portrait photographs, the author is indebted to the great-grandson of James Anderson, Dipl.-Ing. Carl-Artur Bühring in Stuttgart.

daughter Margaret, from his first marriage, helped him with the calculations. Remarried after the death of his wife to the then 17-year-old Anna Watt (see above and *Fig. 57*), he had four more children. Eventually he lost all of his wealth with the bankruptcy of the Bank of Glasgow at the end of 1860. As if to top off the finality of such a disaster, a fire after his death destroyed also his large library with all holdings – among them the correspondence with his friend and benefactor, the Duke of Wellington. So much the greater then is the value of James Anderson's designs for a major bridge across the Firth of Forth in 1818 to be presented here. Quite possibly, they are the only records still existing today of this visionary design.

"Railway Mania"

With the onset of the 1840's, the railway age had come to Scotland – and the call for a bridge across the Forth grew even louder. Fully within the spirit of the new industrial era, the national justification for so great an undertaking was seen in the attempt to serve higher goals than just those of local traffic: "In these days of high pressure, of living and working and eating and drinking at top speed, the saving of an hour or two for thousands of struggling men every day is a point of the greatest importance, and every delay, however excusable and unavoidable, is fatal to enterprise." [29, p. 1] Also, the bridge was considered not just as an achievement for its own sake, but as a portion – "certainly a rather expensive portion" – of a gigantic system of railway lines converging from all directions upon the capital of Scotland. All those participating in this great task were optimistic that courage, foresight and wisdom of the directors of the Forth Bridge Railway Company would be fully confirmed by results which nobody could dream of at the moment. At the same time, the race by the two major railway companies went on as fierce as ever, bent on opening up the industrial areas of northern England and Scotland. After the CR had already succeeded in stalling the erection of the Tay Bridge by 20 years, now the CR and the NBR once again opposed each other in a bitter struggle. Indeed, the history of the Tay Bridge seemed to repeat itself all over again, because the fastest route to northern Scotland still went along the well-managed western line of the competing CR – "whose interests were hostile and in opposition of their own" (the NBR's). Again we see how closely the history of the Forth Bridge is intertwined with the one of the Tay Bridge. Not just the dynamism of "Railway Mania", but even the person of the then still-young Thomas Bouch had charged the excitement of the officials at the Edin-

sein: James Anderson hatte seine erste Berufswahl als Leutnant quittiert, um Zivilingenieur zu werden. Er war sehr vermögend und konnte sich deshalb seine Aufgaben selbst stellen. So verkaufte er viele Erfindungen im Militärwesen und Küstenschutz an die Regierung. Seine Tochter Margeret aus erster Ehe war ihm bei den statischen Berechnungen behilflich.
Nach dem Tode seiner Frau wiederverheiratet mit der 17jährigen Anna Watt (siehe oben und *Abb. 57*) hatte er noch vier weitere Kinder, verlor aber schließlich kurz vor seinem Tode sein gesamtes Vermögen durch den Bankrott der Bank von Glasgow, Ende 1860. Wie um die Finalität solcher Ereignisse komplett zu machen, vernichtete nach seinem Tode auch noch ein Stubenbrand seine große Bibliothek mit allen Hinterlassenschaften – darunter auch die Korrespondenz mit seinem Freund und Gönner, dem *Duke of Wellington*. Um so größer ist der Wert der hier vorgestellten Entwurfszeichnungen von James Anderson für eine Großbrücke über den Firth of Forth von 1818. Möglicherweise sind sie die einzigen heute noch erhaltenen Zeugnisse dieses visionären Vorläufer-Entwurfes.

«Railway Mania»

Mit Anbruch der 1840er Jahre war die Eisenbahn-Epoche nach Schottland gekommen – und der Ruf nach einer Brücke über den Forth wurde immer lauter. So recht im Geiste des neuen Industriezeitalters sah man die nationale Berechtigung für ein so großes Werk in dem Streben, größeren Interessen als lediglich denen des Lokalverkehrs zu dienen: «In diesen Tagen großer Hast, in denen Leben, Arbeiten, Essen und Trinken mit höchster Schnelligkeit vor sich gehen, ist der Gewinn von ein oder zwei Stunden für Tausende arbeitender Männer jeden Tag ein Anliegen größter Wichtigkeit und jede Verzögerung *«fatal to enterprise»* (tödlich fürs Geschäft) – wie entschuldbar oder unvermeidlich auch immer» [29, S. 1]. Auch wird die Brücke nicht als eine Leistung um ihrer selbst willen gesehen, sondern als Teil – «sicherlich ein etwas teurer Teil» – eines gigantischen Systems von Eisenbahnlinien, die aus allen Richtungen auf die Hauptstadt Schottlands zulaufen. Alle an diesem Werk Beteiligten sind zuversichtlich, daß Mut, Weitsicht und Weisheit der Direktoren der *Forth Bridge Railway Company* durch Ergebnisse bewahrheitet werden, die sich noch niemand erträumen kann. Gleichzeitig hielt aber der Wettlauf der wichtigsten Eisenbahngesellschaften um die Erschließung der Industrieregionen Nord-Englands und Schottlands unvermindert an. Nachdem die Caledonian Railway Company schon den Bau der Tay-Brücke um 20 Jahre verzögert hatte,

57
Anna Anderson, nee Watt, 1822–1884

Anna Anderson, geb. Watt, 1822–1884

58 >
Thomas Bouch's revolutionary railway ferries of 1850. In the centre, the sliding platform acting as a "movable cradle" for the trains; to the right, the "Leviathan". Between the large paddle wheels (each driven independently by its own steam engine) rests the low rail bed of the ferry [15, p. 23].

58 >
Thomas Bouchs revolutionäre Eisenbahnfähre von 1850: In der Mitte die Gleitplattform als *«movable cradle»* (fahrbare Wiege) für die Züge, rechts der *«Leviathan»*. Zwischen den großen Schaufelrädern (jedes unabhängig von eigener Dampfmaschine getrieben) liegt das niedrige Schienendeck der Fähre [15, S. 23].

57
James Anderson, 1792–1861
(Photographs in private collection of Carl-Artur Bühring, a great-grandson of James Anderson.)

James Anderson, 1792–1861
(Fotos aus dem Familienbesitz von Carl-Artur Bühring, einem Urenkel von James Anderson.)

burgh and Northern Railway – long before the Tay Bridge had even been started. Bouch had already been working as a railway engineer since he was 17 years of age – which may have led later on to some remarks about his "rather shadowy qualifications" [15, p. 102]. Yet his breadth of imagination and energy – he had already designed a tramway system for Edinburgh on the side – enabled him to embark, for a start, on the construction of very novel railway ferries across the Forth and the Tay *(Fig. 58)*.

Bouch's ingenious idea consisted in bridging the formidable tidal difference of 20 ft at the landings by special sliding platforms on wheels, which were themselves to travel on rails along the loading ramps (slipways), thus being adjustable to any waterlevel. The 60-ft-long (18.29 m) platform was fitted with rails on top, enabling the trains to run from the siding directly onto the rails of the ferries. The famous *Leviathan* of 1850 was the first railway ferry of the world, and it was a great success on the line from Granton to Burntisland. But Bouch's invention had furthered railway traffic to such an extent that the repeated change-overs and interruptions necessary – especially on the long trip to Dundee – were increasingly considered a nuisance. Thus Bouch by his very success with the "floating trains" – but also by his own intervention – only hastened the general demand for a continuous crossing of both estuaries.

standen sich *CR* und *NBR* erneut in erbittertem Kampf gegenüber – ja, die Vorgeschichte der Tay-Brücke schien sich zu wiederholen. Denn der schnellste Weg nach Nord-Schottland lief immer noch über die gutgeführte Weststrecke der konkurrierenden *CR* – «*whose interests were hostile and in opposition to their own*» (der *NBR*).

Es zeigt sich erneut, wie eng die Entstehungsgeschichte der Forth-Brücke mit jener der Tay-Brücke verwoben ist. Nicht nur die gemeinsame Dynamik der «*Railway Mania*», sondern auch die Person des damals noch jungen Thomas Bouch hatte ja – lange vor Baubeginn der Tay-Brücke – bei der *Edinburgh and Northern Railway* für Aufsehen gesorgt. Bouch war schon seit seinem *17. Lebensjahr* als Eisenbahn-Ingenieur tätig gewesen – was allerdings später zu Bemerkungen über seine «*rather shadowy qualifications*» [15, S. 102] Anlaß gab. Aber seine visionäre Kraft und Energie – er hatte nebenbei bereits ein Straßenbahnnetz für Edinburgh entworfen – führten zunächst zum Bau ganz neuartiger Eisenbahnfähren über den Forth und Tay *(Abb. 58)*.

Bouch's genialer Einfall bestand darin, den gewaltigen Tidenhub von 20 Fuß (6,10 m) an den Anlegestellen mit Hilfe einer Gleitplattform auf Rädern zu überbrücken, die selbst wiederum auf den Schienen einer Rampe fuhr und sich so jedem Wasserstand anpassen konnte. Die 60 Fuß (18,29 m) lange Plattform wurde auf ihrer Oberseite mit Schienen versehen, so daß die Züge direkt vom Abfahrtsgleis auf ebenfalls beschiente Fährschiffe laufen konnten: Deren berühmte «*Leviathan*» wurde 1850 zur ersten Eisenbahnfähre der Welt und war ein großer Erfolg auf der Strecke Granton – Burntisland. Aber Bouch's Erfindung förderte den Eisenbahnverkehr derart, daß man das

Bouch's Suspension Bridge Across the Forth

Following the take-over of the Edinburgh and Glasgow Railway by the NBR, the directors in 1863 finally commissioned Thomas Bouch to prepare preliminary plans for a rail bridge across the Forth. But Bouch considered the Queensferry passage too deep for a bridge and so he first chose a wider crossing 5 miles (8 km) upstream, between Blackness to the south and Charlestown in Fife. According to his plans, lattice girders were to cross the two miles (3.2 km) on 61 stone piers, at a height of 150 ft (45.72 m). But the problem was the soil condition, such as a 200 ft-thick (61 m) layer of clay under the riverbed. Bouch thought it well possible to burden this solid layer with the weight of the piers and thus "compact" the soil so that no further setting would occur. But in the end it was not the clay, but the immense stockmarket losses of the NBR which stopped this plan.

Meanwhile Bouch was busy with the Tay Bridge; only in 1873 did he turn his attention towards the new Forth Bridge again. This time, it was to be a gigantic suspension bridge near Queensferry, with a large central pier on Inchgarvie Island *(Fig. 59)*. The towering piers of this bold design were to rise 500 ft (152.40 m) up into the air, with a clearance of 150 ft (45.72 m) above flood level. The central pier measured by itself over 550 ft (167.64 m) in length, which would have brought the foundations down to a depth of over 110 ft (33 m) on the sloping rock of the small island. From there, two large spans of 1,600 ft (487.68 m) each were to reach out to both shores. Those spans were designed to carry two separate tracks of railway at a distance of 100 ft (30.48 m), each pair being supported by strong double girders and braced horizontally by diagonal struts. The piers at South Queensferry and Fife would occupy about the same places as those of today's bridge, with similar approach viaducts extending from the high ground of either shore. The whole structure was to be suspended by an immense chainwork of steel, designed to withstand a maximum tension stress of 10 tons/sq in (1550 kg/cm^2).

Just like before at the Tay, a bridge building company was formed once again, the Forth Bridge Railway Company, FBR for short. Its members were the four railway companies primarily interested in a bridge: The North British, The North Eastern, The Midland and The Great Northern, as they controlled rail traffic along the east coast from London up to Scotland. Together they raised the capital and agreed to send each other so much traffic across the bridge that a yearly dividend of 6 % on the construction cost could be paid out. In July of 1873, the Act of Parliament necessary for

mehrfache Umladen mit den notwendigen Unterbrechungen – besonders auf der langen Fahrt nach Dundee – als immer lästiger empfand. Insofern beschleunigte Bouch durch seinen Erfolg mit den «schwimmenden Eisenbahnen» – aber auch auf eigenes Betreiben – den allgemeinen Wunsch nach einer durchgehenden Überbrückung beider Meeresarme.

Bouch's Hängebrücke über den Forth

Endlich, mit der Übernahme der *Edinburgh and Glasgow Railway* durch die *NBR,* gaben die Direktoren 1863 Bouch den Auftrag, vorläufige Pläne für eine Eisenbahnbrücke über den Forth auszuarbeiten. Bouch schien aber die Queensferry Passage zu tief für eine Brücke und er wählte zunächst eine breitere Stelle 5 Meilen (8 km) stromaufwärts, zwischen Blackness im Süden und Charlestown auf Fife. Nach seinem Plan sollten Fachwerkträger in 150 Fuß (45,72 m) Höhe auf 61 Steinpfeilern die zwei Meilen (3,2 km) überschreiten. Aber das Problem war der Baugrund, nämlich eine 200 Fuß (61 m) starke Tonschicht unter dem Flußbett. Bouch meinte, man könne diese massive Schicht durchaus mit dem Gewicht der Pfeiler belasten und gewissermaßen soweit «komprimieren», daß keine weitere Bewegung auftreten würde. Aber dann stoppten nicht der Ton, sondern die haushohen Börsenverluste der *NBR* diesen Plan.

In der Zwischenzeit war auch Bouch mit der Tay-Brücke beschäftigt und wandte sich erst 1873 wieder der neuen Forth-Brücke zu. Diesmal sollte es eine gewaltige Hängebrücke bei Queensferry sein, mit einem großen Mittelpfeiler auf Inchgarvie. *(Abb. 59)* Die Pfeilertürme des gewagten Entwurfes erhoben sich 550 Fuß (167,64 m) mit einer Durchfahrtshöhe von 150 Fuß (45,72 m) über Hochwasser-Niveau. Der Mittelpfeiler allein war über 500 Fuß (152,40 m) lang, was die Gründung auf dem abfallenden Fels der kleinen Insel auf 33 m Tiefe gebracht hätte; von hier aus sollten zwei große Spannweiten von je 1600 Fuß (487,68 m) zu den beiden Ufern führen. Sie trugen zwei getrennte Schienenstränge in einem Abstand von 100 Fuß (30,48 m), jeder Strang getragen von einem Paar starker Fachwerkträger, und diese horizontal ausgesteift durch Diagonalstreben. Die Pfeiler auf South Queensferry und Fife standen fast an derselben Stelle wie die der heutigen Brücke, mit ähnlichen Auffahrtsviadukten von den beiden Hochufern. Das Ganze sollte getragen werden von einem gewaltigen Kettenwerk aus Stahl mit einer max. Zugbelastung von 10 Tonnen/Quadrat-Inch (1550 kg/cm^2).

Es wurde wieder, wie schon am Tay, eine Brückenbaugesellschaft, die *Forth Bridge Railway Company*

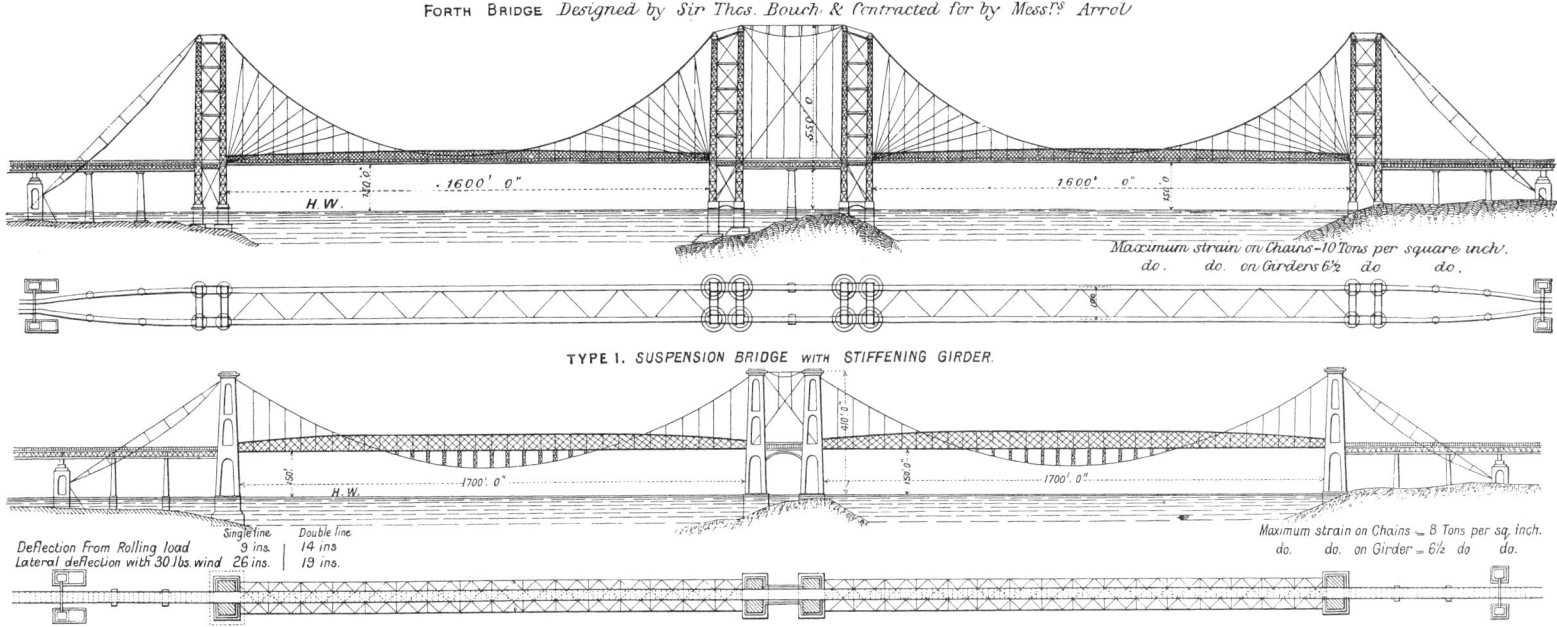

59

Design for a suspension bridge across the Firth of Forth by Sir Thomas Bouch.
"We merely wish it to be understood, that while we raise no objection to Bouch's system, we do not commit ourselves to an opinion that it is the best possible". Thus William Henry Barlow (1812–1902), President of the British Association of Civil Engineers, minced his words in judging this design – and thus also his opinion about this designer ten years younger than himself.

The design is undoubtedly daring, and the arrangement of two separate tracks almost foreshadows modern competition entries for bridges. Yet the parts remain heterogeneous and the whole work seems to be put together out of different structural systems – lacking the "economy of forms" so splendidly illustrated by the final solution. If ever the term "rainbow bridge" is to be applied to any of Bouch's works, then rather to this suspension bridge, than to his Tay Bridge.

Construction began in 1878 under the direction of contractor William Arrol; but in 1880/81 construction was abandoned again under the impact of the Report of Inquiry into the Tay Bridge Disaster of 1879. Not just doubts about Bouch's ability, but also general misgivings regarding the use of suspension bridges for railways had moved the NBR to this decision [29, p. 4].

59

Hängebrücken-Entwurf von Sir Thomas Bouch über den Firth of Forth:
«Wir möchten es lediglich verstanden wissen – selbst wenn wir keine Einwände gegen Bouchs System haben –, daß wir uns nicht auf die Wahl festlegen, dies sei das bestmögliche System.» So formulierte William Henry Barlow (1812–1902), Präsident des britischen Ingenieur-Verbandes, sein Urteil über diesen Entwurf – und damit auch über den 10 Jahre jüngeren Kollegen.

Der Entwurf ist ohne Zweifel kühn und die Führung der beiden getrennten Fahrstränge erinnert durchaus an moderne Brückenwettbewerbe. Aber die Teile sind heterogen und das Ganze scheint aus verschiedenen Tragsystemen zusammengesetzt – ohne die «Ökonomie» der Formensprache, welche die endgültige Lösung auszeichnet: Wenn die Bezeichnung *«Regenbogenbrücke»* auf irgendein Werk Bouchs zutreffen soll, dann eher auf diese Hängebrücke als auf die Tay-Brücke.

Die Bauarbeiten begannen 1878 unter der Leitung des Unternehmers William Arrol, wurden aber 1880/81 infolge des Untersuchungsberichtes über das «Tay Bridge Disaster» von 1879 eingestellt: Nicht nur Zweifel an Bouchs Könnerschaft, auch allgemeine Bedenken gegen Hängebrücken für Eisenbahnen bewogen die *FBR* zu diesem Schritt [29, S. 4].

constructing the bridge was passed and the great contract awarded to W. Arrol & Co. of Glasgow.

Because of the novelty of this design, Chairman John Stirling of Kippendavie – by now familiar to the reader from his active role in the Tay Bridge, and now the new Chairman of the NBR – had asked fellow engineers for their expert opinion of Bouch. Two of them, William H. Barlow (known as well to the reader) and William Pole, drew attention to the expensive materials and complicated stresses; diplomatically they wished it to be understood that while they raised no objection to Bouch's system, they would not commit themselves to the opinion that it was the best possible.

And the Royal Astronomer in Greenwich had given his well-known reply regarding wind pressure, adding how proud he was to have his name associated with such a noble undertaking. Due to lack of funds and the concurrent erection of the Tay Bridge, it still took four more years before, on September 30, 1878, the foundation stone could finally be laid for the great suspension bridge across the Forth – by Bouch's wife Margeret.

Already, workshops and extensive brickyards were being installed at the site. By the spring of 1879, a colossal masonry base was rising up – one of eight bases forming the support for the great Inchgarvie tower at the extreme northwesterly corner of the small island. This lonely base of a pier – it carries a lighthouse today (Fig. 60) – was to be the only trace remaining of "this gigantic undertaking... the crowning triumph in Sir Thomas Bouch's distinguished professional career" [15, p. 27]. With the Tay Bridge Disaster, the tide had turned. At first, the FBR allowed the work to continue, but once the findings of the Rothery Report had become public by the summer of 1880, there arose such a massive storm of criticism among the public and the press against this bridge, and against Sir Thomas Bouch himself, that the FBR very reluctantly abandoned the project in January of 1881.

A New Beginning

The four railway companies were not willing to let the matter rest for long. On February 18, 1881, they commissioned their consulting engineers William H. Barlow, Thomas Harrison and Sir John Fowler, to study the principal possibilities for a bridge across the Forth, and to decide on the type of bridge to be chosen. Fowler had already built two major bridges across the Severn, jointly with his assistant, Benjamin Baker. It was clear which types of bridges would be suitable for this location: 1) one based on Bouch's original design; 2) alternative designs for suspension bridges with

60
At the right foreground rests the large masonry-pier base for Bouch's suspension bridge initiated on Inchgarvie Island. It was started in the spring of 1879 and is topped off today by a lighthouse. It was to remain the only of eight identical masonry piers, which were planned just for the large central tower on Inchgarvie. (Compare *Fig. 59* for Bouch's suspension bridge) [15, p. 64].

60
Vorne rechts der große Ziegel-Pfeilersockel von Bouchs angefangener Hängebrücke auf der Insel Inchgarvie. Er stammt aus dem Frühjahr 1879 und trägt heute einen Leuchtturm. Er sollte der einzige von 8 gleichen Pfeilerfüßen bleiben, die allein für den großen Mittelturm auf Inchgarvie geplant waren (vergl. *Abb. 59* für Bouchs Hängebrücke) [15, S. 64].

61 (Next pages) >
Coloured ink drawing by John Bartholomew, Edinburgh, of the design for a suspension bridge by Sir Thomas Bouch – one of those painterly artist's impressions, featuring popular sea-scape settings right next to the then unfamiliar contraptions of a new technology. Despite its might, the harsh, bloodless ironwork looms somewhat timorously in the setting.

61 (Nächste Doppelseite) >
Kolorierte Federzeichnung von John Bartholomew, Edinburgh, des Hängebrücken-Entwurfes von Sir Thomas Bouch: Eine jener «artist's impressions» der Malerei, in denen z. B. Staffagen aus den beliebten «Seestücken» etwas unvermittelt neben den befremdlichen Apparaturen der neuen Technik auftauchen. Das spröde Gitterwerk steht bei aller Größe eher zaghaft und blutleer in der Gegend.

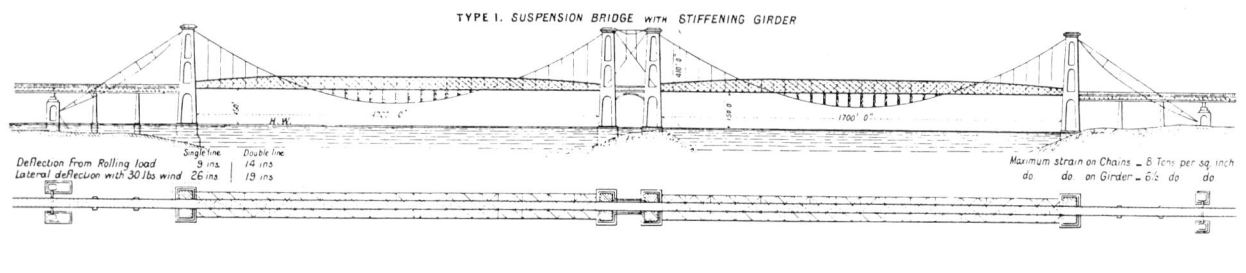
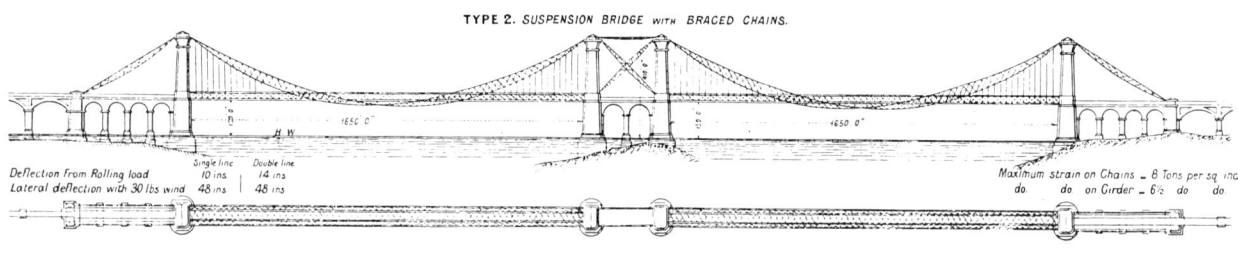
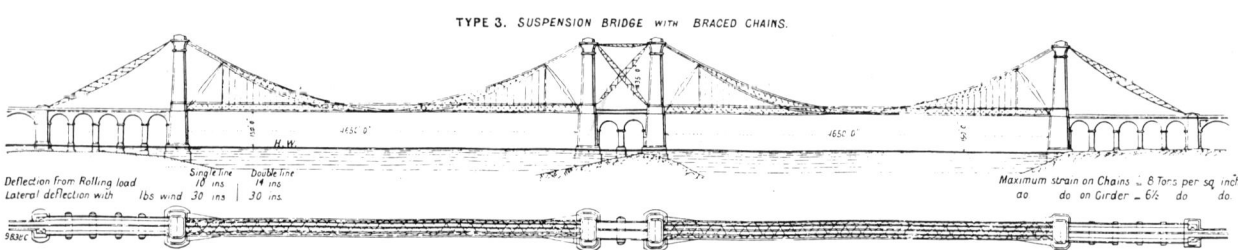

62
Alternative designs for a suspension bridge across the Forth. After construction on Bouch's suspension bridge had been stopped, the engineers Barlow and Harrison, jointly with Fowler and Baker, studied possible design alternatives – resulting in the final design by Fowler and Baker [14, p. 5].

62
Alternative Hängebrücken-Entwürfe zur Forth-Brücke: Nach dem Abbruch der Bauarbeiten an der Bouch'schen Hängebrücke untersuchten die Ingenieure Barlow und Harrison, zusammen mit Fowler und Baker, mögliche Brücken-Varianten – welche schließlich zum endgültigen Entwurf von Fowler und Baker führten [14, S. 5].

stiffening girders and braced chains; and 3) a cantilever bridge *(Figs. 62, 63)*.
The investigations were very thorough, including weight/cost calculations for each type. The information covered not just bridges but also tunnels, both for different locations. Tunnels, however, were quickly discarded because their approaches would have been very long and steep, due to the 200-ft-deep (60 m) shipping channels and the high ground of the shores; they would have required tunnels of many miles in length through unknown soil. Considering all the facts, the Queensferry location was held to be the most suitable for a bridge. And because of the soil condition and the great depth of water, it did not seem advisable to construct a bridge with shorter spans than those indicated by the natural profile of the riverbed; this confirmed in the end the basic assumptions of Bouch's design, as well as those of his predecessor James Anderson. The engineers concluded that a bridge (although not a suspension bridge) would be possible and recommended the preparation of detailed plans.

Neuanfang

Die vier Eisenbahngesellschaften waren nicht gewillt, die Sache lange auf sich beruhen zu lassen. Am 18. 2. 1881 beauftragten sie ihre konsultierenden Ingenieure William H. Barlow, Thomas Harrison und Sir John Fowler, die grundsätzliche Möglichkeit einer Eisenbahnbrücke über den Forth zu prüfen und im positiven Fall zu entscheiden, welchen Brücken-Typ man wählen sollte. Fowler hatte mit seinem Assistenten, Benjamin Baker, bereits zwei große Brücken über den Severn gebaut. Man wußte ziemlich gut, welche Brückenarten für einen solchen Standort in Frage kamen: Bouch's ursprünglicher Entwurf; Varianten von Hängebrücken mit aussteifenden Trägern und verstrebten Ketten; und eine Kragarm- oder Ausleger-Brücke *(Abb. 62/63)*.
Die Untersuchung war sehr umfassend, mit Gewicht/Kosten-Kalkulationen für jeden Typ. Sie schloß nicht nur Brücken ein, sondern auch Tunnel, und beides für verschiedene Standorte. Von Tunneln kam man schnell wieder ab, da wegen der 60 m tiefen Wasser-

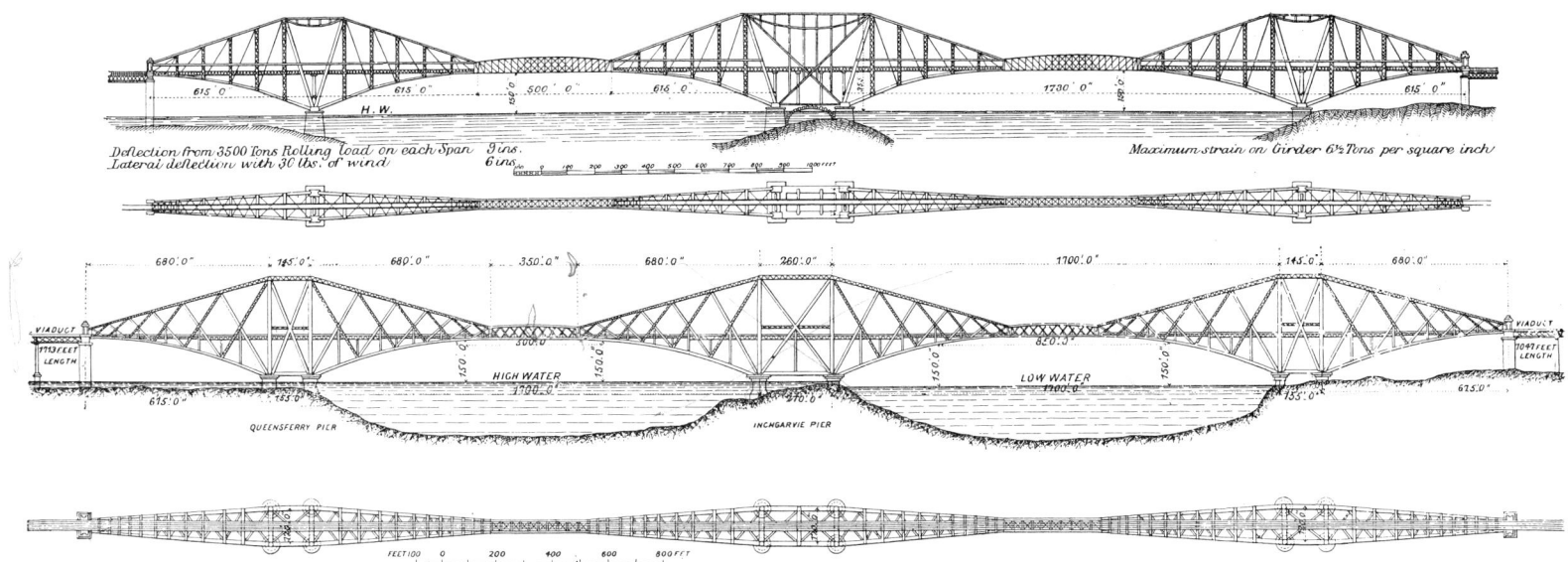

63
Cantilever bridge by Fowler and Baker, original design above, final design below, 1881 [29, p. 5].

63
Auslegerbrücke von Fowler und Baker, oben ursprünglicher Entwurf, unten endgültiger Entwurf, 1881 [29, S. 5].

On September 30, 1881, Fowler and Baker presented their plans to the directors. This was to the day three years after the foundation ceremony for Bouch's abandoned suspension bridge had taken place. Harrison and Barlow had been following the progress of their colleagues enthusiastically, and within two hours the proposals were approved by the railways, with the instructions to prepare the Parliamentary Bill. When the issue was brought before Parliament in the spring of 1882, there was hardly any opposition since the engineers were known to be the best in the country, and the government inspectors together with Fowler and Baker had examined all points of the plan. Even a 13-ft (3.96 m) scale model of the bridge was presented to the government; this stirred so much attention that it continued to be exhibited thereafter. Already by July 12, 1882, the Bill was approved – but not without several severe restrictions: the mistakes made at the Tay Bridge had caused much public anxiety and should most certainly not be repeated here. Thus the Board of Trade was to inspect every phase of construction thoroughly and to present a report four times a year, until the work was finished. The most important stipulation was that the bridge should become known among the public not merely as the largest and strongest, but also as "the stiffest bridge of the world". This implied the greatest stability possible, both vertically under rolling load, as well as horizontally under wind pressure, without the slightest vibration. It meant further

straßen und wegen des Hochufers, die Zufahrten sehr lang und steil geworden wären, mit einem viele Meilen langen Tunnel durch ungewissen Untergrund. Nach Maßgabe aller Dinge hielt man den Queensferry-Standort am geeignetsten für eine Brücke. Und wegen der großen Wassertiefe und Bodenbeschaffenheit riet man davon ab, eine Brücke mit kürzeren Spannweiten als denen zu bauen, die sich aus dem natürlichen Bodenprofil ergaben – womit eigentlich die Voraussetzungen des Bouch'schen Entwurfes, wie auch des früheren von James Anderson, voll bestätigt waren. Die Ingenieure schlossen, daß eine Brücke – allerdings keine Hängebrücke – möglich war und empfahlen, detaillierte Pläne auszuarbeiten.

Am 30. September 1881 legten Fowler und Baker ihre Pläne den Direktoren vor. Harrison und Barlow hatten die Planungsfortschritte ihrer Kollegen verfolgt und waren voller Enthusiasmus. – Innerhalb von zwei Stunden waren die Vorschläge genehmigt, mit dem Auftrag, die notwendige Parlamentsvorlage zu fertigen. Bei der Vorlage im Frühjahr 1882 gab es kaum Widerstand – denn die Ingenieure waren als die besten im Lande bekannt, und die Regierungs-Inspekteure waren vorher mit Fowler und Baker alle Punkte dieses Planwerkes durchgegangen; sogar ein 13 Fuß (3,96 m) langes Modell wurde dem Parlament vorgestellt und erregte solches Aufsehen, daß man es fortan dort ausstellte. Schon am 12. Juli 1882 war die Vorlage genehmigt – allerdings mit mehreren stren-

that the unfinished structure had to be just as stable during each phase of erection, as the finished one. And the steel used had to satisfy the highest requirements by the insurers, Lloyds, as were customary for shipbuilding. Finally, the entire structure was to combine utmost economy with complying to all of the above conditions. – None of all this, according to Westhofen, would have been fulfilled in the suspension bridge, but all stipulations could be met in the cantilever bridge! The great care exercised by the engineers, in conjunction with these stringent requirements, resulted in a bridge that turned out to be at least twice as strong in the end as would have been necessary.

The Principle of the Cantilever Bridge

The choice of a cantilever bridge proved in the end to be the cheapest and best of all solutions. The original design by Fowler and Baker *(Fig. 63)* followed the cantilever-and-suspended-girder principle, but it was altered in a few points to accomodate the differing opinions of the two other engineers. If we compare the original with the final solution, the more harmonious appearance of the first one is noticeable, as is its more daring conception with a longer suspended girder in the centre – meaning more suspended mass – while the cantilevers of the outer piers balance "on one foot" only. This configuration would have offered a few advantages, as we shall see later. The second version, on the other hand, is more down-to-earth, more safety-conscious. In stretching its cantilevers even further, there remains only a relatively short "jeopardized" girder to be suspended in the middle – measuring only two-thirds the length of the central girder in the original design.

Though this principle of bridgebuilding is quite old, so many new problems arose that only after many consultations and renewed calculations could the final form be arrived at. While the term "continuous girder" or "Gerber-Träger" in Germany, was already common for some time[14], many thought of the "cantilever and central girder" as, in fact, a modern invention. Particularly Benjamin Baker was frequently questioned about the "novel" use of cantilevers, whereupon he used to reply that it was nothing but a "bracket" or console – like any balcony or bookshelf. Baker was not

14 Already 23 years before construction of the Forth Bridge, Heinrich Gerber (1832–1912) had fully developed the system of the "console girder" – also called "cantilever" or "girder with free-floating supports". His bridge across the Main River near Hassfurt spanning 37.80 meters was the first bridge of this type.

gen Auflagen versehen: Die Fehler der Tay-Brücke hatten schwere öffentliche Befürchtungen ausgelöst und sollten unter allen Umständen vermieden werden: Danach sollte die IHK jede Bauphase gründlich inspizieren und viermal im Jahr einen Bericht vorlegen, bis das Werk vollendet war. Die allerwichtigste Erwägung dabei war, daß die Brücke in der Öffentlichkeit nicht nur als die größte und stärkste, sondern auch «die steifste Brücke der Welt» bekannt sein sollte. Das bedeutete größtmögliche Stabilität, sowohl vertikal unter rollender Last, als auch horizontal unter Winddruck, ohne die leiseste Vibration; woraus folgte, daß auch die unfertige Brücke in jeder Bauphase so stabil wie die fertige sein mußte; und der verwendete Stahl sollte den hohen Anforderungen von Lloyds entsprechen, die für den Schiffsbau galten. Und schließlich sollte das Ganze von größtmöglicher Wirtschaftlichkeit bei Einhaltung sämtlicher Auflagen sein. – Nichts von alledem, so der Chronist Westhofen, sei in der Hängebrücke erfüllt gewesen, jedoch alles in der Auslegerbrücke! Die Sorgfalt der Ingenieure ergab zusammen mit diesen hohen Anforderungen eine Brücke, die am Ende mindestens zweimal so sicher ausfiel, als es notwendig gewesen wäre.

Prinzip der Ausleger-Brücke

Die Entscheidung für eine Ausleger- oder Kragträger-Brücke erwies sich schließlich als die billigste und beste Lösung vor jeder anderen. Der ursprüngliche Entwurf von Fowler und Baker *(Abb. 63)* folgte ja bereits dem Ausleger- und Mittelträger-Prinzip, wurde aber dann in einigen Punkten geändert, um den abweichenden Meinungen der beiden anderen Ingenieure entgegenzukommen. Vergleicht man ursprüngliche und endgültige Lösung, so fällt die größere Ausgewogenheit der ersten Version auf, aber auch ihre größere Kühnheit mit längerem Mittelträger – also mehr «aufgehängter Masse» – und den «nur auf einem Bein» balancierenden Endauslegern, was, wie sich später zeigte, gewisse Vorteile gehabt hätte. Wogegen der zweite Entwurf der solidere, auf Sicherheit bedachte ist: Er streckt seine erdgebundenen Auslegerarme dermaßen weit vor, daß dazwischen nur ein relativ kurzer, «gefährdeter» Mittelträger übrig bleibt – und nur noch ca. $2/3$ der Länge des Mittelträgers im ursprünglichen Entwurf erreicht! Obgleich dieser Brückentyp im Prinzip sehr alt ist, gab es doch so viele neue Probleme, daß man erst nach langen Beratungen und erneuten Berechnungen zur endgültigen Form kam. Während der Begriff «Durchlaufträger» bzw. «Gerber-Träger» in Deutschland

64
Wandipore Bridge in Tibet, around 1670. This sketch of 1783 is by Lieutenant Davis, member of the embassy at the Court of the Teschu Lama in Tibet. Davies' book published in London in 1800 was very popular at the time, and was translated and republished in Germany; thus British and German engineers got a chance to read the following, and likely the first, description of a "cantilever and central girder bridge" ever published: "The bridge of Wandipore is of a singular lightness and beauty in its appearance. The span measures 112 ft [34.14 m]; it consists of three parts, two sides, and a centre, nearly equal to each other, the sides having a considerable slope raise the elevation of the centre platform, which is horizontal, some feet above the floor of the galleries. A quadruple row of timbers, their ends being set in the masonry of the bank, and the pier supports the sides; the centre part is laid from side to side." Despite the differences of material and dimensions, this structure may well be seen as a prototype of the Forth Bridge [29, p. 6].

64
Wandipore-Brücke im Tibet, ca. 1670. Die Skizze aus dem Jahre 1783 stammt von Leutnant Davis, Botschaftsmitglied am Hofe des Teschu Lama in Tibet. Davis' Buch darüber, 1800 in London erschienen, war damals sehr populär, es wurde in Deutschland übersetzt und neu aufgelegt; englische und deutsche Ingenieure hatten damit Gelegenheit, die folgende und wohl erstmalige Beschreibung einer «Ausleger- und Mittelträger-Brücke» zu lesen:
«Die *Wandipore-Brücke* ist von einzigartiger Leichtigkeit und Schönheit der Erscheinung. Die Spannweite beträgt 112 Fuß (34,14 m). Sie besteht aus drei Teilen: – zwei Seiten- und einem Mittelteil, alle etwa gleich lang. Die Seiten mit ihrer starken Schräge heben das Niveau des ebenen Mittelteiles noch einige Fuß über den Boden der Galerien. Eine vierfache Reihe von Holzbalken, mit den Enden im Mauerwerk der Uferbank verankert, und das Brückentor darüber tragen die Seiten – das Mittelstück ist einfach von einer Seite zur anderen gelegt.» – Berücksichtigt man die Unterschiede in Material und Größe, kann man dieses Werk durchaus als einen Prototyp der Firth-of-Forth-Brücke sehen [29, S. 6].

schon länger bekannt war[14] und in England schlicht «*continuous girder*» hieß, hielten viele seine Umbenennung in «Ausleger und Mittelträger» für eine moderne Erfindung. Besonders Benjamin Baker wurde oft über den «neuartigen» Gebrauch von Kragarmen befragt und pflegte dann zu erwidern, es sei nichts anderes als ein «*bracket*» – nämlich Kragstein oder Konsole, wie bei einem ganz normalen Balkon oder Wandbrett. Baker war über den Begriff «Ausleger» bzw. «Kragträger» oder «Kragarm» *(cantilever)* keineswegs glücklich, wenn man ihn auf seine Brücke bezog:

«Als ich Student war, wurde eine Fachwerkbrücke, die über den Pfeilern einen Obergurt unter Zugspannung und einen Untergurt unter Druck hatte, eine Durchlaufträger-Brücke genannt. Die Forth-Brücke ist ein solcher Typ und ich pflegte sie eine Durchlaufträger-Brücke zu nennen. Aber die Amerikaner bestanden darauf, alle Brücken, die sie nach selbigem Plan bauten, als ‹Ausleger-Brücken› zu bezeichnen. [15, S. 35]

In der Tat war nichts Neues an dem Entwurf, gab es doch schon seit erdenklichen Zeiten Brücken dieser Art in China, Tibet und Indien *(Abb. 64)*. Auch in frühen ägyptischen Tempeln ist eine Kragstein-und-Balken-Kombination gleicher Bauart zu finden *(Abb. 65)*. Ein besonders sinnfälliges Beispiel dieses Prinzips stellte die alte Straßenbrücke in Srinagar, Kaschmir, dar *(Abb. 67)*. In jüngerer Zeit wird der Kräfteverlauf von Durchlaufträgern häufig mit dem eines Mittelträgers verglichen, der von zwei Kragarmen an den Punkten umgekehrter Durchbiegung abgehängt ist. Im Bericht über die *Britannia-Brücke* schlägt Edwin Clark sogar vor, den Träger am Punkt der Kräfteumkehr zu durchtrennen und dann den Mittelteil an den «Halb-Trägern oder Kragarmen» aufzuhängen. Auch John Fowler sprach nicht nur vom Durchtrennen des Trägers an diesem Punkt, sondern probierte die Sache an einem großen Holzmodell auch aus.

1855 gab Barton in einer Abhandlung über den Boyne-Viadukt den entscheidenden Hinweis, daß die Punkte der Kräfteumkehr so gewählt werden können, daß sie mit jedwedem vorher festgelegten Punkt der Trägertrennung zusammenfallen: «Bei großen Spannweiten, wo es entscheidend sein kann, das Gewicht in der

14 Heinrich Gerber (1832 – 1912) hatte bereits 23 Jahre vor dem Bau der Forth-Brücke das System des «Konsolträgers» – auch «Ausleger-Träger» oder «Träger mit freischwebenden Stützpunkten» genannt – zur Baureife entwickelt. Seine Mainbrücke bei Haßfurt mit einer Spannweite von 37,80 m war die früheste moderne Form einer solchen Brücke.

very happy about the term "cantilever" when applied to his bridge:

> "When I was a student a girder bridge which had a top member in tension and the bottom member in compression over the piers was called a continuous girder bridge. The Forth Bridge is of that type, and I used to call it a continuous girder bridge; but the Americans persisted in calling all the bridges they were building on the same plan 'cantilever bridges'." [15, p. 35]

Actually, there was nothing new about his design principle, as bridges of this type had existed since time immemorial in China, Tibet and India *(Fig. 64)*. Even in early Egyptian temples we can find a similar cantilever-and-beam combination *(Fig. 65)*. A particularly telling example of this principle may be seen in the old road bridge at Srinagar in Kashmir *(Fig. 67)*. More recently, the line of stress in continuous girders is frequently compared with that of a central girder being suspended from two cantilevers at the points of contrary flexure. In a report on the Britannia Bridge, Edwin Clark even suggests cutting off the girder at the point of contrary flexure and suspending the central portion from the "semi-beams or cantilevers". Also John Fowler spoke not only of severing the girder at this point, but he even had the thing tested on a large wooden model.

In 1855, Mr. Barton in a treatise on the Boyne Viaduct gave the decisive hint in suggesting that the points of contrary flexure might in fact be chosen to coincide

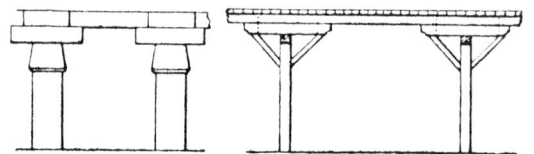

Mitte des Trägers soweit als möglich zu verringern, kann man die Materialmenge in den Ober- und Untergurten, aber auch an den Seiten auf ein Minimum reduzieren, indem man die Punkte der Kräfteumkehr zur Mitte des Trägers hin verschiebt, so daß die große Masse des Gewichtes auf den Stützen zu ruhen kommt.» Das entspricht genau der Trägerausbildung bei der Forth-Brücke. 1859 erhält W. H. Barlow ein Patent für diesen Konstruktionstyp, wobei er als Bauhöhe über den Auflagern das 1,5-fache der Höhe der Feldmitte wählt. Bei unserer Brücke wird dieses Verhältnis allerdings 7 : 1 sein.

1862 verweist Prof. Ritter aus Hannover in seinem Buch «Dach- und Brücken-Constructionen» darauf, daß große Materialersparnis erzielt wird, wenn man einen Durchlaufträger verwendet und die Kontinuität vermittels Angeln unterbricht – und rechnet zum Beweis eine ganze Brücke mit 160 m Spannweite durch *(Abb. 68a)*.

In der zweiten Hälfte des 19. Jahrhunderts kam es zu einer Folge solcher Brückenbauten: 1864 entwarfen Fowler und Baker eine Stahlbrücke von 1000 Fuß

65
Egyptian stone corbel and lintel combination [29, p. 6].

65
Ägyptische Kragstein- und Balken-Kombination [29, S. 6].

67
This old road bridge with shops built along it used to cross the Jehlan River at Srinagar, Kashmir. To be noted are the piers, layered by criss-crossing logs [13, p. 204].

67
Über den Jehlan in Srinagar (Kaschmir) führte diese alte Straßenbrücke mit ihren Verkaufsläden. Beachtenswert die Brückenpfeiler aus kreuzweise aufgeschichteten Baumstämmen [13, S. 204].

66
Five different examples from the treasury of Chinese wooden cantilever bridges [13, p. 39].

66
Fünf verschiedene Beispiele aus dem Schatz chinesischer Holz-Kragträger-Brücken [13, S. 39].

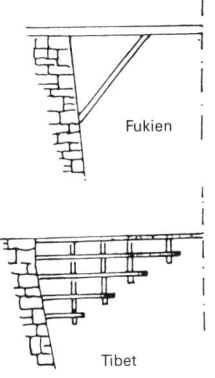
Fukien

Tibet

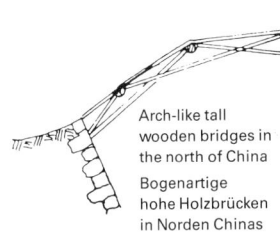
Across ravines in the Tibetan Mountains
Über Schluchten in den Bergen von Tibet

Arch-like tall wooden bridges in the north of China
Bogenartige hohe Holzbrücken in Norden Chinas

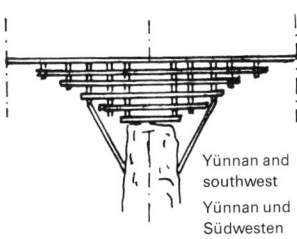
Yünnan and southwest
Yünnan und Südwesten

with any previously determined point of severing the beam: "In very large spans where it may be a matter of great importance to reduce the weight in the middle of the beam as much as possible, the quantity in material in the top and bottom tables as well as of the sides, may be reduced to a minimum by throwing the points of contrary flexure towards the middle of the beam, the great weight of material being placed over the piers." This corresponds exactly to the configuration of girders in the Forth Bridge. In 1859, W. H. Barlow took out a patent for this type of construction, choosing $1^1/_2$ times the height at mid-span for the height of the girder over the piers. In our bridge, however, this ratio would be 1 : 7.

In 1862 Prof. Ritter of Hannover in his book *Dach- und Brücken-Constructionen* demonstrated how a large saving of material could be obtained by using a continuous girder and breaking its continuity by means of hinges; as proof for his conclusions, he calculated in detail the stresses of an entire bridge spanning 160 meters (525 ft) *(Fig. 68a)*.

With the second half of the 19th century, there would be a quick succession of new bridges: In 1864, Fowler and Baker designed on the above system a steel bridge spanning 1,000 ft (304.80 m) for the South Wales and Great Western Direct Railway Bridge across the Severn. And in 1867, Baker emphasized the economic advantages of the continuous girder with varying height in a series of articles on "Long-Span-Bridges". This publication had three editions in Britain and was also translated into German and Dutch. In 1876 nine competition entries were submitted for the proposed New York and Long Island Bridge with spans of 734 ft (224 m) and 618 ft (188 m), two of which used this system. One solution, that by the Delaware Bridge Company, is shown here *(Fig. 68b)*.

In the same year of 1876, the first and till then the only railway bridge of this type, the Warthe Bridge near Posen, was built spanning just 148 ft (45 m), but quite remarkable in its appropriate eloquence and elegance of profile *(Fig. 68c)*. Once again we see that early impulses coming from Britain were being refined and perfected in Central Europe. It may seem strange that the profession had to wait until 1876 before applying such a well-known structural system to railway bridges; there had been thousands of continuous girder bridges, but the engineers had shied away till then from actually severing the bridge at the point of contrary flexure, thereby avoiding girders of varying height.

(304,80 m) Spannweite nach besagtem System für die *South-Wales-and-Great-Western-Direct-Railway*-Brücke über den Severn; und 1867 betonte Baker die wirtschaftlichen Vorteile des Durchlaufträgers mit variierender Höhe in seiner Artikelserie über «Weitgespannte Brücken». Diese Arbeit hatte in England drei Auflagen; sie wurde auch ins Deutsche und Holländische übersetzt. 1876 wurden neun Wettbewerbsentwürfe für die geplante New-York- und Long-Island-Brücke mit Spannweiten von 224 m und 188 m eingereicht, wovon zwei nach diesem System ausgelegt waren: Eine Lösung, die der Delaware Bridge Co., ist hier gezeigt *(Abb. 68b)*.

Im gleichen Jahr 1876 wurde die erste und bislang einzige Eisenbahnbrücke dieses Typs in Deutschland, die Warthe-Brücke bei Posen gebaut, mit zwar nur 45 m Spannweite, aber sinnfälliger Eleganz der Linienführung *(Abb. 68c)*. Es zeigt sich wieder, wie frühe Impulse aus England in Mitteleuropa verfeinert und perfektioniert worden sind. Merkwürdig mag erscheinen, daß man mit einem so wohlbekannten Bausystem bis 1876 wartete; denn es gab Tausende von Brücken mit Durchlaufträgern, aber die Ingenieure hatten sich bis dahin gescheut, die Brücke am Punkt der Kräfteumkehr wirklich zu durchtrennen und Träger von unterschiedlicher Bauhöhe zu verwenden.

Der Entwurf von Fowler und Baker

Auf der Grundlage eines bekannten Brückenbau-Prinzips zeigt sich die Neuartigkeit des Entwurfes von Fowler und Baker in zwei Punkten. Einmal in der beispiellosen Größe des Bauwerkes, und zum andern in der vollkommenen Adaption dieses Prinzips mit den allgemeinen Gesetzen der Festigkeit und den besonderen Bedingungen des Standortes. Die Perfektion ebendieses Durchlaufträger-Prinzips erweist sich nicht nur in der Ansicht, sondern gerade im Plan mit der stark variierenden Breite der Brücke. So ist der Mittelträger gerade breit genug, um zwei Schienenstränge aufzunehmen, während sich die Ausleger bis auf 120 Fuß (36,58 m) an den Pfeilern verbreitern. Um die außerordentliche Höhe des Bauwerkes zu verringern und den Schwerpunkt möglichst weit zu senken, sind die Untergurte der Ausleger gebogen und erheben sich von massiven Mauerwerkspfeilern nur 150 Fuß (45,72 m) über Hochwasser auf die Mittelträger zu, wogegen bei der früheren Hängebrücke die Hauptketten in 550 Fuß (168 m) Höhe aufgehängt waren.

Die Untergurte als Hauptdruckglieder sind als Stahlrohre mit max. 12 Fuß (3,66 m) Durchmesser ausgebildet: Experimente hatten gezeigt, daß «inch for inch» die Rohrform stärker als jede andere ist und so

The Design by Fowler and Baker

Based on a well-known principle of bridge construction, the novelty of the design by Fowler and Baker was demonstrated in two respects: for one, in the extraordinary size of this structure, and for another, in the perfect adaptation of this principle to the general rules of rigidity and the special conditions of the site. The very perfection of the continuous-girder principle is evident not just in elevation but even in the plan, as the width of the bridge varies considerably. The central girder was just wide enough to take a double track of rails, while the cantilevers widened up to 120 ft (36.58 m) at the piers. In order to reduce the enormous height of the structure and to lower the center of gravity as much as possible, the bottom members of the cantilevers were arched, springing from their massive masonry piers to just 150 ft (45.72 m) above water level at the central girders, whereas in the earlier suspension bridge the main chains would have been at a height of 550 ft (168 m).

The bottom members acting as the main compression members were contructed as steel tubes with a maximum diameter of 12 ft (3.66 m). Experiments had shown that "inch for inch" this tubular form was stronger than any other, also reducing the number of bracings and stiffening members to a minimum. The central girder was just an ordinary double-tracked railway bridge spanning 350 ft (106.68 m). During a lecture, Baker demonstrated his famous living model of the bridge – to become known the world over in a newspaper photograph – in order to illustrate the distribution of stresses in the two cantilever arms *(Fig. 69)*.

The distribution of weight chosen had the advantage of concentrating about ¾ of the total load directly upon the piers; thus the smallest possible surface was offered to the wind pressure at those points, where its levering action would be most pronounced – at the extreme ends of the cantilevers. Whereas the wind pressure acting upon the wind-swept Inchgarvie tower was 23 tons per running foot length of the bridge, it was a mere 2 tons at the central girders.

Furthermore, every portion of the structure being fitted served as a subsequent staging for the next portion to be connected, and so on; in rivetting together piece after piece, it was possible to always work from secured positions. Thus, much less temporary staging was re-quired than in any other mode of construction, adding greatly to the confidence of the workmen. Great stability was gained by tapering the sides of the structure as it went up, since the towers had a width of about ⅓ of their height at the base, while they nearly joined up to a point at the top. This made for a strong impres-

auch die Zahl von Aussteifungen und Nebenverstrebungen auf ein Minimum reduziert. Der Mittelträger ist dann nur noch eine ganz gewöhnliche zweigleisige Eisenbahnbrücke von 350 Fuß (106,68 m) Länge. Anläßlich eines Vortrages stellte Baker sein berühmtes lebendes Modell der Brücke vor, – es wurde als Zeitungsfoto in der ganzen Welt bekannt – um die Verteilung der Lasten in den Ausleger-Armen deutlich zu machen *(Abb. 69)*.

Die gewählte Gewichtsverteilung hat den Vorteil, daß etwa ¾ des Gesamtgewichtes unmittelbar über den Pfeilern liegt; und daß dort, wo der Winddruck mit der größten Hebelwirkung ansetzt, nämlich an den Ausleger-Enden, die geringste Angriffsfläche ist. Während so auf dem windgepeitschten Inchgarvie-Turm die Windlast pro laufenden Fuß Brückenlänge 23 Tonnen beträgt, ist sie auf den Mittelträgern nur noch ca. 2 Tonnen. Zudem bietet sich jedes Bauteil nach seiner Montage als Arbeitsbühne für das nächste Teilstück an, d. h. man kann Stück um Stück vernieten und sich so immer von gesicherten Punkten aus vorarbeiten. Dadurch waren viel weniger temporäre Arbeitsbühnen notwendig als in jeder anderen Bauweise, was den Bauleuten große Zuversicht gab. Besondere Standfestigkeit wird durch den seitlichen Anzug der Konstruktion erreicht, denn die Pfeilertürme sind an der Basis ca. ⅓ der Höhe breit und laufen oben fast zur

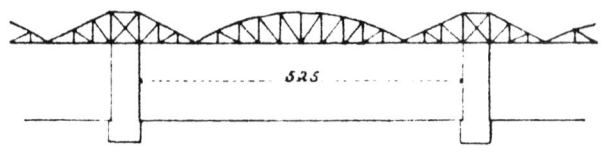

68 a
Prof. Ritter's proposal for a continuous-girder bridge, 1862 [29, p. 6].

68 a
Prof. Ritters Vorschlag einer Durchlaufträger-Brücke, 1862 [29, S. 6].

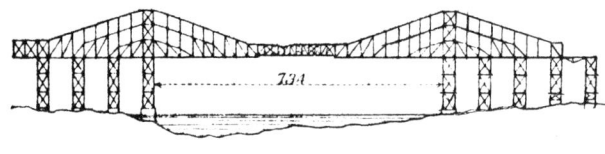

68 b
Competition entry of 1876 for the Long-Island Bridge, following the cantilever principle [29, p. 6].

68 b
Wettbewerbsentwurf 1876 der Long-Island-Brücke nach dem Auslegerprinzip [29, S. 6].

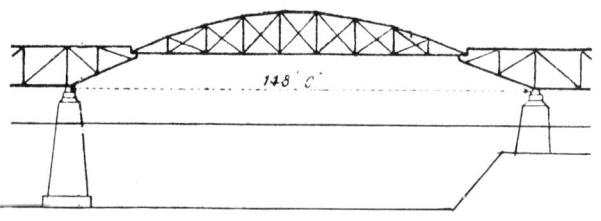

68 c
Bridge across the Warthe near Posen, the first rail bridge built of this type, 1876 [29, p. 6].

68 c
Warthe-Brücke bei Posen, erste gebaute Eisenbahnbrücke dieses Typs, 1876 [29, S. 6].

69
Benjamin Baker's living model of the Forth Bridge. He showed this much publicised illustration for the first time in 1887 during a lecture at the Royal Institution in London, explaining his cantilever principle: "... When a load is put on the central girder by a person sitting on it, the men's arms and the anchorage ropes come into tension, and the men's bodies from the shoulders downwards and the sticks come into compression. The chairs are representative of the circular granite piers. Imagine the chairs one-third of a mile apart and the men's heads as high as the cross of St Paul's, their arms represented by huge lattice steel girders and the sticks by tubes 12 feet in diameter at the base, and a very good notion of the structure is obtained."
The sticks were sawn-off broomsticks, and this model of a Human Cantilever is a telling example of Baker's talent to illustrate candidly matters of technology. The man sitting in the centre – freely suspended between the two human cantilevers – is Kaichi Watanabe, a member of the first generation of Japanese engineers to have come to the West in order to study – and ultimately to surpass – technical achievements. Watanabe's participating in this model was to demonstrate the indebtness of the designers to the Far East, where the cantilever principle had been invented [29, p. 8 revised].

Benjamin Bakers lebendes Modell der Forth-Brücke. Er zeigte dieses vielpublizierte Schaubild zum erstenmal 1887 bei einem Vortrag der Royal Institution in London, zur Erklärung seines Kragträger-Prinzips: «... Wenn eine Last auf den Mittelträger gesetzt wird, kommen die Arme der Männer und die Ankerschnüre unter Zugspannung, und die Körper von den Schultern abwärts sowie die Stäbe unter Druckbelastung. Die Stühle stellen die kreisrunden Granit-Pfeiler dar. Man stelle sich die Stühle ⅓ Meile voneinander entfernt vor, und die Köpfe der Männer so hoch wie das Kreuz von St. Paul's Cathedral, ihre Arme ersetzt durch gewaltige Stahlfachwerke und die Stäbe ersetzt durch Rohre von 12 Fuß Durchmesser – und man erhält einen sehr guten Eindruck von der Konstruktion.»
Die Stäbe waren abgesägte Besenstiele, und das Modell ist ein guter Beweis für Bakers Talent, technische Zusammenhänge anschaulich darzustellen. Der Mann in der Mitte – freischwebend zwischen den menschlichen Auslegern – ist Kaichi Watanabe, ein Angehöriger der ersten Generation japanischer Ingenieure, die zum Studium technischer Errungenschaften in den Westen kamen – und diesen später übertreffen sollten. Watanabes Mitwirkung an dem Modell sollte der Welt die Dankesschuld der Erbauer an Fernost demonstrieren, wo das Kragträger-Prinzip erfunden worden ist [29, S. 8; überarbeitet].

sion of safety against lateral wind pressure. Finally, the method of using cantilevers and central girders allowed for the simplest and most efficient way to accomodate expansion joints; as we shall see later, "this problem is solved here in the happiest manner".

Bridge Esthetics

The task confronting Baker at the Forth was quite similar to the situation Robert Stephenson had encountered at the Menai - with the difference being that the dimensions at the Forth were four-fold. To the Britannia Bridge, the "Britannia Rock" in the middle of the stream had given not just the name, but also a safe support for the central pier. Similarly, the location finally chosen at the Forth was ideal for the purpose, in all its natural features: the level of the surrounding high ground was about as high as was necessary for providing sufficient headroom to the largest vessels of the navy and the merchant fleet. The riverbed consisted of basalt and hard-packed boulder clay at the piers, both very firm and compact, while the soil was no deeper at any point than permissible for pneumatic caissons. Moreover, this site afforded the only narrows along 50 miles (80.5 km) of estuary, reducing the width to just 1 mile and about 150 yards (1759 m), while it was over 2 miles (3.22 km) at all other places *(Fig. 70)*. The rocky promontory to the north further shortened the span and offered sheltered landing sites for a small fleet of barges. About $1/3$ of a mile (536 m) to the south lay the rocky islet of Inchgarvie, and in between the two runs the Main or North Channel for shipping, over 200 ft (61 m) deep. Shipping preferred this channel as it was safer, easier to navigate and also shorter than the South Channel. It is about as wide as the northern one; its southern portion, however, is only 30 ft (9.14 m) deep and extends for another 2,000 ft (609.60 m) to the Queensferry shore. There, about $1/4$ of this channel rises from the water at low tide. In consideration of this natural soil configuration, the position of the three large piers was well chosen, with two large spans of 1,710 ft (521.20 m) each.

The landscape surrounding the bridge is of serene beauty. Westhofen relates that, whatever opinion may be held about the lines of the bridge itself, this or any other bridge must necessarily be "a discordant feature" in a pastoral setting:

> "Standing on Mons Hill in Dalmeny Park, and looking down over its thickly wooded slopes into the broad expanse of the Forth, with the island of Inchgarvie and its old castle breasting the swift current and cutting it into two arms, which below it, unite again in a whirlpool glitter-

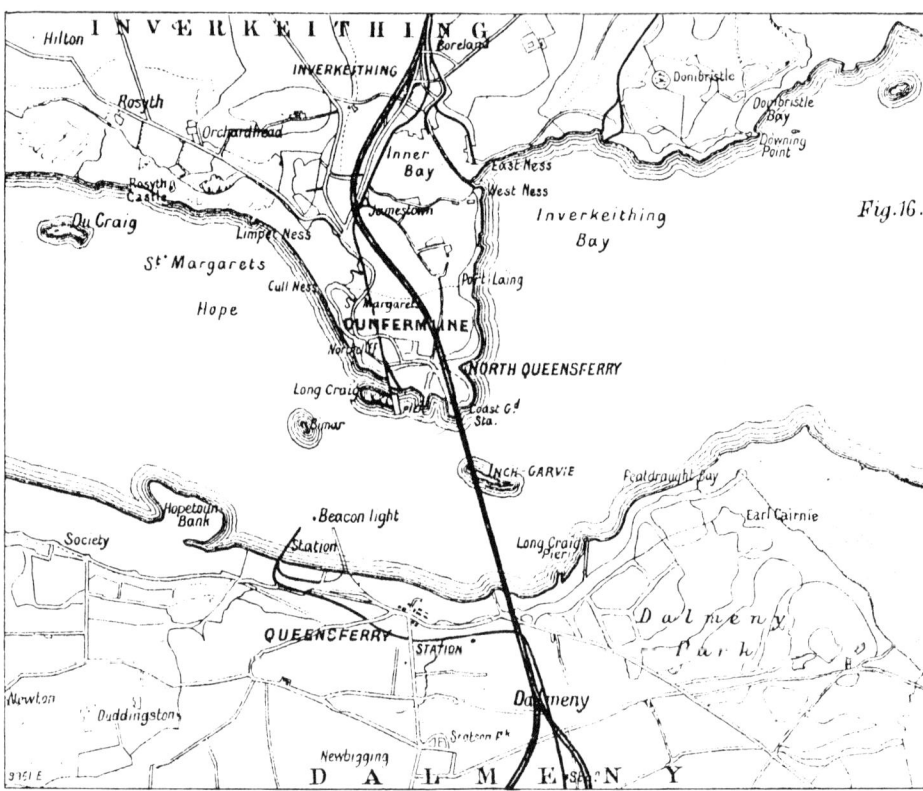

Spitze zusammen: Gerade das vermittelt ein deutliches Gefühl von Sicherheit gegen seitlichen Winddruck. Schließlich erlaubt die Bauweise mit Auslegern und Mittelträgern die einfachste und effektivste Form von Dehnungsfugen – *«this problem is solved here in the happiest manner»* – wie später gezeigt wird.

Brücken-Ästhetik

Die Aufgabe, der sich Baker am Forth gegenübersah ähnelte sehr der Situation Robert Stephensons am Menai – mit dem Unterschied, daß die Dimensionen am Forth das Vierfache betrugen: Die *«Britannia-Brücke»* hatte vom *«Britannia-Felsen»* in Flußmitte nicht nur ihren Namen, sondern auch ein sicheres Auflager für den Mittelpfeiler erhalten. Und so war der schließlich gewählte Standort am Forth in all seinen natürlichen Gegebenheiten ideal für den Bau geeignet: Das Niveau des Umlandes war etwa so hoch wie die nötige Durchfahrtshöhe für die größten Schiffe der Navy und der Handelsflotte; der Flußgrund bestand an den Pfeilerpunkten aus Basalt und hartem Lehm, sehr fest und kompakt; auch lag der Grund bei keinem Fundament tiefer als für pneumatische

70
Site plan of the Forth Bridge around 1890 [29, p. 8].

70
Lageplan der Forth-Brücke um 1890 [29, S. 8].

71
The magnificent landscape at the Firth of Forth – *before* the construction of the bridge: In the foreground, the old landing pier of South Queensferry; to the right, the Castle of Inchgarvie (shown here somewhat more heroically than it is, actually); in the back, the rocky cliffs of North Queensferry. To cast across a bridge at this spot must have seemed an inappropriate, even foolish intervention into nature. – Illustration by Clarkson Stanfield (1793–1867), a renowned landscape and marine painter of the time [15, p. 64 f.].

71
Die großartige Landschaft am Firth of Forth – *vor* dem Bau der Brücke: Vorne die alte Anlegestelle von South Queensferry, rechts die Burg von Inchgarvie (hier heroischer dargestellt als sie ist), dahinter die Felsenklippe von North Queensferry. An dieser Stelle eine Brücke zu schlagen, mußte als ein ungehöriger, ja vermessener Eingriff in die Natur erscheinen. – Nach einer Darstellung von Clarkson Stanfield (1793–1867), einem bekannten See- und Landschaftsmaler seiner Zeit [15, S. 64 f.].

ing in all the colours of the rainbow, the whole backed by the Fifeshire Hills, the Ochills, and the great peaks of Dumbarton, Stirling, and Pertshire, is a view hard to be excelled in any part of the world. Hardly less fine and perhaps more grand still is the view down the estuary into the limitless ocean, from the grounds round Hopetoun House." *(Fig. 71)* [29, p. 7]

Contemporaries view the scene differently once the bridge has been built. When looking downstream from Hopetoun House, the horizon is now dissected by the girders' strict geometry of triangular and square shapes – harmonising in no way with the soft lines of the landscape. Therefore the best view of the landscape would be from the bridge itself, as it would then no longer be seen ("because the disturbing element is left out"!); reversely, by far the best view of the bridge would be obtained from the river, at a distance of a mile or so, with the structure rising to a great height against the sky: "Thus viewed, its simple lines, its well-proportioned parts, its impressive air of strength and solidity and yet of lightness and grace, never fail to strike the mind of the beholder. Four-square to the wind and immovable it stands!" *(Fig. 72)* [29, p. 7]

The general appreciation of major engineering structures already began to change at that time. Whereas a generation before there was nothing but praise for the Britannia Bridge, or for Paxton's Crystal Palace in London, now the artists at least became increasingly critical of such feats of technology. Benjamin Baker, too, was caught up in the fire and had to defend his bridge; William Morris even condemned it as the "supremest specimen of all ugliness" [23, pp. 194, 195]. He declared flatly that there would never be an architecture of iron, every improvement in machinery being uglier and uglier, whereupon Baker replied:

"Probably Mr. Morris would judge the beauty of a design from the same standpoint, whether it was for a bridge a mile long or for a silver chimney ornament. It is impossible for anyone to pronounce authoritatively on the beauty of an object without knowing its function. The marble columns of the Parthenon are beautiful where they stand, but if we took one and bored a hole through its axis and used it as a funnel for an Atlantic liner it would, to my mind, cease to be beautiful, but of course Mr. Morris might think differently."

This was the beginning of a battle between art and technology, still going on today. Baker did have a deep feeling for the past, just as much as Morris did; Baker himself had restored some of the works of his great predecessor, Telford. Baker was justified as an en-

Caissons zulässig. Schließlich liegt hier die einzige Engstelle entlang der 50 Meilen (80,5 km) des Meeresarmes – und zwar nur 1 Meile und 150 Yards (1759 m) breit, anstatt über 2 Meilen (3,22 km) an allen anderen Stellen *(Abb. 70)*. Der Felsvorsprung im Norden verkürzt die Spannweite und bietet geschützte Landungsstellen für die kleine Flotte von Lastkähnen. Etwa $1/3$ Meile (536 m) südwärts liegt die Felseninsel Inchgarvie, dazwischen der Haupt- oder Nord-Kanal mit 200 Fuß (61 m) Tiefe. Die Schiffahrt benutzt fast nur diese Fahrrinne, da sie sicherer und leichter zu navigieren und außerdem kürzer als die Südrinne ist. Diese ist etwa gleich breit und tief wie der Nord-Kanal, ihre Südflanke liegt allerdings 30 Fuß (9,14 m) unter Wasser und erstreckt sich noch 2000 Fuß (609,60 m) bis zum Queensferry-Ufer, wovon bei Ebbe $1/4$ aus dem Wasser taucht. Aus diesem Bodenprofil ergibt sich die Stellung der drei großen Stützen mit den zwei Spannweiten von 1710 Fuß (521,20 m).

Die Landschaft in unmittelbarer Umgebung der Brücke ist von herber Schönheit. Westhofen meint, welcher Meinung auch immer man über die Linienführung der Brücke selbst sein mag, daß diese oder jede andere Brücke notgedrungen «*a discordant feature*» in pastoraler Landschaft sein muß: «Steht man auf Mons Hill im Dalmeny Park und blickt hinunter über dicht bewaldete Hänge auf die breite Fläche des Forth mit Inchgarvie und seiner alten Burg, wie sie den schnellen Strom in zwei Arme teilt und dahinter in einem Strudel in allen Farben des Regenbogens glitzernd wieder vereint, das Ganze abgeschlossen vom Fifeshire Hügelland, den Ochills und den großen Gipfeln von Dumbarton, Stirling und Perthshire – dann ist das ein Anblick ohnegleichen in der Welt. Kaum weniger großartig ist der Blick die Mündung hinab zum offenen Meer, vom Gelände um Hopetoun House herum» *(Abb. 71)* [29, S. 7].

Das ändert sich aber im Urteil der Zeitgenossen, seitdem es die Brücke gibt: Beim Blick stromabwärts von Hopetoun House überschneidet sich nun der Horizont mit der strengen geometrischen Regelmäßigkeit von Drei- und Vierecken der Brückenträger – die in keiner Weise mit den weichgeschwungenen Linien der Landschaft harmonisieren können: Daher sei der beste Blick auf die Landschaft von der Brücke aus, da man sie selbst dann ja nicht sehen würde (!) – «*because the disturbing element is left out*» – und umgekehrt sei die weitaus beste Ansicht der Brücke vom Fluß aus, etwa 1 Meile (1,6 km) entfernt, so daß das Bauwerk sich zu großer Höhe gegen den Himmel reckt: So gesehen verfehlten seine einfachen Linien, seine gutproportionierten Teile, seine eindrucksvolle Aura von Kraft und Festigkeit und dennoch Leichtigkeit und Grazie nie

72
Looking at the profile of the southern cantilever tower from the west. To the left, the southern central girder, to the right, the approach viaduct at South Queensferry.

72
Profil-Ansicht des südlichen Auslegerturmes von Westen. Links der südliche Mittelträger, rechts der Anfahrtsviadukt bei South Queensferry.

ihre Wirkung auf den Betrachter. «Vierbeinig fest und unverrückbar gegen den Wind steht es da!» *(Abb. 72)* [29, S. 7].

Die allgemeine Einstellung gegenüber technischen Großbauwerken begann sich damals bereits zu ändern. Hatte man eine Generation vorher z. B. nichts als Lob für die *Britannia-Brücke,* oder für Paxtons Kristallpalast in London, so wurden nun zumindest die Künstler immer kritischer gegenüber Werken der Technik. Auch Benjamin Baker mußte seine Brücke in Schutz nehmen, und William Morris verurteilte sie gar als *«the supremest specimen of all ugliness»* (das äußerste Exemplar aller Häßlichkeit, oder kurz: den Gipfel an Scheußlichkeit) [23, S. 194, 195]. Er erklärte rundweg, daß es niemals eine Architektur aus Eisen geben

gineer in insisting on functional honesty, but Morris, the artist, was justified as well in pointing out the growing destruction of the magnificent British landscape; it had become quite clear to him how fast man was to become a stranger to his very own environment, by his increasing mastery of technology.

Today we may rightly experience the Forth Bridge as part of a grand natural spectacle itself. The real impact of this structure within a magnificent setting can only be comprehended by someone who has been there in person – who has sensed the fulfilled silence and expectancy of this place, who has witnessed how packs of clouds passing overhead throw entire sections of the bridge into darkness, while the far end stands out again brightly in the sunlight; this is underlined, though hardly noticably, by the steady high-pitched singing of the wind high up in the meshwork of bracings, periodically interrupted by a strange bang ringing up, which may be interpreted as friction thrusts of the steel colossus under tension *(Fig. 73)*.

Tides, Wind and Weather

The tide at Queensferry may reach 18, sometimes 22 ft (5.50 to 6.70 m). Due to the narrows of the Firth, the current is quite rapid at this spot, especially in the North Channel. In erecting the overhanging Inchgarvie north cantilever, it was therefore quite difficult to lift all material directly from the steam barges below, right in the middle of the current. The combined forces of tide, wind and current made it almost impossible to keep the barges in place long enough to fasten the hook of the crane – even with the most experienced and best skipper at the helm.

The prevailing wind direction was southwest, the next frequent was an unpleasant easterly wind, whereas the westerly winds usually were warmer due to the Gulf Stream, but brought frosty weather in the winter. Frost was indeed a problem; because much hydraulic equipment was being used, many of the pipes had to be disconnected at night to drain dry. This almost caused a disaster when in 1889 during a furious February gale, a fire broke out on Inchgarvie which was on the verge of threatening the steelwork before the pumps could be activated. During three or four days of the year, the storms blew so fiercely that not even the big paddlewheeler could set out and all work had to rest since even the cranes could not be operated. On the average, only 22 to 23 full working days remained per month. Frequently, no more than an hour or two need have been lost mornings due to heavy rain – which drenched the men sending them back to their quarters, but even when the weather cleared up for the

werde, da jeder Fortschritt in diesen «Apparaturen» immer nur häßlicher und häßlicher werde. Worauf Baker erwiderte:

«Wahrscheinlich würde Mr. Morris die Schönheit eines Entwurfes vom selben Standpunkt aus beurteilen, gleich, ob es dabei um eine Brücke von einer Meile Länge oder um ein silbernes Schornstein-Ornament geht. Es ist unmöglich für irgend jemand, sich mit Autorität über die Schönheit eines Objektes zu äußern, ohne seine Funktion zu kennen. Die Marmorsäulen des Parthenon sind schön, wo sie stehen, aber wenn wir eine nähmen und ein Loch durch ihre Achse bohrten, und sie als Schornstein für einen Atlantikdampfer nützten, würde sie meiner Ansicht nach aufhören, schön zu sein; aber natürlich mag Mr. Morris anders darüber denken.»

Das war der Anfang eines Kunst/Technik-Streites, wie wir ihn bis heute kennen. Baker hatte ein ebenso tiefes Gefühl für die Vergangenheit wie Morris – hatte er doch selbst einige der Bauwerke seines großen Vorgängers Telford restauriert. Baker hatte recht, als Ingenieur auf funktionaler Ehrlichkeit zu bestehen – und Morris als Künstler wies mit Recht auf die zunehmende Verunstaltung der wunderbaren englischen Landschaft hin; ihm war klargeworden, wie rasch der Mensch durch die wachsende Beherrschung der Technik sich seiner eigenen Naturwelt entfremdete.

Heute erlebt man die Forth-Brücke selbst schon als eine Art Naturereignis: Von der wirklichen Erscheinung dieses Bauwerkes in grandioser Natur kann sich nur derjenige ein Bild machen, der selbst dort war – der jene erfüllte Lautlosigkeit und Wartestimmung des Ortes gespürt hat, auch gesehen hat, wie vorüberziehende Wolkenfelder ganze Partien der Brücke ins Dunkel tauchen und gleichzeitig am anderen Ende die Konstruktion wieder hell im Lichte steht; dazu jenes beständige, kaum hörbare Singen des Windes hoch oben im Gewirr der Verstrebungen, periodisch unterbrochen von einem seltsam metallisch aufklingenden Schlag, hinter dem man Spannungsstöße des Stahlgebirges verstehen mag *(Abb. 73)*.

Gezeiten, Wind und Wetter

Der Tidenhub bei Queensferry beträgt 18, manchmal auch 22 Fuß (5,50 – 6,70 m). Wegen der Engstelle des Firth ist die Strömung hier sehr schnell, besonders in der Nordrinne. Beim Bau des überhängenden Inchgarvie-Nordauslegers war das besonders schwierig, wenn mitten in voller Strömung alles Material direkt von den Dampfkähnen in die Höhe gehievt werden

73
Forth Bridge as seen from the landing pier at South Queensferry.

73
Forth-Brücke vom Pier bei South Queensferry.

mußte: Die vereinten Kräfte von Gezeitenschub, Wind- und Flußströmung machten es so gut wie unmöglich, die Kähne gerade lange genug an Ort und Stelle zu halten, um wenigstens den Kranhaken anzubringen – selbst mit dem erfahrensten und geschicktesten Steuermann am Ruder.

Die Hauptwindrichtung ist Südwest, zweithäufig ein unangenehmer Ostwind, während die Westwinde wegen des Golfstroms wärmer sind, im Winter allerdings Frostwetter bringen. Frost war ein Problem, und da viel hydraulisches Gerät verwendet wurde, mußte man die Leitungen abends abkuppeln und leerlaufen lassen. Das rächte sich 1889, als in einem schweren Februar-Sturm ein Feuer auf Inchgarvie ausbrach und bereits die Stahlkonstruktion bedrohte, bevor die

rest of the day, "no power of persuasion was great enough to bring them back to work again".

Furthermore, sea fogs in winter and spring caused much anxiety during work shifts, as every morning and night many hundreds of people had to be ferried from the shores to their job sites. The foggy banks came creeping up the Firth like a solid wall of dazzling white cotton wool, leaving just the peaks of the towers standing out clearly in the bright sunlight, while down below darkness set in.

Wind Pressure Experiments

The collapse of the Tay Bridge had taught the profession to take wind pressure seriously. The Board of Trade had already raised the permissible wind pressure in exposed positions to 56 lbs/sq ft (273 kg/m^2) – and this twice across the whole surface of the girder, whereby the resistance to this pressure had to be provided by the deadweight of the structure itself. Baker took charge of the matter with much attention and foresight before construction commenced. A wind gauge installed atop the old Inchgarvie Castle was still of simple construction, recording only maximum pressures ranging from 27 to 41 lbs/sq ft (130 to 200 kg/m^2), but at least it proved that maximum wind pressures occurred most often during sudden and localised gusts, rather than in a steady and even pressure extending over a large surface. Once the central towers had reached their full height, Baker installed improved wind gauges at the top and he found that the wind was acting differently upon the various columns, with a variance of up to 12 lbs/sq ft (58 kg/m^2).

Since so little was really known about the actual effect of wind pressure on level or curved surfaces, Baker embarked on a sequence of remarkable experiments *(Fig. 74)*. But even more important for him was to obtain exact data for an oblique wind action upon braced box girders, as the wind pressure might be multiplied according to the angle of impact. In the case of two box girders, one positioned close behind the other, such as in the top member, the surfaces lying further back would be covered by the ones in front, given a rectilinear impact of the wind; while under an oblique impact, the wind pressure would be quadrupled! Since there already existed some information about diagonal wind pressure on cubic shapes, Baker first confirmed these findings by his model. But in the case of covered surfaces, he proved that the respective surface exposed to wind pressure was in no case more than 1.8 times the direct front surface of box girders. This figure then was doubled as a margin of safety.

Pumpen eingesetzt werden konnten. An drei oder vier Tagen des Jahres waren die Stürme von solcher Heftigkeit, daß nicht einmal der große Raddampfer übersetzen konnte und alle Arbeit ruhen mußte, denn auch die Kräne waren nicht zu bedienen. So blieben im Durchschnitt nur 22 bis 23 volle Arbeitstage im Monat. Häufig hätten nur eine oder zwei Stunden wegen starken Morgenregens ausfallen müssen, der die Männer durchnäßte und in ihre Unterkünfte schickte – aber selbst wenn das Wetter für den Rest des Tages wieder schön wurde, «waren keine Überredungskünste gut genug, die Männer wieder an die Arbeit zu bringen». Schließlich machte Nebel im Winter und Frühjahr große Sorgen bei Schichtwechsel, da man jeden Morgen und Abend viele hundert Leute von den Ufern zu ihren Arbeitsstätten bringen mußte: Die Nebelbänke kamen den Firth heraufgezogen wie eine massive Wand blendend weißer Watte und ließen oben nur die Pfeilerspitzen im Sonnenlicht stehen, während es unten dunkel wurde.

Winddruck-Experimente

Der Einsturz der Tay-Brücke hatte gelehrt, die Windkräfte ernst zu nehmen. Das *Board of Trade* selbst hatte die zulässige Windbelastung in exponierten Lagen inzwischen auf 56 Pfund/Quadratfuß (273 kg/m^2) festgelegt – und das zweimal über die ganze Trägeroberfläche, wobei der Widerstand gegen den Winddruck nur vom Eigengewicht der Konstruktion geleistet werden mußte. Baker nahm sich der Sache noch vor Baubeginn mit großem Einsatz an. Ein Winddruckmesser auf der alten Burg Inchgarvie war noch von einfacher Bauweise und verzeichnete nur Höchstwerte zwischen 27 und 41 Pfund/Quadratfuß (130 und 200 kg/m^2). Immerhin zeigte sich, daß die maximalen Winddrücke eher bei plötzlichen und lokalisierten Böen auftraten als bei stetigem Druck auf eine größere Fläche. Nach Errichtung der Mittelpfeiler zu voller Höhe installierte Baker dort oben verbesserte Drehwindmesser und fand, daß der Wind die einzelnen Stützen unterschiedlich stark mit bis zu 12 Pfund/Quadratfuß (58 kg/m^2) Differenz belastete.

Die dürftigen Kenntnisse über den wirklichen Effekt von Winddruck auf ebene oder gekrümmte Flächen veranlaßten Baker zu einer Reihe bemerkenswerter Versuche *(Abb. 74)*. Doch darüber hinaus war es für Baker noch wichtiger, genaue Daten für *schrägen* Windaufprall in verstrebten Kastenträgern zu erhalten, da sich hier je nach Winkelstellung der Winddruck vervielfachen konnte: Bei zwei nahe hintereinander liegenden Kasten-Fachwerkträgern, wie z. B. im Obergurt, wären bei genau rechtwinkligem Windeinfall die

A page about the wind

Eine Seite über Wind

Gustave Eiffel – the predecessor:
Already during the 1860's and 1870's, Eiffel had designed tall railway viaducts for the ore and coal trains in the *Massif Central*. He had not only used advanced methods for calculating the loading and bending moments, but he had also studied for the first time the influence of wind pressures on bridges, in detail. In Britain, the wind loads had been studied, too, at one or two bridges, such as for the Britannia Bridge, but not all bridge-builders were as careful as Eiffel.

Gustave Eiffel – der Vorläufer:
Eiffel hatte bereits in den 1860er und 1870er Jahren hohe Eisenviadukte für die Erz- und Kohle-Züge im Massif Central gebaut. Dabei benutzte er nicht nur fortschrittliche Berechnungsmethoden für die Last- und Biegemomente, sondern studierte auch erstmalig im Detail den Einfluß der Windkräfte auf Brücken. – In Großbritannien hatte man zwar auch bei ein oder zwei Brücken die Windkräfte untersucht, so bei der Britannia-Brücke, aber nicht alle Brückenbauer waren so sorgfältig wie Eiffel.

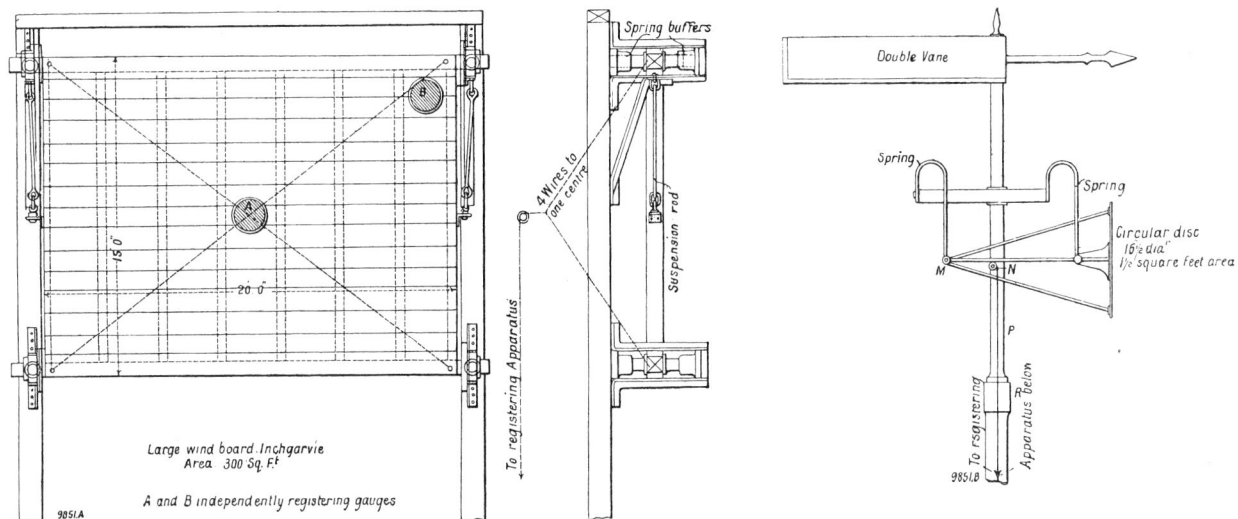

74
Baker's wind gages at Inchgarvie. The wind board (to the left) consisted of an area of 300 sq ft (27.87 m²), with two holes for the wind gages to be inserted, with connecting wires to the recording instrument. To the right, we see a freely rotating weather vane with an added disk exposed to the wind. The instruments were not very precise, but they gave an approximation of the wind pressures to be expected – such as a maximum pressure of 41 lbs/sq ft [29, p. 9].

74
Die Windmeßgeräte Bakers auf Inchgarvie. Das Windbrett (links) hat eine Fläche von 300 Squarefoot (27,87 m²) mit zwei Löchern für eingesetzte Windmesser und Verbindungsdrähten zum Aufzeichnungsinstrument. Rechts eine frei rotierende Wetterfahne mit angesetzter Winddruckscheibe. Die Geräte waren nicht sehr genau, gaben aber eine Annäherung der zu erwartenden Windlasten – so einen Höchstwert von 41 Pfund/Squarefoot [29, S. 9].

Baker's Testing Model:
Benjamin Baker's wind experiments were the first of their kind in Britain. More complicated was the measuring of wind pressures upon curved or honeycombed surfaces, such as lattice girders. Due to the difficulty of obtaining exact results under varying speed or direction of the wind by the use of models, Baker simply reversed the matter by *letting the wind be stationary and making the model movable*. This experimental model – a masterpiece of common sense – consisted merely of a light wooden stick suspended in the center from the ceiling and balanced horizontally. On one end of the stick, Baker fastened a cardboard model of the surface to be tested – be it part of a tube, bracing, top member or the entire cantilever. At the other end, he fastened a sheet of cardboard in the same parallel po-

Bakers Versuchsmodell:
Benjamin Bakers Windexperimente waren die ersten dieser Art in Großbritannien. Komplizierter war die Ermittlung des Winddruckes auf gekrümmte oder durchbrochene Flächen, wie z. B. Gitterträger: Wegen der Schwierigkeit, bei wechselnder Windgeschwindigkeit und -Richtung mit Modellen exakte Resultate zu bekommen, drehte Baker einfach den Spieß um, *indem er den Wind stationär sein ließ* und *das Modell beweglich* machte. Dieser Versuchsapparat – ein Meisterstück an «*common sense*» – bestand aus einem leichten Holzstab, der in der Mitte an einer Schnur von der Decke hing und sich waagrecht ausbalancierte. An einem Ende des Stabes befestigte Baker ein Papp-Modell der zu testenden Oberflächenform – gleich ob Rohrteil, Seitenstrebe, Obergurt oder der ganze Ausleger.

sition as the model. The surface of this cardboard could be enlarged or reduced by another sheet, which could slide in and out of the first at will. Once the stick was now pulled close and let go swinging parallel, it showed – "if this was properly done" – that the swinging stick kept its position, provided the model and the sheet of cardboard opposite were of equal weight and surface. As soon as one surface was larger than the other one, it would be turned back by the resistance of the air pressure. If this was repeated often enough, the effective surface for the wind pressure was obtained in the end.

These new findings were extremely important, because with large bridges, the wind pressures produce a much greater moment than rolling loads do, like trains for instance. That is to say, the areas of impact for wind pressure should, for the sake of economy, be reduced to a minimum.

Final Structure

The actual structure of the Forth Bridge consists of two double cantilevers and two connecting central girders. To these are added the two approach viaducts with 10 spans in the south and five in the north. Each double cantilever consists of a central tower resting on four circular masonry piers. The two outer cantilever-ends to the north and south are supported additionally by the large stone piers of the viaducts, while the central tower on Inchgarvie is free-standing and extra strong (*Fig. 63*, fig. 5).

The total length of the bridge (without viaducts) measures 5,330 ft (1624.58 m). This consists of the central tower on Inchgarvie of 260 ft (79.25 m) length, the two Fife and Queensferry towers of 145 ft (44.20 m) each, the two central girders of 350 ft (106.68 m) each and the six cantilevers of 680 ft (207.26 m) each. If we add the two viaducts to this, a total length of 8,296 ft (2528.62 m) is obtained – with two main spans of 1,710 ft (521.21 m) each.

The large diagonal trusses between the top and bottom members cross each other in pairs in the center, whereby the compression members are constructed as tubes, while the tension members are built as open lattice girders, thus demonstrating the interplay of stresses. Originally it was intended to continually decrease the slant of the side walls toward the cantilever ends. But this would have resulted in very complicated connections and in a twisted top member, whereas the continuous slant with a ratio of 1 : 7.5 adds significantly to the overall impression of great rigidity against the wind *(Fig. 75)*.

Ans andere Ende setzte er ein Pappschild in gleicher paralleler Ausrichtung wie das Modell. Dieses Schild konnte mittels eines zweiten herausschiebbaren Pappstückes in seiner Oberfläche nach Wunsch vergrößert oder verkleinert werden. Zog man nun diesen Stab zu sich heran und ließ ihn parallel schwingen, zeigte sich – «if this was properly done» –, daß der schwingende Stab seine Lage beibehielt, vorausgesetzt, das Modell und die Pappscheibe gegenüber hatten dasselbe Gewicht und die gleiche Oberfläche. Sobald eine Fläche größer als die andere war, wurde sie vom Luftdruck zurückgedreht. Wiederholte man das oft genug, erhielt man schließlich die effektive Winddruck-Fläche.

hinteren Flächen von den vorderen verdeckt, würden aber bei schrägem Windaufprall den Druck vervierfachen! Da man bereits Kenntnis über diagonalen Winddruck auf kubische Formen hatte, bestätigte Baker mit seinem Modell zunächst diese Regeln. Im Falle verdeckter Flächen aber wies er nach, daß die betroffene Winddruckfläche keinesfalls größer war als das 1,8-fache der direkten Frontalfläche von Kastenträgern. Dieser Wert wurde dann zur statischen Sicherheit verdoppelt. Die neuen Erkenntnisse waren außerordentlich wichtig, da bei großen Brücken die Windkräfte ein viel größeres Moment erzeugen als beispielsweise rollende Lasten; d. h. die Angriffsflächen für Winddruck sollten aus Gründen der Wirtschaftlichkeit auf ein Minimum reduziert werden.

Endgültige Konstruktion

Die eigentliche Konstruktion der Forth-Brücke besteht aus drei Doppelauslegern (Kragträgern) und zwei verbindenden Mittelträgern. Dazu kommen die Auffahrtsviadukte mit 10 Trägerfeldern im Süden und fünf im Norden. Jeder Doppelträger besteht aus einem Mittelturm, der auf vier kreisrunden Mauerwerkspfeilern ruht. Die beiden äußersten Ausleger-Enden im Norden und Süden haben je ein zusätzliches Auflager in den großen Steinpfeilern der Viadukte: Der mittlere Auslegerturm auf Inchgarvie dagegen steht frei mit verstärktem Mittelturm (*Abb. 63*, Fig. 5).

Die Gesamtlänge der Brücke (ohne Viadukte) beträgt 5330 Fuß (1624,58 m). Sie setzt sich zusammen aus dem Mittelturm auf Inchgarvie von 260 Fuß (79,25 m), den beiden Fife- und Queensferry-Türmen zu je 145 Fuß (44,20 m), den zwei Mittelträgern zu je 350 Fuß (106,68 m) und sechs Auslegern von je 680 Fuß (207,26 m). Nimmt man die beiden Viadukte hinzu, erhält man eine Gesamtlänge von 8296 Fuß (2528,62 m) – mit zwei Hauptspannweiten von je 1710 Fuß (521,21 m).

Die großen Diagonalstreben zwischen Unter- und

75
View of the Forth Bridge from northwest near North Queensferry. The slant of the structure with a ratio of 1 : 7.5 can clearly be seen along the entire length of the bridge.

75
Nordwest-Ansicht der Forth-Brücke von North Queensferry. Der Anzug von 1 : 7.5 ist über den Gesamtverlauf der Konstruktion deutlich zu erkennen.

Obergurt kreuzen sich paarweise in der Mitte, wobei die Druckstreben rohrförmig und die Zugstreben als offene Gitterträger ausgebildet sind und so den Kräfteverlauf anschaulich machen. Ursprünglich war geplant, die Seitenneigung zum Ausleger-Ende hin stetig zu verringern, um so parallele Mittelträger zu erhalten: Das hätte aber zu sehr komplizierten Verbindungen und einem verdrehten Obergurt geführt; wogegen der durchgehende Anzug im Verhältnis von 1:7,5 ganz wesentlich zum Gesamteindruck großer Windsteifigkeit beiträgt (Abb. 75).

Der interne Eisenbahnviadukt läuft auf eigener schmaler Brückenkonstruktion durch die drei Doppelausleger hindurch. Für das Gleichgewicht des ganzen Werkes war es notwendig, die beiden äußeren Ausleger-Enden im Norden und Süden – die ja keine Mittelträger zu halten hatten – jeweils mit dem halben

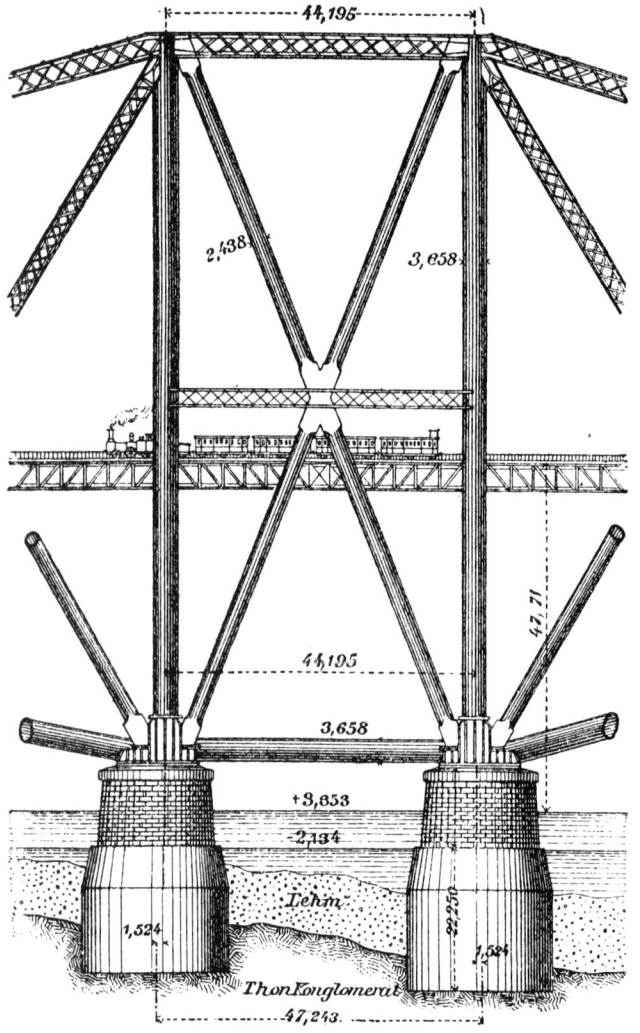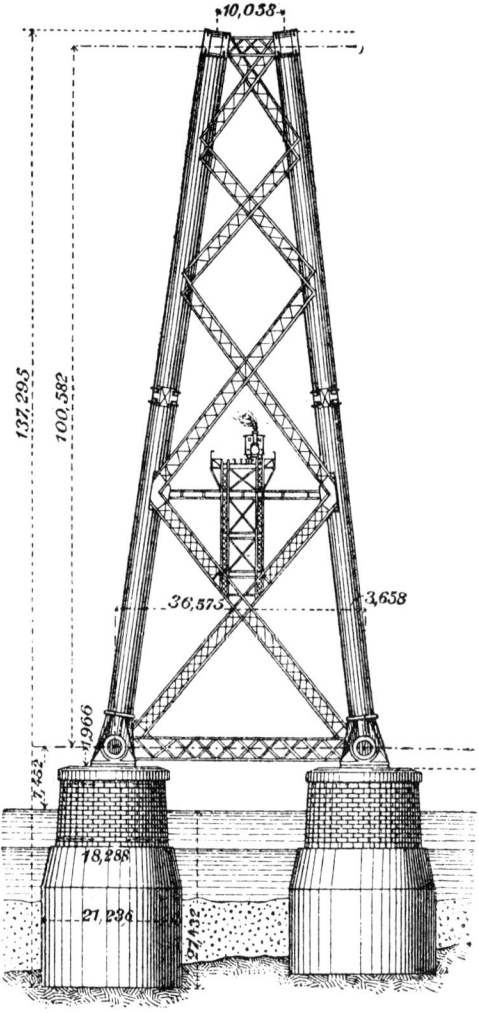

76
The position of the railway viaduct inside the bridge. A train passing through is necessary to impact a sense of scale, so that one can appreciate the true height of the structure [5, p. 24].

76
Die Lage des inneren Eisenbahnviaduktes in der Brücke. Erst ein durchfahrender Zug gibt einen Maßstab für die Größe des Bauwerkes [5, S. 24].

The internal railway viaduct runs on its own narrow bridge right through the three double cantilevers. For the balance of the whole structure, it was necessary to load down the two outer ends of the cantilevers to the south and north, by half the weight of the central girders, as they did not have to support a central girder. This was done by placing 1,000 tons of scrap iron into steel boxes inside the tall end-piers of the viaducts. Thus, equal loading conditions were created for all six cantilevers in the state of rest. In order to compensate for the additional weight of a train entering the opposite end of the cantilever, however, additional weights were imposed on the fixed cantilevers, so the free ends would not be able to sag. By these anchorages at the ends via additional weights, all conceivable maxi-

Gewicht des Mittelträgers zu belasten. Das geschah mit je 1000 Tonnen Eisenschrott in Stahlkästen innerhalb der beiden großen Viadukt-Endpfeiler; so waren gleiche Lastbedingungen für alle sechs Ausleger im Ruhezustand geschaffen. Um aber nun das Gewicht eines einfahrenden Zuges im gegenüberliegenden Ausleger-Ende aufzufangen, wurde zusätzliches Gewicht auf die festen Ausleger gesetzt, so daß sich die freien Enden nicht mehr senken konnten. Durch diese Endverankerungen mit Zusatzgewicht wurden alle erdenklichen Höchstlasten berücksichtigt. Nur im Inchgarvie-Doppelausleger konnte noch der Fall eintreten, daß sich zwei Züge gleichzeitig am Ausleger-Ende treffen und damit das Gleichgewicht empfindlich stören würden: Aber auch hierfür wurde vorgesorgt, wie

77
"Skewback" at the Fife pier near North Queensferry.

77
«Skewback» (Schrägwiderlager) am «Fife-Pfeiler» in North Queensferry.

mum stresses were being considered. Only at the double cantilever of Inchgarvie, could there be the case of two trains meeting simultaneously at the cantilever ends, thus offsetting the balance considerably. But even for this case, precautions were made, as shown in Fig. 78b: By virtue of many and strong diagonal bracings, this Inchgarvie tower is of enormous strength and great weight. It is thus well equipped to take up the enormous loads of dead load, live load and wind pressure.

All of these forces must ultimately come together in the "skewbacks" of the four masonry piers of each tower. Here, five tubular girders and five lattice girders have to be linked up in giant junctions, resting with their base upon "bed plates" *(Fig. 77)*. The function of

Abb. 78b zeigt. Durch vielfache und kräftige Diagonalverstrebungen ist dieser Inchgarvie-Mittelturm von immenser Kraft und großem Gewicht und damit gut imstande, die enormen Belastungen von Eigengewicht, Nutzlast und Winddruck zu bewältigen.
Alle diese Kräfte aber müssen letztlich zusammenkommen in den *«skewbacks»* (Schräg-Widerlagern) oder Hauptauflagern auf den vier Mauerpfeilern: Hier treffen sich je fünf Rohr- und fünf Gitterträger in gewaltigen Knotenpunkten, die an der Basis auf *«bed plates»*, den Lagerplatten, aufsitzen *(Abb. 77)*. Die Funktionen dieser Lagerplatten – nämlich Widerstand und doch begrenztes Nachgeben gegen Winddruck sowie Dehnungen bzw. Schrumpfungen der Mitteltürme – sind so wichtig, daß man ihnen viel Sorgfalt

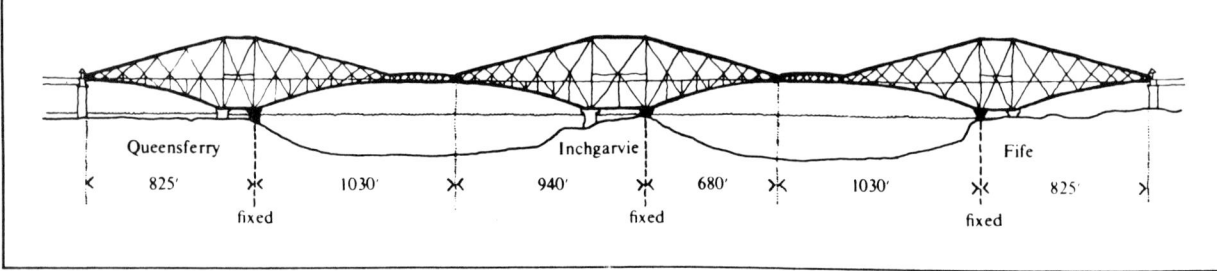

78 a
Schematic drawing of the fixed and movable supports of the piers: Only one out of four supports of the three cantilever towers is fixed, whereas the others are somewhat movable in order to take up expansion or torsion by the action of the sun or by wind pressure [1, p. 116].
Firmly fixed are
1) the southeastern support of the Fife pier;
2) the northeastern support of the Inchgarvie pier; and
3) the northeastern support of the Queensferry pier.

78 a
Schema der festverankerten und beweglichen Pfeilerfüße. Nur 1 von 4 Pfeilerfüßen der 3 Pfeilertürme ist festverankert, die anderen sind in begrenztem Maße beweglich, um Dehnungen bzw. Verdrehungen durch Sonneneinstrahlung oder Winddruck aufzufangen [1, S. 116].
Festverankert sind
1) der Südost-Fußpunkt des Fife-Pfeilers
2) der Nordost-Fußpunkt des Inchgarvie-Pfeilers
3) der Nordost-Fußpunkt des Queensferry-Pfeilers

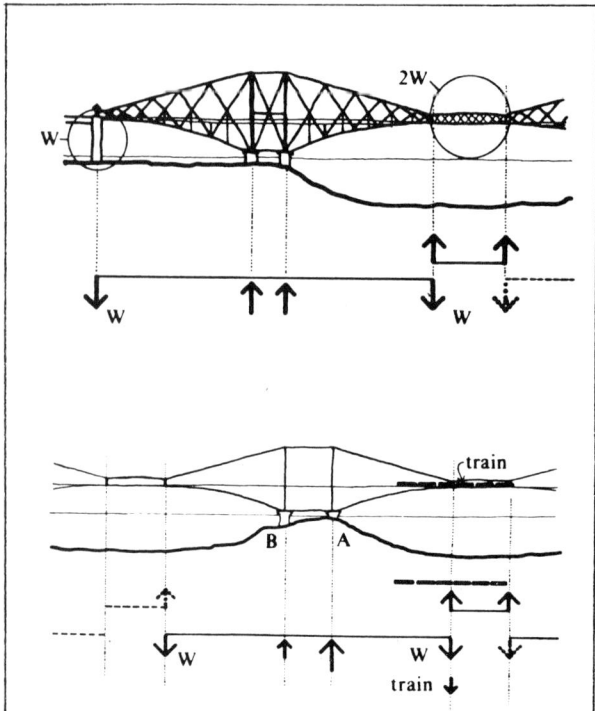

78 b
Schematic drawing of the distribution of loads:
The Fife and Queensferry piers are fixed at the last piers of the viaducts, each carrying half of the central girders at their free ends (upper sketch).
The central Inchgarvie pier supports half the weight of the two central girders at each cantilever, thus being balanced as far as dead load is concerned (lower sketch).
However, once a train enters, or two trains meet at the far end of a cantilever, the balance is upset. The tower would tend to rotate around point A and lift point B, which is being held down by the anchoring bolts. But since they were not intended to ever take up tension stresses, the distance between points A and B was widened to about double that of the other towers. This would prevent the central tower from toppling even under the heaviest of trains, transferring compression stresses to the foundations instead [1, p. 118].

78 b
Schema der Lastverteilung:
Die Fife- und Queensferry-Pfeiler sind jeweils am Viaduktendpfeiler festverankert und tragen an ihren freien Enden je die Hälfte der aufgehängten Mittelträger (obere Skizze).
Der zentrale Inchgarvie-Pfeiler trägt an jedem Ausleger je das halbe Gewicht der beiden Mittelträger und ist damit ausbalanciert, sofern es nur um Eigengewicht geht (untere Skizze).
Kommt aber ein Zug, oder treffen sich gar zwei Züge am Ende eines Auslegers, ist dieses Gleichgewicht gestört: Der Pfeilerturm würde um Fußpunkt A kippen und Fußpunkt B anheben wollen, was durch die Ankerschrauben verhindert wird; da man sie aber keinen Zugkräften aussetzen wollte, wurde der Abstand zwischen Fußpunkt A und B soweit auseinandergelegt – ca. auf das Doppelte der beiden anderen Pfeiler –, daß der Mittelturm selbst unter dem schwersten Zug nicht mehr kippen kann und nur noch Druckkräfte auf die Fundamente überträgt [1, S. 118].

these bed plates – affording resistance and still some movement under wind pressure, as well as allowing for some expansion or shrinkage of the central towers – is so important that much care and thought was put into their design. It is a curious fact that these problems would not have arisen in the original design by Fowler and Baker (*Fig. 63,* fig. 4). They are the result of the alterations requested previously by two associate engineers. At first it had been intended to fix the main supports firmly to the masonry piers. But in the end it was decided to anchor firmly just one of the four main supports, and to allow the three remaining ones to react in a limited way to the forces of temperature, as well as to lateral bending of the bridge, as might be caused by the sun's radiation to one side only, and by wind pressure acting on the cantilevers *(Fig. 78a).*

It may be noted that the tubular bottom members of the cantilevers are not bent like arches, but run polygonally from junction to junction. Aside from the greater rigidity which straight tubes provide, arched tubes would have required forming every steel plate in two directions – a process made more difficult by their diameter decreasing from 12 to $6^1/_2$ ft (3.66 to 1.98 m). The top members, being designed solely for tension, run straight without interruption from the top of the towers down to the last strut of the cantilevers. The viaduct girders for the rail bed inside are firmly connected to the cantilevers, adding greatly to the lateral stiffness of the bottom members.

Sequence of Construction

Once the contract had been awarded to the firm of Tancred, Arrol & Co of Glasgow on December 21, 1882, preparatory work got started at South Queensferry. Immediately, the "No. 1 Shed" of Bouch's suspension bridge was used again as the first job office. A careful triangulation, with an observatory especially built at the south shore and with 20 surveying stations, determined the exact location of the main piers. Soon, an entire town of over 100 barracks and 16 stone houses for workers' families and engineers sprang up, as well as canteens, reading rooms, shops etc. To these came extensive workshops including a large drafting shop with a blackened floor, whereupon the plans in 1:1 scale and wooden models of entire building segments could be assembled, in full size. When the narrow shores got too small for all these works, the ill-fated Tay Bridge was recalled once again by using two of its girders for bridging the old train cutting. In the busy years from 1886 to 1889, 24 hectares of goods yards were finally covered up with the bridge's prefabricated girders. Immense amounts of material were brought

und Denkarbeit gewidmet hat. Pikanterweise hätte es diese Probleme im ursprünglichen Entwurf von Fowler und Baker (*Abb. 63,* Fig. 4) so nicht gegeben; sie sind eine Folge der gewünschten Änderungen. Zunächst war beabsichtigt, die Hauptauflager völlig fest mit den Mauerpfeilern zu verbinden. Schließlich entschied man aber doch, nur je ein Hauptauflager von vieren fest zu verankern, und die übrigen drei in begrenztem Maße den Temperatureinflüssen nachgehen zu lassen, ebenso den seitlichen Durchbiegungen durch einseitige Sonnenerwärmung und durch Winddruck auf die Ausleger *(Abb. 78a).*

Es fällt auf, daß die rohrförmigen Untergurte der Ausleger nicht als reine Bögen gekrümmt sind, sondern polygonal mit einem leichten Knick von Feld zu Feld laufen. Abgesehen von der größeren Steifigkeit gerader Rohre, hätte bei gekrümmten Rohren jede einzelne Stahlplatte in zwei Richtungen extra geformt werden müssen – erschwert noch durch den abnehmenden Rohrdurchmesser von 12 auf 6,5 Fuß (3,66 auf 1,98 m). Die Obergurte sind durchweg auf Zug beansprucht und laufen vom Scheitelpunkt der Mitteltürme bis zur Endstrebe der Ausleger ohne Unterbrechung durch, mit abnehmendem Querschnitt. Die inneren Viadukt-Träger des Schienenweges sind fest mit den Auslegern verbunden und tragen wesentlich zur Seitenaussteifung der Untergurte bei.

Bauablauf

Nach der Auftragserteilung an die Firma Tancred, Arrol & Co. aus Glasgow am 21. Dezember 1882 wurde in South Queensferry mit den Vorarbeiten begonnen. Sogleich bediente man sich des «*No. 1 Shed*» von Bouch's Hängebrücke als erster Bauhütte. Eine sorgfältige Vermessung mit einem eigens gebauten «*observatory*» am Südufer und 20 Meßstationen bestimmte die genaue Lage der drei Hauptpfeiler. Bald entstand dort eine ganze Stadt mit über 100 Baracken und 16 Steinhäusern für Arbeiterfamilien und Ingenieure, sowie Kantinen, Leseräumen usw.; dazu große Werkstätten samt einer 20 × 70 m großen Zeichenhalle mit schwarzem Boden, auf dem in Kreide die Pläne im Maßstab 1:1 und naturgroße Holzschablonen ganzer Bauteile hergestellt werden konnten. Als der Uferstreifen für all diese Arbeiten zu schmal wurde, entsann man sich noch einmal der unglückseligen Tay-Brücke und benutzte zwei ihrer Träger zur Überbrückung des alten Bahneinschnittes. In den geschäftigen Jahren 1886 bis 1889 waren dort schließlich 24 Hektar Lagerfläche mit vormontierten Brückenträgern belegt. Immense Materialmengen wurden ans Hochufer und von dort über Schrägschienen zum Uferstreifen geschafft.

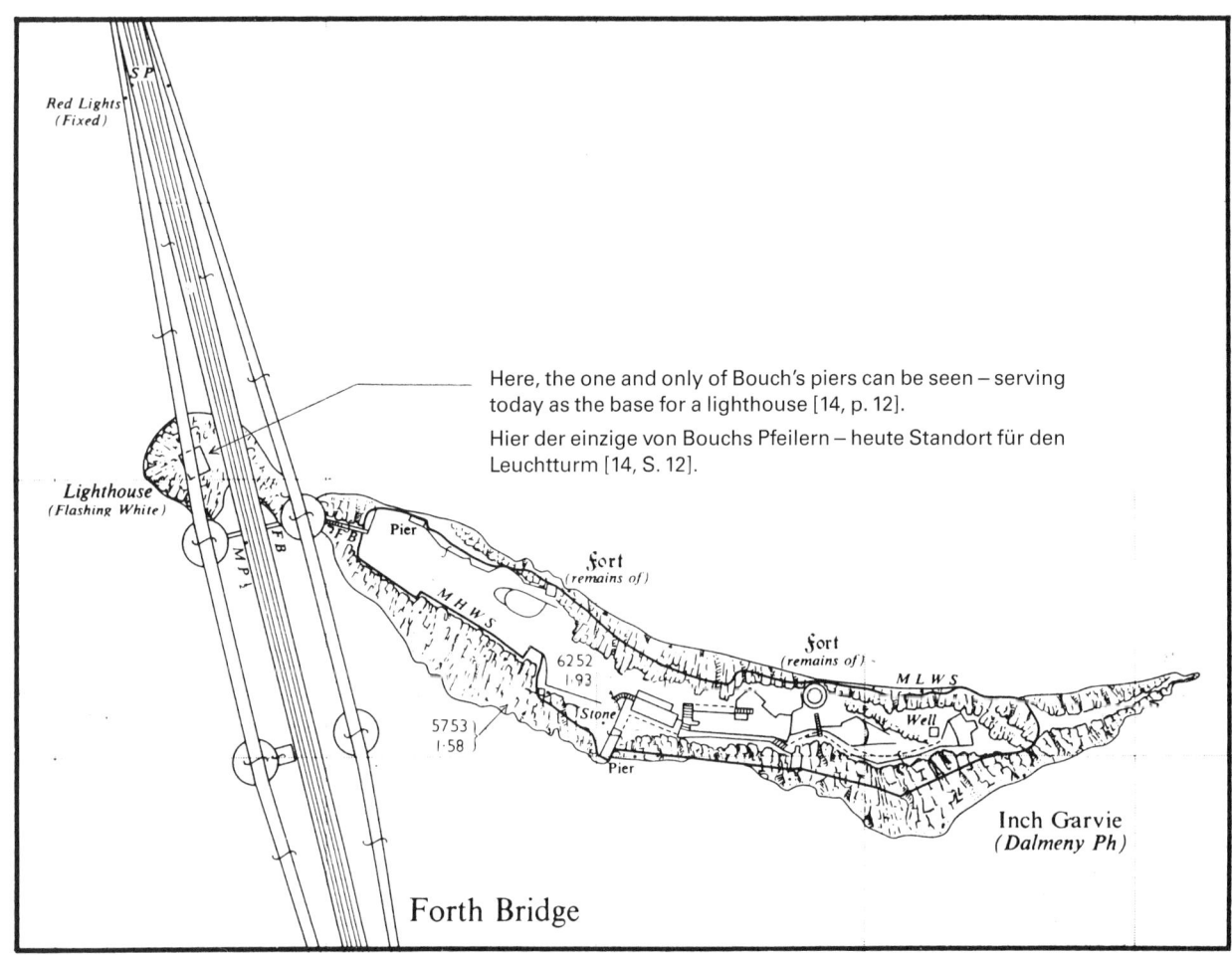

79
The rocky islet of Inchgarvie in the middle of the Firth of Forth. At the beginning of construction, the old Dundas family's ancient castle of 1490 still occupied this place. In former times, the island was called "Inchgarde" – its garrison offering refuge in waters infested by pirates. It posessed the right to levy dues from shipping, and the story goes that old Lord of Dundas not only kept soldiers, but he protected pirates as well so they would chase the ships up the Forth for him to collect the dues. And now this island was to become the mere support for a bridge – indeed, the castle's keep, of all places, should be crowned by Baker's new wind gauges!

79
Die Felseninsel *Inch-Garvie* in der Mitte des Firth of Forth: Hier stand bei Baubeginn noch die Stammburg der alten Dundas-Familie von 1490. Früher hieß die Insel «*Inchgarde*» und war mit ihrer Garnison ein Refugium in Piraten-verseuchten Gewässern. Sie besaß das Recht, Schutzzölle von der Schiffahrt zu erheben – wobei erzählt wird, der alte Lord of Dundas habe nicht nur die Soldaten gestellt, sondern auch die Piraten geschützt, damit sie ihm zum Abkassieren die Schiffe den Firth hinaufjagten. Und nun sollte die Insel das Auflager für eine Brücke werden – ja, ausgerechnet auf dem Bergfried wollte man die neuen Windmeßgeräte Bakers aufpflanzen!

from the high ground via inclined rails down to the shore.

But on Inchgarvie – "which providence has so kindly placed in the middle of the Firth" – things were different *(Fig. 79)*. When Baker's wind gauges were installed on the old keep of Inchgarvie Castle in 1882, peace was upset already. Once foundation work for the piers got started, more working area was required, after the western portion of the island had already been obtained by Parliamentary Act, for the erection of the four masonry piers. When finally the castle's entire grounds had to be covered over with shops and lodgings for 90 foreign workmen, the patience of the old Lord of Dundas came to an end. Court proceedings and compulsory expropriation were to follow, awarding to the Lord in the end 1,500 Pounds for the "right to erect piers" and another 2,800 Pounds for the rest of the island. – One way or other, this was itself an act of piracy!

Further progress was hampered by the fact that all work had to be carried out from a tiny island. The granite stones for the base of the piers were transported by ship from Aberdeen, directly to large stagings built according to the rock formation across 10,000 sq yds (9140 m²) of the island's western part. All steel was preassembled at the south shore and transported by steam barges. A task in its own right was the transport of the workmen. In order to ferry them to and from their many job sites, a paddlewheel steamer with 450 seats was specially built. To this came a whole fleet of covered lighters and small boats, as well as one lifeboat each positioned underneath the 6 cantilevers, manned by two expert watermen, to save anyone who may have fallen off. In this way, at least 8 lives would be saved during the entire time of construction.

Electrical Matters

The night shifts were particularly dangerous; although there was electrical lighting, it was still of very low wattage. One tried using the bright but fragile carbon arc lights along with the so-called Lucigen lamps (fed by a mixture of creosote oil and compressed air) and with incandescent lights. Yet even with the best of lighting, the night shift did not exeed 50 % of the work done by the day shift. As the main piers were not lit uniformly (as the circumstances demanded), or were not lit at all, there was a great deal of confusion to the shipping during those Scottish winter nights. Once, the captain of a tug-boat coming down the river with his barge in tow, mistook the lights of the Fife Tower, in a slight mist, for those of Inchgarvie and held directly north toward the village of North Queensferry. While

he managed to back out his tug at the last moment, the barge went on, causing much damage to the landing jetty. Thereafter, a lighthouse was erected – of all places, atop the single masonry base which was the only permanent piece of work remaining of Bouch's suspension bridge *(Fig. 60)*. With those hundreds of arc lamps, the bridge was a fascinating sight at night: "Three pillars of fire – a truly wonderful and unique spectacle."

Novel Caissons

The work under water for erecting the circular masonry piers was only possible by using caissons, just like on the Tay. On Inchgarvie, the site for the two northern piers *(Fig. 63)* was permanently under water during high tide, and partly during low tide. Because they were quite exposed on the narrow cliff, it was decided to use temporary open caissons. They could only be used during low tide, and the water pouring in over the top had to be pumped out following every high tide. Thus excavation work at the northeastern pier lasted 10 months, until the actual masonry work could be started atop the concrete footing. Despite the greater depth of the northwestern pier, the work there lasted only 8 months thanks to the lessons gained from past experience. But these were just two of twelve foundations of piers for the whole bridge.

80 >
Pneumatic caisson with men working inside the compression chamber below. The caissons of 70 ft (21.34 m) diameter were far bigger than those used at the Tay. Pre-assembled at the Dalmarnock Ironworks in Glasgow, these double-shelled containers of wrought-iron were taken apart for transport, to be rivetted up again in South Queensferry. Upon a cylindrical lower section, a conical upper section was placed. Then this stationary caisson was bolted to a temporaray one, positioned on top. The "diving-bell" or compression chamber of only 7 ft (2.13 m) in height was equipped with a circular "cutting edge" running all around the bottom. The air-tight ceiling above the chamber was reinforced by strong girders to withstand the internal air pressure. The Inchgarvie caissons had two air shafts equipped with air-locks, one for excavated material, the other for access and exit of the workmen. At South Queensferry though, the caissons had several shafts, as it was possible there to dig across the entire surface soon after hitting the ground. The equalisation of air pressure inside the air-locks took about two minutes by hand valve, lest one had to suffer caisson disease in coming up from down below too fast. Still, there were many complaints about severely aching joints, especially when returning to normal air pressure; this stopped immediately upon returning into the high pressure zone – so many men actually preferred to rather stay below on Saturdays and Sundays, only to come up in emergencies [29, p. 22].

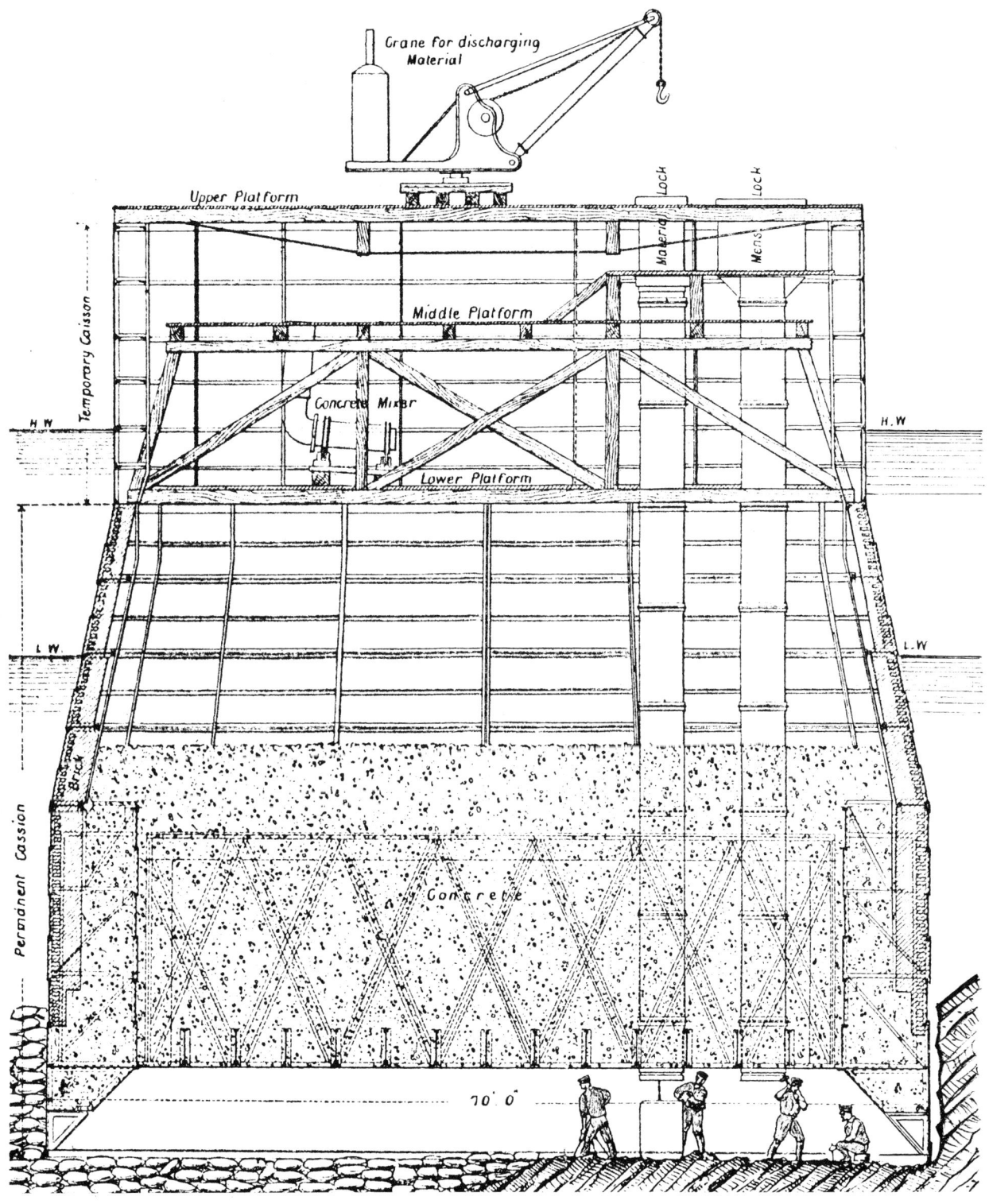

Caisson Details

Caisson-Details

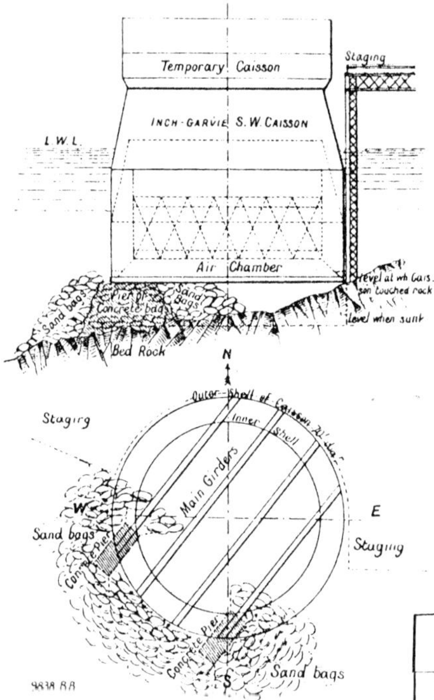

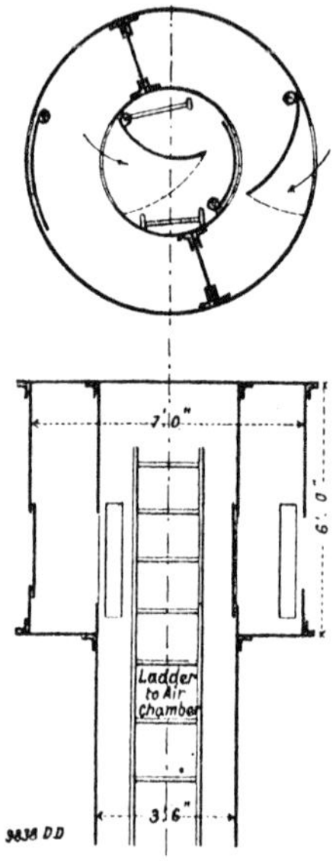

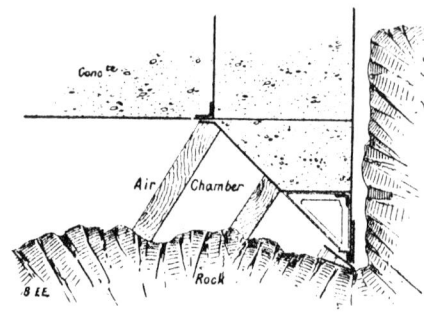

81 b
Air-lock with compression chamber and access shaft [29, p. 23].

81 b
Luftschleuse mit Druckkammer und Einstiegsschacht [29, S. 23].

81 c
Temporary supports of the cutting edge on bed-rock [29, p. 30].

81 c
Abstützung der Schneidekante auf Felsgrund [29, S. 30]

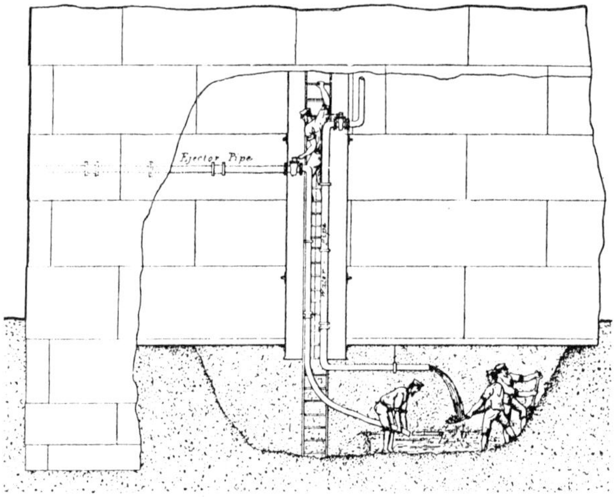

81 a
Inchgarvie caisson on sloping bed-rock with temporary supports [29, p. 16].

81 a
Inchgarvie-Caisson auf schrägem Fels mit temporärer Unterstützung [29, S. 16].

81 d
Excavation at riverbed, using a mixture of water and mud. Baker feared the caisson work just as much as his men did who were literally digging away the ground on which they were standing. If the caisson dropped too fast, they could easily be smothered between the sea-bed below and the many thousand tons of steel and concrete above them. Just a few years before, this had happened under the waters of the Neva in St Petersburg in 1876, when a caisson suddenly dropped 18 inches. Of the 28 men in the air chamber, 9 remained locked up inside. Only two were taken out alive. All the others had been smothered to death by the mud. And during construction of New York's Brooklyn Bridge and the St Louis Bridge, 119 of 600 workmen suffered from "caisson disease", causing the death of 16 men. Naturally, this syndrome was not unknown at the Forth; it was called "the bends" by the men and they were proud not to have suffered a single casualty [29, p. 27].

81 d
Aushub am Meeresgrund mit Wasser-Schlamm-Gemisch. Baker fürchtete die Caisson-Arbeiten ebenso wie seine Leute es taten – denn sie gruben sich buchstäblich den Boden weg, auf dem sie standen. Sank der Caisson zu schnell, konnten sie leicht zwischen Meeresboden und den vielen Tausend Tonnen Stahl und Beton über sich zermalmt werden. Wenige Jahre vorher, 1876, war das unter den Wassern der Newa in St. Petersburg geschehen, als ein Caisson plötzlich 18 Inch (45,7 cm) absackte; von den 28 Männern in der Kammer blieben 9 darin eingeschlossen, und nur 2 konnten gerettet werden. Die übrigen waren im Schlamm erdrückt worden. Beim Bau von New York's Brooklyn Bridge und der St. Louis Brücke erlitten 119 von 600 Arbeitern die Taucherkrankheit, woran 16 starben. Natürlich war die «Caisson-Krankheit» auch am Forth nicht unbekannt – von den Arbeitern *«the bends»* genannt –, aber man war stolz, daß es deshalb keine Toten gab [29, S. 27].

For the two southern Inchgarvie caissons, only enclosed pneumatic caissons could be used, due to the depth and rapid tidal current. Before they were floated to the site, a large number of bags filled with sand and concrete were placed on the sloping ground, after soundings of the ground profile were taken opposite the spot where the cutting edge of the caisson would first hit the rock. Additional guide posts, appearing in a semi-circle above the water, were used in order to let the caissons float to the correct location. It was quite a task every time to tow those gigantic cylinders weighing 3,000 to 4,000 tons through wind, waves and current to their places, to hold them there and to finally set them safely and exactly onto the ground, through all types of soil. Despite the hazards, these large caissons caused fewer problems than did the open ones used for the other foundations *(Fig. 80)*.

The caissons were assembled on launching cradles, just like in ship-building. There, the caissons were filled up with concrete, from 4 to 6 ft (1.22–1.83 m) above the ceiling of the air chamber, in order to increase stiffness and weight. The launching was always performed during high tide in order to get the caissons afloat fast, with their draught of 9 to 10 ft (2.74–3.05 m) of water, and away from the shallow fore-shore to the final resting place. The two Inchgarvie caissons each weighed 500 tons when floated; once they had reached deep water, they were loaded with the temporary caisson, with machines, concrete, brick and a crane, in order to save the double journey. So they weighed 2,870 in the end, causing a draught of 31 ft (9.45 m) – "two floating monsters!"

The sinking of the six pneumatic caissons had been entrusted to Monsieur L. Coiseau of Paris. He had already been engaged at the Suez Canal and for regulating the Danube in Vienna, and had also built the great harbour docks at Antwerp with a large staff of experienced men. At this point, our patriotic Mr. Westhofen does away with one of the prejudices of his time, that foreign workers, mostly northern Italians, but also a few Frenchmen, Belgians, Austrians and Germans would be better equipped to withstand the high atmospheric pressure in the sticky air inside the caissons. On the contrary, many British workmen and helping hands had frequently been sent down, many for the first time and yet they had suffered no harm, it being merely a matter of habit, good health and – "moderation in taking strong spirituous liquors". Indeed, some of Monsieur Coiseau's own men were the ones suffering from too much whisky and sudden changes of temperature. Still, all six caissons were finally grounded after a mishap with one tilted caisson was eventually dealt with too. Their depths ranged

Für die beiden südlichen Inchgarvie-Fundamente kamen wegen der Tiefe und starken Gezeitenströmung nur geschlossene pneumatische Caissons in Frage. Bevor man diese anschleppen konnte, legte man nach Ausloten des Grundprofiles eine große Zahl von Sand- und Betonsäcken auf den schrägen Grund – gegenüber der Stelle, wo die Schneidekante des Kastens den Fels zuerst berühren würde. Dazu kamen Führungspfähle, die oben halbkreisförmig aus dem Wasser ragten, um den Senkkasten an die richtige Stelle anschwimmen zu lassen. Es war jedesmal ein großes Stück Arbeit, diese gewaltigen Zylinder mit je 3000 – 4000 Tonnen bei Wind, Wellen und Strömung an ihren Platz zu schleppen, dort festzuhalten und schließlich durch alle Bodenarten hindurch sicher und genau auf Grund zu setzen. Trotzdem verursachten die großen Caissons weniger Probleme als die offenen der übrigen Gründungen *(Abb. 80)*.

Der Zusammenbau der Caissons erfolgte wie beim Schiff auf einer Helling, wo die Kästen zwecks Aussteifung und Ballast 4 bis 6 Fuß (1,22 – 1,83 m) über der Luftkammerdecke mit Beton vorgefüllt wurden. Der Stapellauf geschah immer bei Flut, damit die Caissons mit ihrem Tiefgang von 9 – 10 Fuß (2,74–3,05 m) rasch vom seichten Ufer zum Senkplatz geschleppt werden konnten. Die beiden Inchgarvie-Caissons wogen bei Stapellauf je 500 Tonnen, wurden dann aber, sobald sie tieferes Wasser erreicht hatten, noch mit dem temporären Caisson, mit Maschinen, Beton, Ziegeln und Kran belastet, um den doppelten Transportweg zu sparen. So wogen sie schließlich je 2870 Tonnen bei 31 Fuß (9,45 m) Tiefgang – *«two floating monsters!»*.

Das Absenken der sechs pneumatischen Caissons hatte man Monsieur L. Coiseau aus Paris anvertraut, der schon am Suez-Kanal und bei der Wiener Donauregulierung tätig gewesen war und mit seiner erfahrenen Arbeitsgruppe die Hafenanlagen von Antwerpen gebaut hatte. Hier räumt der patriotische Mr. Westhofen mit einem Vorurteil der Zeit auf, daß nämlich ausländische Arbeiter, größtenteils Nord-Italiener und einige Franzosen, Belgier, Österreicher und Deutsche, besser als britische Arbeiter imstande gewesen wären, den hohen atmosphärischen Überdruck in feuchtheißer Caisson-Luft auszuhalten: Im Gegenteil, viele britische Fach- und Hilfsarbeiter seien häufig da unten eingesetzt worden, viele zum erstenmal und ohne Gesundheitsschäden zu erleiden; das sei hauptsächlich eine Frage der Gewohnheit, aber auch guter Gesundheit und – *«moderation in taking strong spirituous liquors»*. In der Tat waren es einige von Coiseau's eigenen Leuten, denen zuviel Whisky und die krassen Temperaturunterschiede nicht be-

from 64 to 89 ft (19.20–27.13 m), with an average working time of 78 days for each – without a single death being due to atmospheric pressure. Although two men died during this period, they had actually been ill already before, and "another man became insane, and had to be sent back to his own country".

The regular procedure for sinking the South Queensferry caissons was carried out by mixing the layers of mud with water, just as it had been done before at the Tay. By holding the ejector hose into this mixture, the air pressure inside the chamber hurled the mud up to the surface of the water (Fig. 81 d). Still, this was dangerous for the men locked inside during low tide, since only the natural buoyancy was preventing the caisson from suddenly sinking into the mud, smothering the men below; although the colossus once dropped by 7 ft (2.13 m), filling the chamber and access shaft with mud, not a single life was lost – quite unlike the scenes at many other construction sites of that time. When finally the very hard boulder clay was reached, even dynamite did not help much. At that stage, Mr. Arrol thought of an old principle: steady drops hollow any stone, and thus he invented the hydraulic spade (Fig. 82).

When at least a depth of 89 ft (27.13 m) had been reached, the outside water pressure by itself was enough to totally seal off the lower edge, enabling the work to be carried on under normal pressure; despite the much greater depths here, work was much easier now than further up. In any case, the amount of excavation work achieved was enormous, in view of the simple means available then. During 24 hours of work in very hard soil, and with a full shift of 27 men in the compression chamber using four hydraulic spades, every 5 minutes a bucket of excavation material was sent up, corresponding to 3.65 m³/per man/per 12-hour shift. Thus the time from floating the caisson to its final founding on the ground was shortened from 96 to 69 days in the end.

On Inchgarvie, the bed-rock had to be dynamited from below the cutting edge once the men had gotten out to safety via the access shaft.

Under the lower edge, pieces of evenly spaced rock were left standing as supports; only in the end, during low tide, were they dynamited away, causing the caisson to come down with a tremendous roar. In these two caissons, the air pressure had to be increased with every sinking, since the cutting edge never closely rested on the rock and allowed water to flow through; but by the same token the air was always fresh, as large volumes of compressed air were being forced outside, rising steadily like a coat of bouncing bubbles. Here the work went on around the clock, to

kamen. Dennoch wurden alle sechs Caissons sicher auf Grund gesetzt, nachdem man auch einen umgekippten Senkkasten wieder aufrichten konnte. Deren Tiefen betrugen 64–89 Fuß (19,20–27,13 m), bei einer durchschnittlichen Arbeitszeit von je 78 Tagen –, ohne daß ein einziger Todesfall dem atmosphärischen Überdruck anzulasten war. Zwar starben zwei Mann während der Zeit – «die waren aber schon vorher krank, und ein dritter Mann wurde wahnsinnig und mußte in sein Heimatland zurückgeschickt werden».

Der normale Senkvorgang der South-Queensferry-Caissons lief so ab, daß man wie schon am Tay die Schlammschichten mit Wasser versetzte und den Ausstoßschlauch so in das Gemisch hielt, daß der innere Luftdruck es bis zur Wasseroberfläche schleuderte (Abb. 81 d). Allerdings war das für die eingeschlossenen Männer bei Ebbe gefährlich, da nur der natürliche Auftrieb den Kasten davon abhielt, plötzlich in den seichten Schlamm zu sinken und die Männer zu erdrücken – und obwohl einmal der Koloß 7 Fuß (2,13 m) absackte und Kammer samt Schacht mit Schlamm füllte, ging kein Mann verloren, im Gegensatz zu vielen anderen Baustellen jener Zeit. Als man endlich auf den allerhärtesten eiszeitlichen Geschiebe-Lehm stieß, half selbst Dynamit nicht mehr viel. Da besann sich Mr. Arrol eines alten Prinzips – steter Tropfen höhlt den Stein – und erfand den hydraulischen Spaten (Abb. 82).

Schließlich war man auf 89 Fuß (27,13 m) Tiefe angelangt, so daß der Wasserdruck allein genügte, die untere Fuge völlig abzudichten und wieder mit Normaldruck zu arbeiten, d. h. trotz großer Tiefe war es jetzt leichter als weiter oben. Jedenfalls war die Arbeitsleistung angesichts der damaligen Mittel enorm: Bei 24-stündiger Arbeit auf härtestem Grund – mit einer vollen Besatzung von 27 Mann in der Druckkammer und vier hydraulischen Spaten – wurde alle fünf Minuten ein Korb mit Aushub nach oben geschickt, was 3,65 m³ pro Mann je 12-Stundenschicht entsprach. So konnte man zuletzt die Zeit vom Stapellauf bis zur endgültigen Gründung von 96 auf 69 Tage verkürzen.

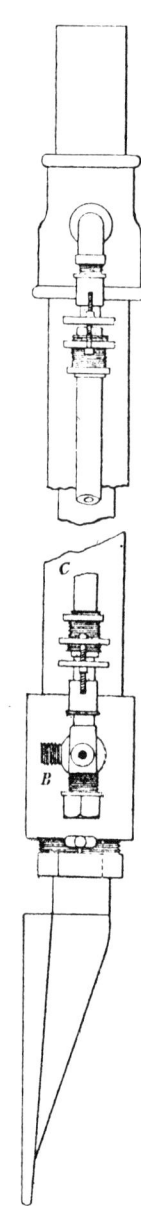

82 >
Der «hydraulische Spaten» – eine der vielen Erfindungen von William Arrol. Dieser Spaten war eigentlich eine Ramme, vorne mit einer Schaufel versehen und einer Wasserdüse mit 1 000 Pfund/Quadrat-Inch (340 kg/cm²) Druck. Der Spaten wurde von zwei Mann so gehalten, daß er mit dem oberen Ende an der Kammerdecke ein Widerlager fand und der dritte Mann das Schaufelende und den Wasserhahn bediente. So konnte man flache Lehmstücke in langen Furchen vom Boden spalten – «a most simple yet a most efficient tool to grapple with the problem». [29, S. 27].

be suspended only from Saturday 6 p.m. to Sunday 12 p.m. In the beginning, there were 8-hour shifts with 8 hours off, later these were 6-hour shifts with 10 hours off, and in the end under the highest air pressure, there were 4-hour shifts with 8 hours off. Finally, air chamber and shafts were totally filled up with concrete, whereafter also the large hollow of the caisson was filled with concrete up to low-water level. Upon these solid foundations, the subsequent erection of the 12 circular masonry piers was mere routine work. The piers started at 18 ft (5.49 m) below the high-water mark and had a diameter of 48 ft (14.63 m) on top, with a tapering profile of 1 : 10.5.

Raising the Viaducts

The final height for all the stone piers supporting the viaducts was to be 130$\frac{1}{2}$ ft (39.78 m) above high-water level. Since the work was to go on without scaffolding, the steel girders were first assembled in an easily accessible working position near the ground, then to be lifted hydraulically and simultaneously with the masonry piers as the masonry work on the piers progressed. Altogether, 15 viaduct spans of 168 ft (51.21 m) had to be raised, each weighing 200 tons and made of regular double girders. From these girders supported only temporarily, staging was suspended around each pier; with the successive lifting of the girders, the staging thus always maintained its appropriate working height for the masonry work to follow suit. This step-by-step process was somewhat reminiscent of Baron von Münchhausen's "pulling oneself by the scalp out of the mud". Furthermore, instead of lifting the whole mass of girders linked to each other at once, as was planned originally, the girders were lifted piece by piece in small increments of 3 inches (76 mm), until a continuous level of 3$\frac{1}{2}$ ft (1.07 m) had been reached throughout, allowing for yet another two courses of masonry to be added. In the end, the viaducts doubled as transport means for wagons drawn to and fro by Shetland ponies, between the various piers and two hoists for material at the far ends. 15 months passed

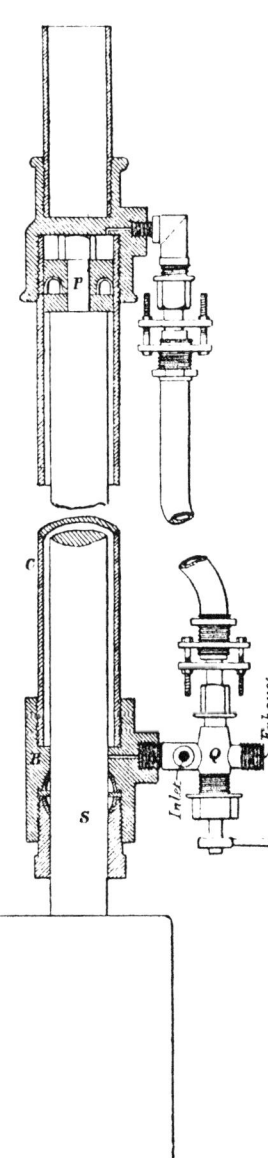

< 82
The "hydraulic spade" – one of William Arrol's many inventions. This spade actually was a ram, equipped with a spade in front and a water jet of 1000 lbs/sq inch (340 kg/cm^2) pressure. The spade would be held by two men in such a way that it would rest with its upper end at the ceiling of the air chamber, while the third man would operate the working end of the spade with the water jet. Thus shallow pieces of mud could be cut off the ground in long furrows – this spade being "a most simple yet a most efficient tool to grapple with the problem" [29, p. 27].

Auf Inchgarvie mußte man den Fels unter der Schneidekante wegsprengen, nachdem sich die Männer durch den Luftschacht in Sicherheit gebracht hatten. Dabei ließ man unter der Schneidekante in regelmäßigen Abständen Felsblöcke als Auflager stehen: Erst zuletzt und bei Ebbe wurden sie weggesprengt, so daß der Senkkasten mit gewaltigem Getöse in die Tiefe fuhr. In diesen beiden Caissons mußte auch der Luftdruck mit jedem Senkschritt erhöht werden, da die Schneidekante auf dem Fels nie dicht aufsaß und ständig Wasser durchkam; allerdings war die Luft so immer frisch, da große Luftmengen nach draußen drangen und ständig wie ein Mantel brodelnder Blasen aufstiegen. Hier ging die Arbeitszeit rund um die Uhr und ruhte nur von Samstag 18.00 Uhr bis Sonntag 24.00 Uhr. Anfangs gab es 8-Stundenschichten mit 8 Freistunden, später 6-Stundenschichten mit 10 Freistunden und zuletzt im höchsten Luftdruck 4-Stundenschichten mit 8 Freistunden. Am Schluß wurden Druckkammer und Schächte völlig zubetoniert und erst dann der große Hohlraum der Caissons bis hinauf zum Ebbe-Niveau mit Beton gefüllt. Auf diesen soliden Fundamenten war dann der Bau der 12 kreisrunden Mauerpfeiler nur noch Routine-Arbeit. Sie begannen 18 Fuß (5,49 m) unter Flut-Niveau und hatten oben einen Durchmesser von 48 Fuß (14,63 m), mit einem Anzug von 1 : 10,5.

Viadukt-Anhebungen

Die Endhöhe aller Viaduktpfeiler sollte 130,5 Fuß (39,62 m) über Hochwasser betragen. Da man ohne Gerüst auskommen wollte, wurden die Stahlträger zuerst in günstiger Arbeitshöhe zusammengebaut und dann im Gleichklang mit dem Hochmauern der Pfeiler systematisch angehoben. Insgesamt waren 15 Viaduktfelder zu je 168 Fuß (51,21 m) Stützweite und 200 Tonnen Gewicht aus normalen Doppel-Gitterträgern anzuheben. Man hing also von den – selbst nur temporär aufgesetzten – Trägern um jeden Pfeiler herum Arbeitsbühnen ab, die mit dem Anheben der Träger immer in der richtigen Arbeitshöhe für das nachgezogene Mauerwerk standen: Im taktweisen Verfahren war das wie Münchhausens «sich selbst am Schopf aus dem Sumpf ziehen». Anstatt die ganze Masse der miteinander verbundenen Träger auf einmal anzuheben, wie zuerst geplant, hob man sie hintereinander in kleinen Schüben von 3 Inch (76 mm) an, bis jedesmal eine durchgehende Höhe von 3,5 Fuß (1,07 m) erreicht war und man so zwei volle Mauerwerkschichten durchmauern konnte. Zuletzt wurden die Viadukte noch als Transportwege für Förderkarren benutzt, welche Shetland-Ponies zwischen den einzel-

from the first to the last lifting and completion of the viaducts, without accident and despite uncommon building practices *(Fig. 83)*.

The Steel

The Forth Bridge signalled the end of the cast-iron age. Considering the novelty of the design, the tremendous size and the extraordinary stresses during construction, only this new material, steel, could be applied here – "and this choice was a happy one". Aside from the iron balancing weights atop the two end-piers of the viaducts, cast-iron was no longer being used here. The steel withstood many test runs, "and one could not have wished for a more uniform and homogeneous material"; the trust by the workmen in its strength under maximum stresses was extraordinary:

> Mr. Westhofen himself tested the quality of the novel material. A piece of steel scrap was placed under an ordinary diamond-headed drill about 1 inch (25,4 mm) in diameter; a hole was drilled about $^3/_4$ inch (19 mm) deep and the result was one single corkscrew shaving, about 1 yard in length, starting from the moment the drill touched the steel and attached yet by the end to the steel plate, out of which it was bored. A truly sensational working material!

The stipulations by the Board of Trade regarding the quality of steel were scanty at the time, coming down to the rule-of-thumb that the maximum working stress should not exceed $^1/_4$ of the ultimate breaking strain of steel. With that, no difference was made between tensile and compressive stresses, nor between live load and dead load, not to talk even about changes in the nature of these stresses themselves. Thus the engineers, in collaboration with the Board of Trade, established new rules in regard to the stresses admissible under varying conditions: for tension members, the steel was to have an ultimate resistance of not under 30 tons/sq. inch (4.650 kg/cm^2); for compression members not under 34 tons/sq. inch (5.270 kg/cm^2). Under conditions of varying stresses due to changing maximum and minimum loads, 20 tons/sq. inch (3.100 kg/cm^2) were to be used; for tensile stresses alternating with compressive ones, the ultimate stress was to be 10 tons/sq. inch (1.550 kg/cm^2), with $^1/_3$ of the ultimate stress to be considered the working stress. By comparison, cast-iron possessed, despite its much greater weight, a maximum tensile strength of only 22 tons/sq inch (3.410 kg/cm^2) – just $^2/_3$ that of steel. The original estimate for the amount of steel required for the cantilever bridge had been 42,000 tons of steel. A

nen Pfeilerbaustellen und zwei Materialwinden an den Enden hin und her bewegten. 15 Monate vergingen von der ersten bis zur letzten Hebung und Vollendung der Viadukte, ohne Unfall, trotz ungewohnter Baumethoden *(Abb. 83)*.

Der Stahl

Die Forth-Brücke besiegelte das Ende der Gußeisenzeit. Angesichts der Neuartigkeit des Entwurfes, der gewaltigen Größe und der außerordentlichen Belastungen während der Bauzeit kam hier nur das neue Material Stahl in Frage – *«and this choice was a happy one»*. Außer den Gegengewichten auf den beiden Viadukt-Endpfeilern wurde kein Gußeisen mehr verwendet. Der Stahl bestand lange Testreihen «und man hätte sich kein gleichmäßigeres, homogeneres Material wünschen können»; das Vertrauen der Arbeiter in seine Stärke unter höchsten Belastungen war außerordentlich:

> Mr. Westhofen stellte selbst die Qualität dieses neuen Materials auf die Probe: Ein Stück Stahlabfall wurde mit einem normalen Diamantbohrkopf von 1 Inch (25,4 mm) Durchmesser etwa 3/4 Inch (19 mm) tief angebohrt und ergab einen einzigen Korkenzieher-Span von Anfang bis Ende des Bohrvorganges, 1 m lang und das Ende immer noch fest mit der Stahlplatte verbunden: Ein wahrhaft sensationelles Baumaterial!

Die *Board-of-Trade*-Bestimmungen über Stahlgüte waren damals dürftig und erschöpften sich in der Faustregel, daß die maximale Belastung 1/4 der äußersten Bruchbelastung nicht überschreiten sollte. Dabei wurde weder nach Zug- bzw. Druckkräften unterschieden, noch nach Nutz- oder Eigenlast, ganz zu schweigen von Veränderungen in der Art dieser Kräfte selbst. Die Ingenieure legten deshalb in Zusammenarbeit mit der IHK neue Regeln fest, für die zulässigen Lasten unter verschiedenen Bedingungen: Für Zug-Glieder mußte der Stahl eine Festigkeit von nicht weniger als 30 Tonnen/Quadrat-Inch (4650 kg/cm^2) haben, für Druck-Glieder nicht unter 34 Tonnen/Quadrat-Inch (5270 kg/cm^2). Bei unterschiedlichen Kräften wegen wechselnder Höchst- und Niedrigstbelastung galten 20 Tonnen/Quadrat-Inch (3100 kg/cm^2), für abwechselnde Zug- und Druckkräfte war die Höchstbelastung 10 Tonnen/Quadrat-Inch (1550 kg/cm^2) – wobei immer 1/3 der Höchstwerte als *«working stress»*, also normale Arbeitslast galt. Zum Vergleich: Gußeisen hatte bei viel höherem Gewicht nur eine Zugfestigkeit von maximal 22 Tonnen/Quadrat-Inch (3410 kg/cm^2), also nur etwa 2/3 der von Stahl. Der ursprüngliche

83
The lifting of the viaducts by hydraulic presses, as the masonry piers were being heightened simultaneously. The staging was suspended from the girders above, and was thus elevated at regular intervals to keep pace with the rest of the work in progress. Photograph of September 17, 1885 [8, p. 139].

83
Die Hebung der Viaduktträger durch hydraulische Pressen und sukzessives Hochmauern der Pfeiler. Die Arbeitsbühnen hingen an den Trägern und hoben sich immer gleichzeitig mit dem Baufortschritt in die Höhe. Aufnahme vom 17. September 1885 [8, S. 139].

large quantity of this material was supplied by the Steel Company of Scotland, and the rest by the "Mssrs. Siemens" of Swansea in South Wales. The steel there was manufactured by the newly-invented Siemens-Martin process. All plates, angles and other sections were readily cut to length and shape as ordered and delivered directly from the steel works. Sub-

Mengenanschlag für die Auslegerbrücke belief sich auf 42 000 Tonnen Stahl. Einen Großteil davon lieferte die *Steel Company of Scotland* und den Rest «die Herren Siemens» aus Swansea, South Wales: Der Stahl wurde nach dem neu-erfundenen Siemens-Martin-Verfahren hergestellt. Alle Platten und Profile kamen nach Länge und Form fertig zugeschnitten von den

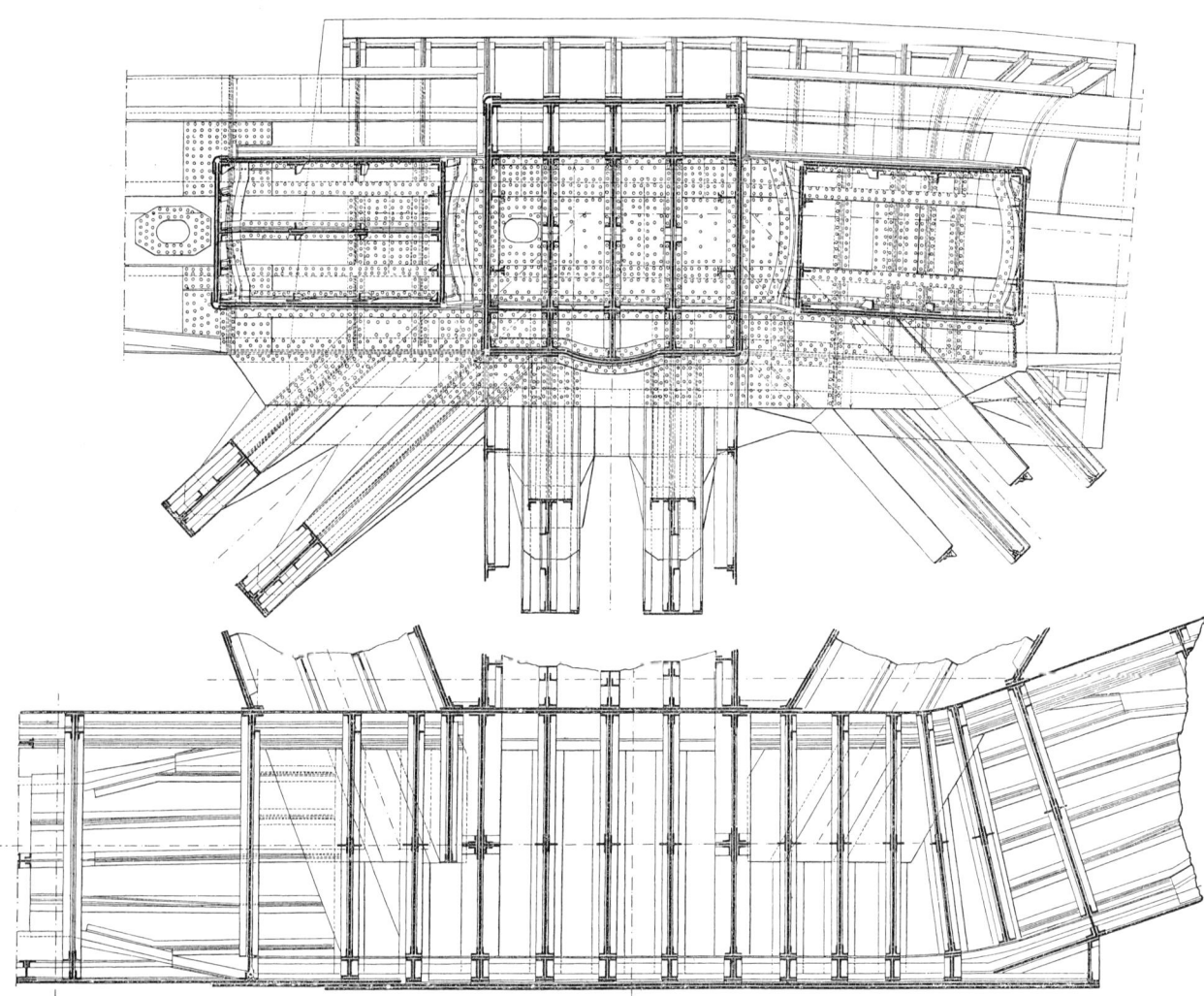

84
Lower juncture of primary members, called "skewback" – seen from top (above) and in section (below). Hardly any other point of the structure equals this impression of controlled force and massive solidity. Five large tubular girders of 12 ft (3.66 m) in diameter join up with five lattice girders in a single point. Mr. Biggart, foreman of the drawing offices, shops and yards, reported in a paper he read to the Institute of Engineers and Shipbuilders in Scotland, the "skewbacks" were the most complicated parts of the entire bridge's superstructure: "... The difficulties are mostly on account of the various angles at which the different parts lie in relation to one another and the skewback" [29, p. 42].

84
Unterer Hauptknotenpunkt oder «skewback» (Schräg-Widerlager) – oben Draufsicht, unten Längsschnitt: Wohl kein anderes Konstruktionsteil gibt einen vergleichbaren Eindruck der kontrollierten Kraft und Massigkeit dieses Bauwerkes: Fünf große Rohrträger von 12 Fuß (3,66 m) Durchmesser vereinen sich mit fünf Gitterträgern in einem Punkte. Mr. Biggart, Leiter des Zeichenbüros und der Werkstätten, schilderte in einem Vortrag beim Verband der Ingenieure und Schiffsbauer Schottlands, daß die «skewbacks» der komplizierteste Teil des gesamten Brücken-Oberbaues waren: «Die Hauptschwierigkeit liegt in den unterschiedlichen Neigungswinkeln, mit denen die verschiedenen Bauteile zueinander und im Verhältnis zum ‹skewback› zusammenlaufen» [29, S. 42].

sequent alterations required some additional 16,000 tons of steel, to which one needed to add 3,200 tons for the viaducts and 4,200 tons just for the rivets!

Bed Plates and "Skewbacks"

The actual superstructure of the bridge commenced once the bed plates were on the masonry piers. These plates measuring 17 × 37 ft (5.18 × 11.28 m) and rivetted together in several layers, had a total thickness of 4 inches (102 mm), weighing 44 tons at each support. Every one of these elaborate steel packages was anchored to the masonry piers by 24-ft-long (7.32 m) bolts. Since there was no bitumen-membrane in those days for moisture insulation, two layers of sailcloth – well soaked in red lead – were laid crosswise between the lower cement bed and the bed plate. The upper bed plates rested in recesses of the lower ones, thus able to slide in a bed of oil. Only one of the four legs of each tower was anchored firmly, while the others were thus movable in a limited way, both laterally and longitudinally. Considering the size of the structure, extensive movements due to tension stresses had to be provided for. In the end, some 58,000 tons of total weight rested on all 12 supports (4 at each tower) of the bridge. (Compare plan in *Fig. 63*).

The "skewbacks" at the base of the three main piers transferred all vertical and lateral loads from the superstructure down to the masonry piers. Five large tubular girders joined here in a single point, – transferring about 3,000 tons of dead load and another 3,000 tons of wind load from each of these "skewbacks" onto the foundations *(Fig. 77)*. The skewbacks, rising box-like from the base, change their configuration on the way up into the various tubular sections; the drawings give only a scant impression of the complicated transition geometry of the major junctions, each joining five tubular members of 12 ft (3.66 m) diameter, with five braced lattice girders *(Fig. 84)*. The skewbacks were pre-assembled at the shore and their joining was so complicated that on certain days only 6 rivets could be closed. The speed of "closing the rivets" had become an indicator for general progress. Later on, in erecting the central piers, a record of closing 800 rivets per day was achieved *(Fig. 85a + b)*!

The Tubular Girders

Baker intended to combine the utmost rigidity with the least amount of weight for all compression members. He compared the combined load of the compression members, resulting from live load, wind load and dead load, with the total weight of an ocean liner. The

Lagerplatten und «Skewbacks»

Der eigentliche Brücken-Oberbau begann mit den Lagerplatten auf den Mauerpfeilern. Diese Platten, 17 × 37 Fuß (5,18 × 11,28 m) groß und mehrschichtig vernietet, ergaben eine Gesamtstärke von 4 Inch (102 mm) mit einem Gewicht von 44 Tonnen je Auflager. Jedes dieser aufwendigen Stahlpakete wurde mit 24 Fuß (7,32 m) langen Schrauben in den Mauerpfeilern verankert. Da es damals noch keine Dachpappe zur Feuchtigkeits-Isolierung gab, legte man zwischen unterem Zementbett und Lagerplatte zwei kreuzweise Lagen Segeltuch ein – «gut in Mennige getränkt». Die oberen Lagerplatten griffen in Aussparungen der unteren ein und konnten dort in einem Ölbett gleiten. Nur einer von vier Füßen je Pfeilerturm war fest verankert, die anderen auf diese Weise in Längs- und Querrichtung begrenzt beweglich: Bei der Größe des Bauwerkes war mit erheblichen Spannungsbewegungen zu rechnen. – Letztlich ruhen auf den insgesamt 12 Fußpunkten der Brücke (4 je Pfeilerturm) immerhin 58 000 Tonnen Gesamtgewicht (vergl. Planzeichnung in *Abb. 63*).

Die unteren Hauptknotenpunkte oder *«skewbacks»* (wörtlich: «Schräg-Widerlager») am Fuße der drei Pfeilertürme übertragen alle Vertikal- und Lateral-Kräfte aus dem Oberbau auf die Mauerpfeiler: Fünf große Rohrträger treffen hier in einem Punkte zusammen – wobei ca. 3000 Tonnen Eigengewicht und weitere 3000 Tonnen Windlast von jedem dieser *«skewbacks»* auf die Fundamente übertragen werden *(Abb. 77)*. Die Knotenpunkte beginnen unten kastenförmig und gehen nach oben in die verschiedenen Rohrquerschnitte über: Die Zeichnungen geben nur einen unvollkommenen Eindruck von der komplizierten Durchdringungs-Geometrie dieser Knoten, mit fünf Rohrträgern von je 12 Fuß (3,66 m) Durchmesser und fünf Aussteifungs-Gitterträgern *(Abb. 84)*. Die *«skewbacks»* wurden am Ufer vormontiert und ihr Zusammenbau war so kompliziert, daß man an manchen Tagen nur um sechs Nietlöcher vorankam: Die Schnelligkeit *«of closing the rivets»* (Nietlöcher zu schließen) wurde zum Maßstab des Baufortschrittes: Später, beim Bau der Mitteltürme, erreichte man einen Rekord von 800 Nietverbindungen pro Tag *(Abb. 85a und b)!*

PORTABLE HYDRAULIC RIVETTERS.

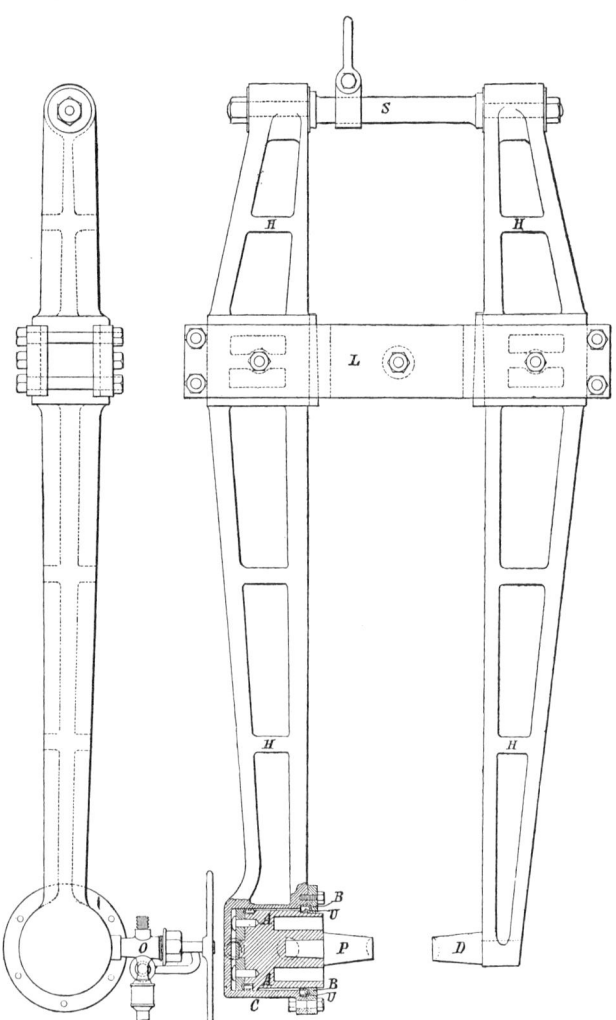

<<
Rivetting Work

85 a
Hydraulic rivetting machine. William Arrol had previously invented such time-saving devices for the construction of his bridges across the Clyde. During assembly, this movable device could reach every point of the tubular girders. With a pressure of 40 tons, "rivets were thus pressed and closed as if they were made of clay" [29, p. 44].

<<
Nietverbindungen

85 a
Hydraulische Nietmaschine. William Arrol hatte solche arbeitssparenden Geräte bereits für den Bau seiner Clyde-Brücken entwickelt. Bei der Montage konnte man mit dem transportablen Gerät jeden Punkt der Rohrträger erreichen. Mit einem Druck von 40 Tonnen wurden so «Nieten gepreßt und verschweißt, als wären sie aus Ton» [29, S. 44].

85 b
Rivetting cage. In order to rivet the large tubular girders in place, special rivetting cages were constructed. They were manned by 6 men – 3 working outside at the surface of the tubes, 3 working inside the tubes – who in turn where helped by 2 youths pre-heating the rivets and bringing them up. The cages travelling like large cylinders along the tubes were equipped with a rotating beam for supporting the movable rivetting machine. The rivets were pre-heated to yellow glow in small stoves and closed by hydraulic pressure from inside and out. As soon as a 18 ft (5.49 m) tubular section had been joined, the cage was moved ahead to the next section; thus, an average of 600 to 700 rivets could be closed per day [29, plate X].

85 b
Vernietungs-Käfig. Für die Vernietung der großen Rohrträger vor Ort baute man eigene «rivetting cages» (Vernietungs-Käfige): Sie hatten 6 Mann Besatzung – 3 draußen an der Rohroberfläche, 3 im Rohrinneren, und dazu noch 2 Jungen für das Vorglühen und Heranholen der Nieten. Die Käfige schoben sich zylinderförmig über die Rohre und hatten einen Schwenkarm mit beweglicher Nietmaschine. Die Nieten wurden in kleinen Öfen zur Gelbglut erhitzt und mit hydraulischem Druck von außen und innen vernietet. Sobald ein 18 Fuß (5,49 m) langes Rohrstück vernietet war, schob man den Käfig zum nächsten Teilstück voran und konnte so durchschnittlich 600–700 Nieten pro Tag verarbeiten [29, Tafel X].

tubular members were assembled totally out of steel plates of ³/₈ (9.5 mm) and 1¹/₄ inch (31.8 mm). They first had to be heated to red glow in stoves at the shore. Then they had to be formed by hydraulic presses and pre-drilled by special multiple drilling machines in order to obtain their tubular form. The large drilling machine was more like a "drilling wagon" that travelled on its own tracks along a "drilling road" on the shore; it was equipped with 10 drilling spindles – all driven simultaneously by one steam engine *(Fig. 86)*.
The large tubular members, whether vertical support or diagonal strut, all shared the same method of as-

Die Rohrträger

Baker wollte für alle Druck-Glieder größte Festigkeit mit größtmöglicher Leichtigkeit verbinden: Er verglich die kombinierte Belastung der Druckrohre durch Nutzlast, Winddruck und Eigengewicht mit dem Gesamtgewicht eines Ozeandampers. Die Rohrträger wurden zur Gänze aus 3/8 Inch (9,5 mm) und 1 1/4 Inch (31,8 mm) starken Stahlplatten zusammengebaut. Dazu mußten sie erst in Öfen am Ufer rotglühend erhitzt, in hydraulischen Pressen gekrümmt und mit speziellen Mehrfach-Bohrmaschinen zur Rohrform

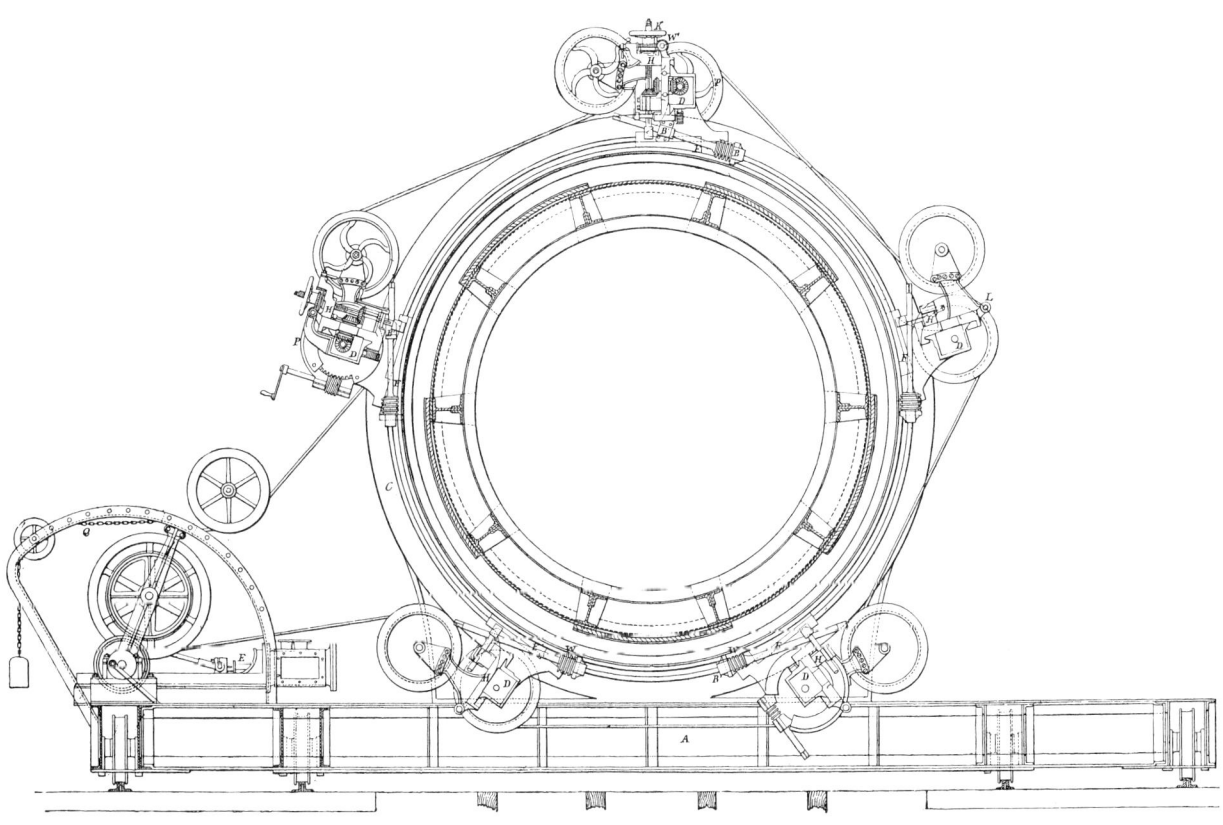

86
The large drilling machine for pre-assembling the tubular girders. The machine enclosed the tubes like a cylinder while drilling, travelling on tracks along the shore. In this drawing, the tubular section is shown in the center (compare also *Fig. 87*). All tubular sections were first joined provisionally at the shore and held in place by bolts for their correct fit, before they were finally rivetted together on the bridge in their exact location. Which prompted *The Scotsman* in 1886 to the following observation: "It thus comes about that the whole of the bridge will have been twice built ere it is finally erected" [29, p. 36].

86
Die große Bohrmaschine für die Vor-Montage der Rohrträger: Das Gerät umgriff die Rohre zylindrisch beim Bohren und fuhr auf eigenem Schienenweg am Ufer entlang: In der Zeichnung erscheint der Rohrquerschnitt in der Mitte (vergl. auch *Abb. 87*). Alle Rohrteile wurden probeweise auf Paßform aneinandergefügt und zunächst mit Bolzen gehalten, damit man sie später auf der Brücke in genauer Lage und endgültig vernieten konnte. Was den «*Scotsman*» 1886 zu folgender Betrachtung veranlaßt: «Daraus ergibt sich, daß die ganze Brücke praktisch zweimal gebaut sein wird, bevor sie endlich fertig da steht» [29, S. 36].

sembly, despite differing diameters. The steel plates were an even 16 ft (4.88 m) long, every tubular circumference being composed of 10 overlapping plates with staggered joints. The overlappings were stiffened inside longitudinally by I-sections of steel, which in turn were joined laterally every 8 ft (2.44 m) by an inner ring. *(Fig. 87)* In order to keep to an exactly circular cross-section, heavy dividing diaphragms were placed inside at larger intervals, with a man hole left in the centre to crawl through. How necessary this was became evident as the largest horizontal members flattened by almost 3 inches (76 mm) due to their own weight, before they could finally be rivetted up. In this manner extremely strong and rigid tubular members were obtained. The diagonal members in the central bay had a flattened tubular shape in order to facilitate their connections and intersections.

vorgebohrt werden. Das große Bohrgerät war eher ein «Bohrwagen», der auf eigenen Schienen am Ufer seine «Bohrstraße» entlangfuhr und zehn Bohrspindeln besaß – alle gleichzeitig von einer Dampfmaschine angetrieben *(Abb. 86)*.

Die großen Rohrträger, gleich ob Vertikal-Stütze oder Diagonal-Strebe, hatten bei unterschiedlichem Durchmesser alle denselben Aufbau: Die Stahlplatten waren einheitlich 16 Fuß (4,88 m) lang, jeder Rohrumfang bestand aus zehn überlappten Platten mit versetzten Stößen. Die Überlappungen waren in der Längsrichtung innen mit I-Profilen verstärkt und diese wiederum alle 8 Fuß (2,44 m) mit einem inneren Ring verbunden *(Abb. 87)*. In größeren Abständen legte man – zur Wahrung des genauen Kreisquerschnittes – schwere Scheidewände mit nur einem Kriechloch in der Mitte ein; wie nötig dies war, zeigten gerade die

87
The yard at South Queensferry showing the tubular sections, awaiting preliminary assembly to the large tubular girders. Photograph of August 5, 1889 [8, p. 63].

87
Das Werkgelände bei South Queensferry mit den Teilstücken für die Vor-Montage der großen Rohrträger. Aufnahme vom 5. August 1889 [8, S. 63].

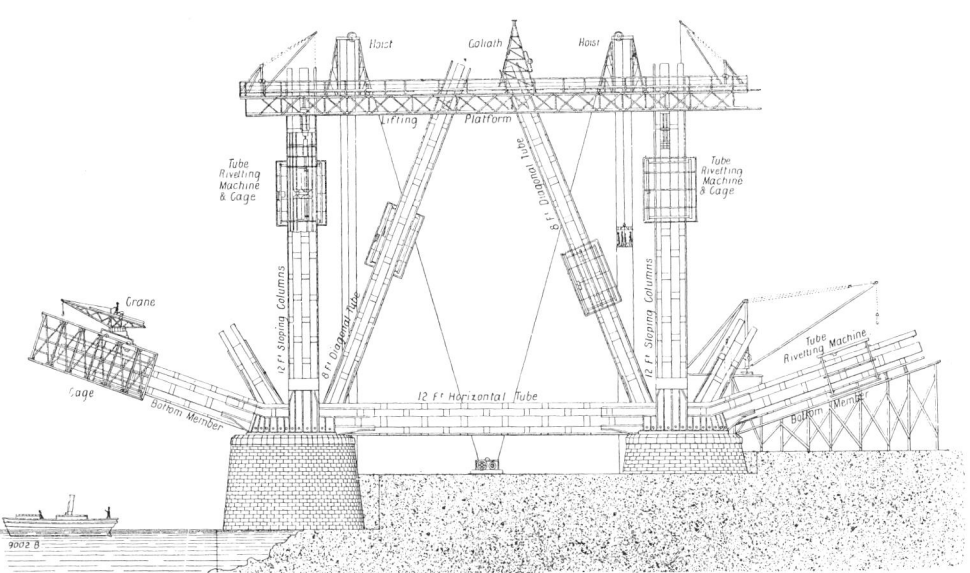
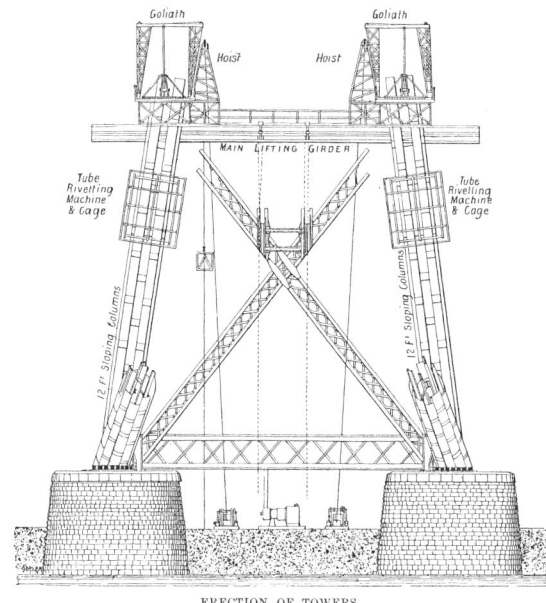

88
Assembly of a central tower with working platforms; to either side of the tower, work on the first bay of the cantilever is being started. Side view to the left, cross section to the right [29, p. 46].

88
Montage eines Pfeilerturmes mit Hebebühne; auf beiden Seiten Beginn des ersten Ausleger-Feldes. Links Seitenansicht, rechts Querschnitt [29, S. 46]

Erection of the Towers

The superstructure was begun by erecting steam derrick cranes on special staging, 30 to 40 ft (9.14–12.10 m) above the piers. Bit by bit, the skewbacks, parts of the diagonal struts, the vertical columns and the compression members of the first bay of the cantilevers could thus be put together. The assembly of these parts up into the air was difficult, as the members were actually inclined in two angles, set not only to an inclination towards the point of intersection, but also to follow the uniform taper of the towers, and had therefore a strong tendency to lean inward to the center of the towers. When the vertical columns and struts had reached a height of 50 ft (15.24 m), special lifting platforms were built (Fig. 88). For this, 190-ft to 350-ft-long (57.91–106.68 m) lattice girders were used; later on, they were to become the large tensile members for the first bay of the cantilevers, whereas they were now being "borrowed" so to speak, for this temporary task. By using the parts already assembled, together with the temporary ones to be fixed later, the structure thus reared itself upward.

For transporting men and material, the "Goliath" was used, a hydraulic "traveller" running on tracks across the entire length of the platforms (Fig. 88). The water pressure of the lifting pumps of 35 cwt/sq.inch (11 470 kg/cm²) often was too much for the leather washers in the hydraulic rams; this and frozen pipes caused much delay, until the men had gotten more ex-

stärksten Horizontalrohre durch ihre Abflachung von fast 3 Inch (76 mm), durch Eigengewicht, bevor sie fertig vernietet waren. So erhielt man außerordentlich stabile und biegesteife Rohrträger. Die Diagonalstreben im Mittelfeld hatten eine abgeflachte Rohrform, was die Anschlüsse und Kreuzungspunkte erleichterte.

Montage der Pfeilertürme

Der Oberbau begann mit der Errichtung von Dampfkränen auf eigenen Arbeitsbühnen, 30 – 40 Fuß (9,14 – 12,19 m) über den Auflagern. Man konnte so nacheinander die «skewbacks», Teile der Diagonalstreben, der Vertikalrohre und der Druckstreben des 1. Auslegerfeldes zusammenbauen. Die Montage der Teile in der Luft war schwierig, da sie wegen des starken Neigungswinkels der Streben und des gleichzeitigen Schräganzuges der Pfeilertürme doppelt geneigt waren und sich nach innen zur Pfeilermitte lehnten. Als die Vertikalrohre und Streben 50 Fuß (15,24 m) Höhe erreicht hatten, baute man spezielle Hebebühnen (Abb. 88). Dazu wurden vier 190 – 350 Fuß (57,91 – 106,68 m) lange Gitterträger verwendet, die später zu Teilen der großen Zugstreben des 1. Auslegerfeldes werden sollten und nun sozusagen «ausgeliehen» wurden. Unter Verwendung der bereits gebauten Teile und Vorwegnahme der zukünftigen baute man sich so selbst in die Höhe.

Für den Material- und Personentransport diente der

perienced. The first lifting procedure on Inchgarvie in January of 1887 took a full 18 days, whereas the last one only took 5 hours.

All work had to be carried on within a structurally incomplete system; the diagonal struts for example were hanging for a long time one-sidedly and unconnected in the air, and had to be pushed apart again hydraulically into their correct position, following every second lifting (every 30–40 ft) (9.14–12.19 m). Due to their two-fold inclination, they had deflected from their true alignment under their great weight, downward and sideways. By using a triangulation along the center line of the bridge, the vertical columns were pulled time and again into position; with their heavy load they were comparatively unconnected and vulnerable to lateral winds, while the rivetting of the columns had stiffened them evermore against such "corrections". Another problem was overcome again half way up the towers, when at a level of 200 ft (60.96 m) the crossing of the large diagonal struts had been reached. On Inchgarvie, 80 tons of steel came together here on either side, as 6 directions of members joined at one junction. Thus a very strong framework was obtained in order to close at last the lower half of the towers. In having all the vertical columns fixed in their final positions, a new start upwards could be made – just as the previous one at the main piers below.

At last, the platforms had reached their highest point, so they could be combined forming a single platform (Fig. 89). For the extensive work at the upper junctions, they were enclosed like houses with wooden walls, windows and electric illumination, so work could go on there day and night, protected from the weather. Particularly for the ends of the diagonal struts changing from tubes to squares, precise templets and fittings had to be made up. An impression of the extent of this work may be gained by the size of the superimposed "sky houses" – four to each tower – each with three stories and perched 360 ft (109.73 m) above the water level. (Fig. 107) The top junctions take up two tubular columns and five lattice girders. The transmission of tresses is effected by very large web plates in direct continuation of the webs of the top members, extending by so-called "horns" into the various other sections (Fig. 90).

The Cantilevers

The first bay of the cantilevers to either side of the towers was the largest and most difficult to erect, because of the length of members. Once the lifting platforms had arrived at the point of intersection of the towers' diagonal struts, the bottom members were

«Goliath», ein Brückenkran, der als hydraulischer «traveller» auf Schienen über die ganze Länge der Bühnen fahren konnte (Abb. 88). Der Wasserdruck der Hebepumpen betrug 11 470 kg/cm^2 und war derart groß, daß häufig die Lederdichtungen der Hydraulik nachgaben; dazu kamen gefrorene Leitungen und es ging viel Zeit verloren, bis die Männer genügend Erfahrung damit hatten: Die erste Hebung auf Inchgarvie im Januar 1887 dauerte volle 18 Tage, die letzte nur noch 5 Stunden.

Da man ständig innerhalb eines unfertigen statischen Systems arbeitete – weil z. B. die Diagonalstreben lange Zeit einseitig und ohne Halt in der Luft hingen, mußte man sie nach jeder zweiten Hebung (alle 30 – 40 Fuß [9,14 – 12,19 m]) wieder in der richtigen Lage hydraulisch auseinanderspreizen; sie waren wegen der zweifachen Neigung unter ihrem großen Gewicht immer wieder von den vorgeschriebenen Achsen seitlich und nach unten abgewichen. Mit einer Hauptvermessungslinie in der Brückenachse zog man auch die Vertikalrohre immer wieder «auf Kurs», denn sie waren mit ihrer großen Auflast relativ freistehend und anfällig gegen Seitenwind, dabei aber auch wegen fortschreitender Vernietung immer steifer gegen solche «Korrekturen» geworden. Schwierig wurde es wieder in halber Höhe, als man 200 Fuß (60,96 m) hoch am Kreuzungspunkt der großen Diagonalrohre angelangt war. Auf Inchgarvie kamen hier an jeder Seite 80 Tonnen Stahl zusammen, wo sich 6 Trägerachsen in einem Raumknoten trafen. Man erhielt so aber ein ausgesprochen stabiles Gefüge und konnte endlich die untere Hälfte der Pfeilertürme statisch abschließen. Damit waren auch alle Vertikalrohre in ihrer endgültigen Lage fixiert und ein neuer Anfang gemacht – wie vorher mit den unteren Auflagern.

Zuletzt hatten die Hebebühnen den höchsten Punkt erreicht und wurden zu einer einzigen Plattform vereint (Abb. 89). Wegen der umfangreichen Arbeiten an den oberen Knotenpunkten wurden sie hausartig mit Holzwänden, Fenstern und Arbeitslampen umkleidet, so daß man dort wettergeschützt Tag und Nacht arbeiten konnte. Besonders für die Enden der Diagonalrohre mit ihrem Übergang von Rundung auf Rechteck mußten genaue Schablonen und Paßstücke gefertigt werden. Einen Eindruck vom Umfang dieser Arbeiten vermittelt die Größe der aufgesetzten «Lufthäuser» – vier Stück je Pfeilerturm – mit je drei Stockwerken und 360 Fuß (109,73 m) über dem Wasserspiegel (Abb. 107). Die oberen Knotenpunkte vereinen in sich zwei Rohrstützen mit fünf Gitterträgern. Die Kraftübertragung erfolgt durch große Stegplatten in Fortsetzung der Obergurt-Stege, die mit sogenannten «Hörnern» in die verschiedenen Querschnitte eingreifen (Abb. 90).

89
View from northwest of the central tower on Fife. At the bottom, near the water, the initial assembly of the cantilevers' bottom members [29, plate XII].

89
Nordwest-Ansicht des Pfeilerturmes auf Fife. Unten Beginn des Vorbaues der Ausleger-Untergurte [29, Tafel XII].

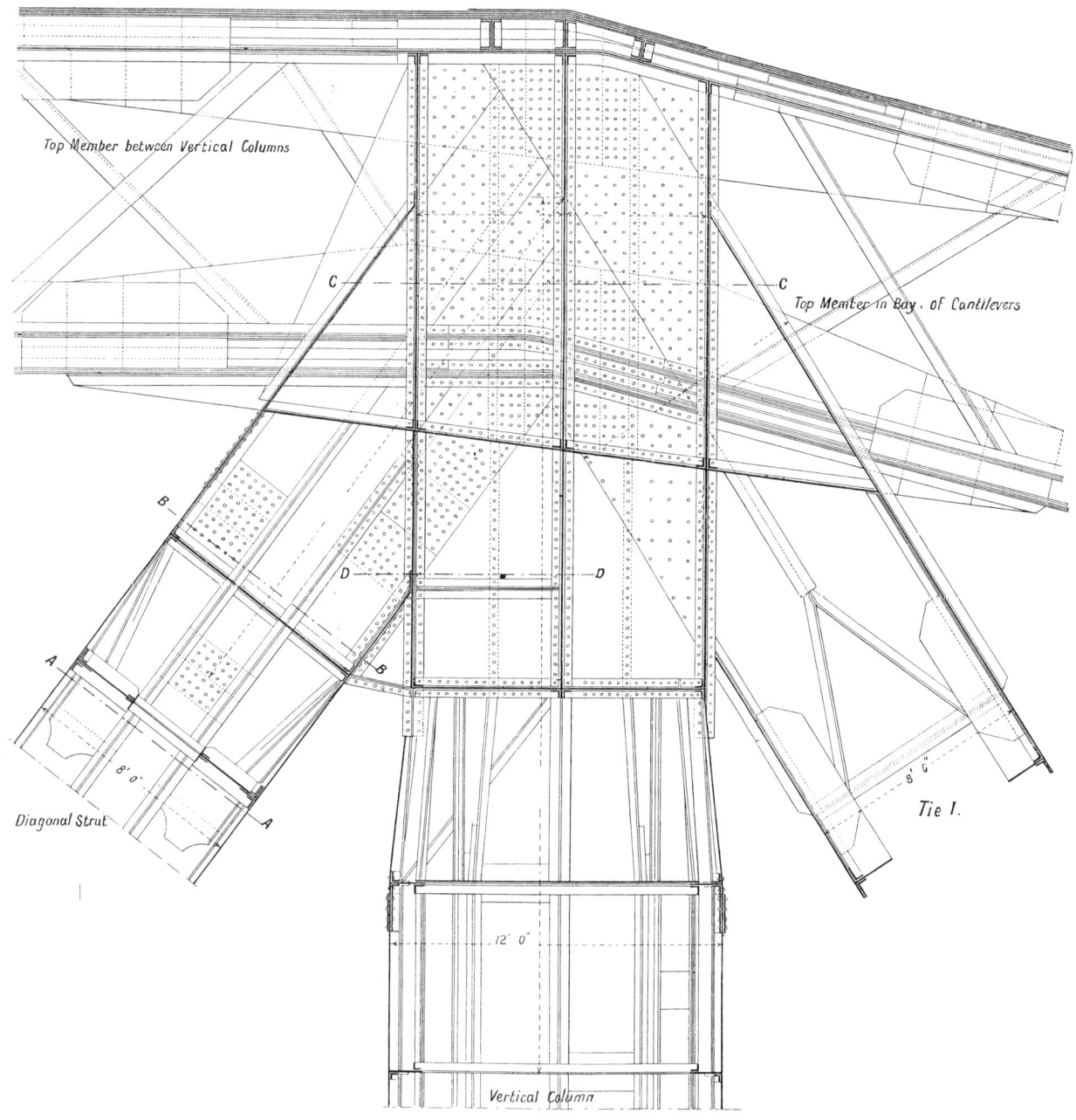

90
Configuration of the top main junction [29, p. 29].

90
Ausbildung des oberen Hauptknotenpunktes [29, S. 29].

built out simultaneously in both directions, using this time a rectangular sliding cage, equipped with a hydraulic crane on top *(Fig. 88).* Inside the forward part of the cage, the platers were building up the tube, while the hydraulic crane swung round all material from behind to the front. For a length of 64 ft (19.51 m) the bottom member extending unsupported could thus be

Die Ausleger

Das erste Ausleger-Feld beidseits der Mitteltürme war das größte und wegen der langen Bauteile schwierig zu erstellen. Sobald die Hebebühnen das Diagonalkreuz der Pfeilertürme erreicht hatten, trieb man auf beiden Seiten den Untergurt-Rohrträger voran, dies-

91
This period engraving of April 15, 1887, illustrates the first assembly stages of the freely extended bottom members and diagonal struts. The crane can be seen atop the rectangular sliding cage [28].

91
Die zeitgenössische Darstellung vom 15. April 1887 zeigt die Anfänge der Arbeiten an den freitragenden Untergurten und Diagonalstreben. Auf dem rechteckigen Gleitkäfig ist der Kran zu sehen [28].

rivetted up *(Fig. 91)*. When the cage had gone beyond the center of the cantilevers' first bay, a temporary plate-tie was inserted, running from the cantilever up to the vertical columns and back down again at the other end to the opposite cantilever; thus a state of balance between the bottom members was achieved *(Fig. 92)*. Also this tie was "borrowed" from the main webs of the future top members of the second bay.

Once the towers had been completed, the lifting platforms were no longer required; instead, 3-ton cranes were put atop each of the four vertical columns. The viaduct girders were also extended in a similar way. They were strong enough to carry themselves, overhanging for a length of 100 ft (30.48 m) into both cantilevers, and even carrying a 3-ton crane at the far end. As soon as the diagonal ties from above were intersecting with the diagonal struts from below, new fixed points were thus secured. By now the "Jubilee Crane" – as it was popularly called, coinciding with Queen Victoria's Jubilee of 50 years' reign in 1887 – could be erected on the top member. The towers had now reached their full height, precisely at this historic occasion, and there were celebrations galore. This ingenious crane *(Fig. 93)* travelled on slides down the top member's incline of 1:4. With its jib reaching 34 ft (10.36 m) it could slew around by 220° and it could be worked independently from the ground by a pair of reversible steam engines and its own boiler. Suspended from below it carried along its own working platform

mal mithilfe eines rechteckigen Gleitkäfigs, auf dem ein hydraulischer Kran stand *(Abb. 88)*. Im Vorderteil des Käfigs bauten die Plattner den Rohrträger zusammen, während der Kran alles Material von hinten nach vorne schwenkte. So konnte man den Untergurt im freien Vorbau auf einer Länge von 64 Fuß (19,51 m) fertig vernieten *(Abb. 91)*. Als der Käfig über die Mitte des ersten Ausleger-Feldes hinausgelangt war, baute man ein temporäres Zugband ein, das vom Ausleger zum Mittelpfeiler und wieder hinunter zum Gegenausleger lief und so für Gleichgewicht der Untergurte sorgte *(Abb. 92)*. Auch dieses Zugband war «geborgt» aus den Hauptstegen für den späteren Obergurt des zweiten Feldes.

Mit Vollendung der Pfeilertürme konnte man auf Arbeitsbühnen verzichten und dafür jedem der vier Pfeiler-Rohre einen 3-Tonnen-Kran aufsetzen. So wurden auch die Viaduktträger vorgebaut; sie waren stark genug, 100 Fuß (30,48 m) weit in die beiden Ausleger vorzukragen und am fernen Ende noch einen 3-Tonnen-Kran aufzunehmen. Sobald die Zugdiagonalen von oben die Druckdiagonalen von unten kreuzten, hatte man einen neuen Festpunkt gewonnen und konnte auf dem Obergurt den «*Jubilee* Kran» errichten – so genannt, weil er mit dem 50-jährigen Regierungs-Jubiläum von Queen Victoria, 1887, zusammenfiel: Die Pfeilertürme hatten genau zu diesem historischen Datum ihre volle Höhe erreicht und es gab Feiern allerorten. Der sinnreiche Kran *(Abb. 93)* fuhr mit Gleitlagern auf der Obergurt-Neigung von 1:4 entlang. Er konnte mit 34 Fuß (10,36 m) Ausladung um 220 Grad herumschwenken und war durch zwei umkehrbare Dampfmaschinen mit eigenem Boiler vom Boden unabhängig. Nach unten abgehängt brachte er auch seine eigene Arbeitsbühne mit sich, die immer mitfuhr. Für den Obergurt hievte man die 24 Fuß (7,32 m) langen Profile vom Viadukt nach oben und baute sie auf besagter Bühne zusammen, worauf der Kran 24 Fuß weiterfuhr, usw. Er wog 64 Tonnen und stand schließlich 80 Fuß (24,38 m) von den Pfeilerköpfen entfernt, was eine erhebliche Belastung für die beiden frei auskragenden Obergurte bedeutete und –

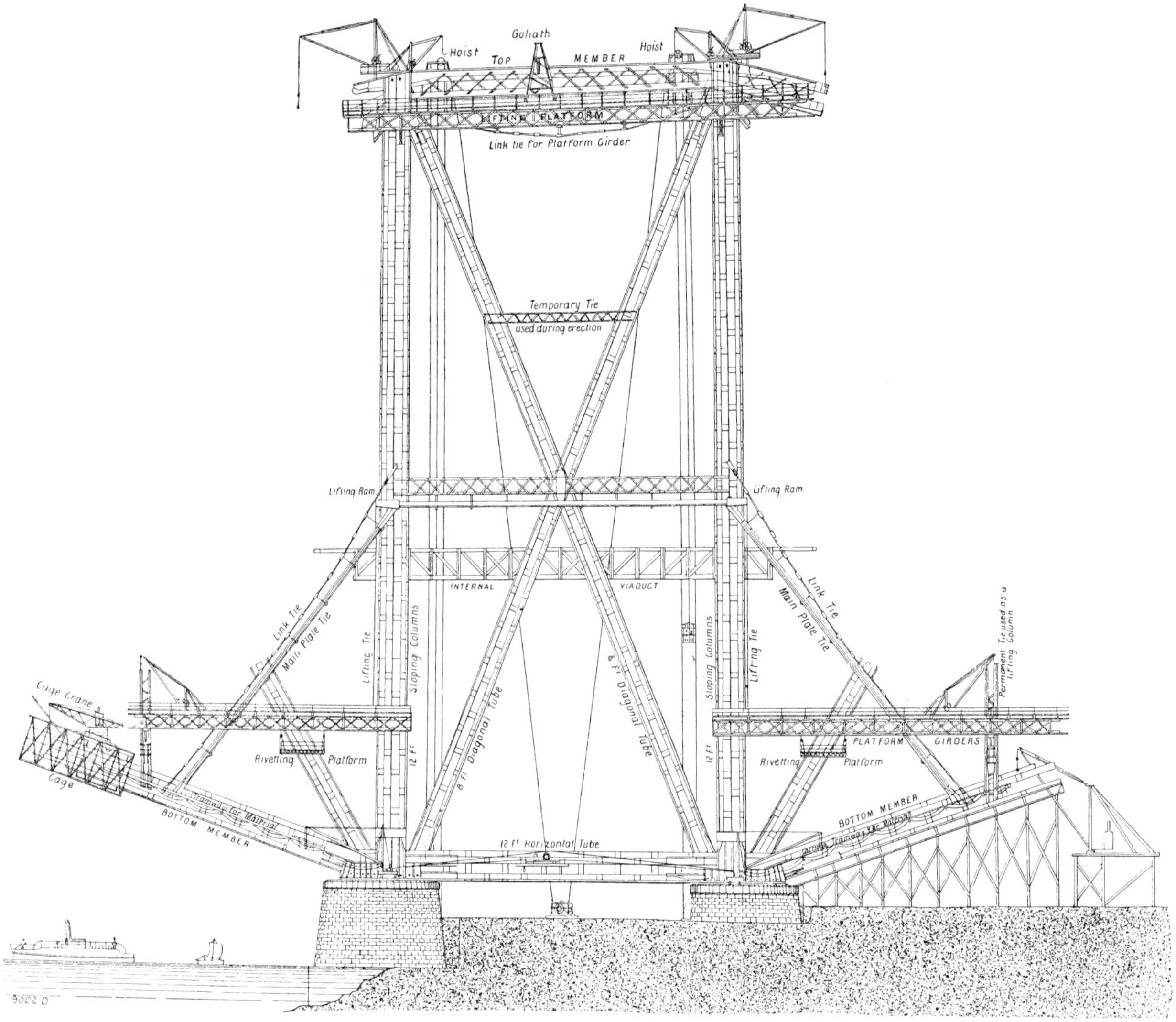

92
The central tower almost completed. Across the intersection of the diagonal struts, the plate-tie – "borrowed" from parts of the future top members – can be seen supporting temporarily the freely extended bottom members of the cantilevers to either side [29, p. 52].

92
Mittelturm fast vollendet. Am Kreuzungspunkt der Diagonalstreben dient das Zugband – «geborgt» aus Teilen des späteren Obergurtes – zur vorläufigen Entlastung der freitragenden Untergurte [29, S. 52].

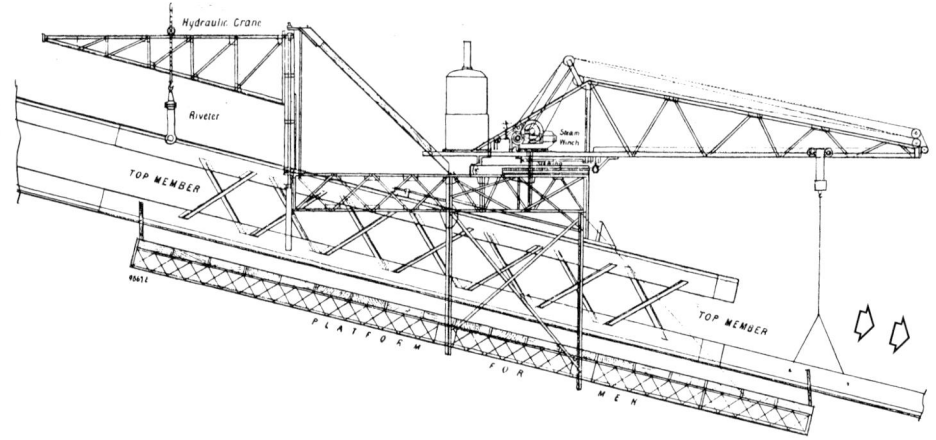

93 >
This curious couple consisting of the "Jubilee Crane" (to the right) and the rearward rivetting crane (to the left) travels downward to the right along the top member, while assembling continuously its very own slide-way, piece by piece. The working platform being fixed to the cranes' platform always travels along. The platform enables two work gangs – platers and rivetters – to work in sequence and simultaneously [29, p. 54].

93 >
Das kuriose Doppelgespann aus «Jubilee-Kran» (rechts) und rückwärtigem Vernietungskran (links) fährt nach rechts abwärts auf dem Obergurt entlang und montiert sich dabei selbst seinen fortlaufenden Gleitweg aus den Teilstücken. Die Arbeitsbühne ist mit der Kran-Plattform fest verbunden und fährt immer mit. Auf der Bühne können zwei Trupps – Plattner und Vernieter – im Taktverfahren gleichzeitig arbeiten [29, S. 54].

94
This photograph of the Inchgarvie tower taken April 6, 1888, illustrates very well the freely extending assembly of the two bottom members, by the use of sliding cages. In this advanced phase, the components are being lifted directly out of the steam barges in the water [8, p. 155].

94
Diese Aufnahme des Inchgarvie-Pfeilerturmes vom 6. April 1888 zeigt deutlich den freien Vorbau der Untergurt-Rohrträger mit Hilfe von Gleitkäfigen. In diesem fortgeschrittenen Stadium werden die Werkstücke direkt aus den Dampfkähnen im Wasser gehievt [8, S. 155].

95
View of the Forth Bridge from the northwest on March 22, 1889. In this seventh and final year of construction, at last the parts are beginning to form a whole and the final shape can almost be imagined [8, p. 97].

95
Nordwest-Ansicht der Forth-Brücke am 22. März 1889: In diesem siebten und letzten Baujahr beginnen sich endlich die Teile zum Ganzen zu schließen – die fertige Form der Brücke wird zumindest vorstellbar [8, S. 97].

readily travelling along. In building the top members, the 24-ft-long (7.32 m) sections were lifted from the viaduct below and joined at once on the said platform, whereupon the crane moved on for another 24 ft (7.32 m), etc. It weighed 64 tons and finally came to rest at a distance of 80 ft (24.38 m) from the top junctions, resulting in a tremendous strain upon the two top members still unsupported; it was quite a threatening sight – causing much anxiety in this first and largest bay. As the top members were never designed for such a load, the side bracings were stiffened for the duration of the work.

Press Commentaries

As the cantilevers could now be seen growing from afar, also public interest rose by the day. The press and any conceivable periodicals served their readers with regular reports. *The Scotsman* described that strange progress by which the bridge acted as a scaffolding for its very own construction – quite peculiar at times

höchst bedrohlich anzusehen – gerade in diesem ersten und größten Auslegerfeld einige Sorge bereitete. Da die Obergurte nie für diese Last geplant waren, verstärkte man die seitlichen Gitterstreben für die Dauer der Arbeiten.

Presse-Kommentare

Mit dem weithin sichtbaren Vorbau der Ausleger wuchs auch das Interesse der Öffentlichkeit von Tag zu Tag. Die Presse und alle möglichen Wochenblätter bedienten ihre Leser mit regelmäßigen Lageberichten. *The Scotsman* erläuterte jenen sonderbaren Vorgang, bei dem die Brücke selbst als ihr eigenes Baugerüst diente – sehr ungewöhnlich in Tagen, wo zusätzliche Hilfsgerüste aus Holz die Regel waren – im März 1888:

«Die Arbeiter stehen heute praktisch auf ihrer Arbeit von gestern. Sobald ein neuer Ring von Stahlplatten an die Rohre gefügt oder ein weiterer Träger-Abschnitt mit den Obergurten ver-

when temporary wooden scaffolding was the rule of the day – in March of 1888:

> "The workers of today are practically standing upon their labours of yesterday. As soon as a fresh round of steel plates is added to the tubes, or an additional girder section rivetted to the top arms, the platforms with their freight of men and cranes and other mechanical appliances are slid out correspondingly and a new piece of work is begun which again, when completed, will give the necessary standing support for a further extension.... Every piece of work done becomes the basis of another advance, and the Forth Bridge men labour much in the same way as the Esquimaux who ascends the ice-cliff by cutting steps, one after the other in its face."

The further the cantilevers stretched out across the water, the clearer the still-open gaps could be perceived between them – already, the final shape of the bridge could be imagined. Crowds of curious people watched in awe, as the large girder arms were being built outward by tiny men and machines suspended between sky and sea. Sir Robert Purvis exclaimed in his Memoir to Sir William Arrol *(Fig. 95)* [14, p. 16]:

> "More and more was the amazement as, week by week, the columns were perceived to be throwing out enormous, far-reaching growths on either side. Each of these was ever increasing in weight and altering in shape but ever in perfect balance."

The Builder described to its readers in May, 1888, how the steel cantilevers had to be extended out in either direction at the same pace, in order to hold the whole thing in equilibrium:

> "The whole work, as we saw it from the engineers' steam launch, has a most remarkable appearance at present. The three piers (cantilever towers) suggesting the idea of three enormous ships with the tubes projecting like gigantic bowsprits over the water, and everything at present hanging out into the air unsupported."

Corrective Liftings – "A Trifle Beyond"

What appeared to be so light and airy from afar, was a cumbersome iron mass from nearby. The long top members had deflected about 10 inches (25.4 cm) and had to be lifted again by hydraulic rams, beyond the point of intersection intended; only in this manner could one allow for the probable compression to follow. But the massive tubes of the bottom members had also deflected due to their own weight, the rivetting cages and various machinery. Now they could no

nietet ist, werden die Plattformen mit ihrer Last von Menschen, Kränen und anderem Gerät im Gleichschritt nach vorne geschoben und ein neuer Arbeitsabschnitt begonnen, der bei Fertigstellung wieder die notwendige Standfläche für die folgende Verlängerung bietet... Jedes fertige Stück Arbeit wird zur Basis eines erneuten Vorbaues – und die Männer der Forth-Brücke arbeiten ganz ähnlich einem *Eskimo,* wie er sich eine Eisklippe hinaufarbeitet und dabei eine Stufe nach der anderen in die Oberfläche schlägt.»

Je weiter die Ausleger über das Wasser hinauswuchsen, desto meßbarer wurde auch die noch freie Strecke zwischen den Pfeilertürmen – man konnte sich bereits die fertige Form der Brücke vorstellen. Scharen von Neugierigen verfolgten staunend, wie die großen Trägerarme von winzigen Menschen und Maschinen vorgebaut wurden, die frei zwischen Himmel und Wasser schwebten. Sir Robert Purvis rief in seiner Denkschrift für Sir William Arrol aus *(Abb. 95)* [14, S. 16]:

> «Mehr und mehr stieg die Verwunderung, als man Woche für Woche sehen konnte, wie die Säulen enorme, weitreichende Auswüchse an jeder Seite von sich reckten. Jeder von ihnen gewann ständig an Gewicht und veränderte seine Form – aber immer in vollkommenem Gleichgewicht.»

The Builder beschreibt seinen Lesern im Mai 1888, wie die Stahl-Kragarme nach beiden Richtungen im gleichen Rhythmus gebaut werden müssen, um das Ganze im Gleichgewicht zu halten:

> «Das ganze Werk hat, wenn man es von der Dampfbarkasse der Ingenieure aus betrachtet, derzeit ein höchst bemerkenswertes Aussehen. Die drei Pfeiler (Kragarm-Türme) suggerieren das Bild von drei ungeheuren Schiffen, die mit ihren Rohren wie gigantische Bugspriete über das Wasser ragen, und alles hängt derzeit ohne Unterstützung frei in die Luft hinaus.»

Korrektive Hebungen – «a trifle beyond»

Was aus der Ferne so leicht und schwerelos erschien, war aus der Nähe eine schwerfällige Eisenmasse: Die langen Obergurte waren um ca. 10 Inch (25,4 cm) abgesunken und mußten erst mit hydraulischen Pressen über den geplanten Verbindungspunkt hinaus angehoben werden: Nur so konnten sie die spätere Kompression vorwegnehmen. Aber auch die massigen Rohre der Untergurte hatten sich durch Eigengewicht, Nietungskäfige und Maschinenlast gesenkt und konn-

August 1884

May 1886

April 1887

October 1887

June 1888

October 1888

June 1886

96

Construction sequence of erecting the Forth Bridge. To this, the words by Sir Robert Purvis seem fitting: "From year to year the wonder grew, as the mighty piers slowly arose out of the sea and the ascending columns climbed ever higher and higher" [14, pp. 24, 25].

96

Abfolge des Baufortschrittes der Forth-Brücke. Dazu noch einmal Sir Robert Purvis: «Von Jahr zu Jahr wuchs das Wunderwerk, als sich die mächtigen Pfeiler langsam aus dem Meer erhoben und die aufsteigenden Säulen immer höher und höher kletterten» [14, S. 24, 25].

Early 1888

June 1888

June 1889

The bridge complete   Die vollendete Brücke

97
This photograph taken May 25, 1889, looking at Inchgarvie from the northwest, shows the considerable extension of the cantilevers – without using any supporting scaffolding, but working platforms suspended from the bottom members. The strong current made it very difficult to keep the boats in place while hoisting the materials [8, p. 16].

97
Die Aufnahme vom 25. Mai 1889 mit Blick auf Inchgarvie von Nordwest zeigt die beträchtliche Ausladung der Pfeilertürme – ohne Hilfsgerüste, dafür mit abgehängten Arbeitsbühnen an den Untergurten. Bei der starken Strömung war es schwierig, die Boote zum Hochhieven der Werkstücke auf Stelle zu halten [8, S. 16].

longer be joined with the large diagonal ties descending already from above. In order to lift the tubes, a kind of "cradle" was built around the bottom members, using steel angles in continuation of the diagonal struts. By means of hydraulic rams placed inside this cradle, the bottom members were pushed up again – "a trifle beyond" the intended point of intersection, as nobody knew exactly to what extent such liftings had to be excessive. Only then were the junctions finally closed.

By way of these liftings the absolutely correct position of the first bottom junctions of the cantilevers was obtained; moreover, some compensation was effected for the likely elongation of the diagonal ties, proportionate to the weight of the completed structure. Since little was known in those days about the plasticity of steel, the extent of these liftings may well be de-

ten nun nicht mehr mit den großen Diagonal-Zugstreben, die bereits von oben herabkamen, verbunden werden: Zur Hebung baute man aus Stahlwinkeln, in Verlängerung der Diagonalstreben, eine Art «Schaukel» um den Untergurt herum und drückte ihn hydraulisch wieder nach oben – «a trifle beyond» den geplanten Punkt: Denn wieviel diese «Überfahrt» bei den Hebungen genau zu sein hatte, wußte niemand; erst dann wurden die Knoten endgültig verbunden.

Mit den Hebungen erreichte man einmal die absolut genaue Lage der ersten unteren Ausleger-Knotenpunkte; und zum anderen einen Ausgleich für die spätere Belastung der Diagonalstreben: Weil man damals über die Plastizität von Stahl noch wenig wußte, darf man wohl das Maß dieser Hebungen als intuitiv bezeichnen. Immerhin waren so zum erstenmal außerhalb der Pfeilertürme – und frei über dem

scribed as intuitive. Be that as it may, there were now established, outside the central towers – and freely above the sea – two fixed points, capable of sustaining their full design load. In further extending the top members, the "Jubilee Crane" was successively moved ahead, facilitating also the extension of the diagonal struts; once their top junctions were rivetted up, yet another pair of fixed points was thus secured. – Despite all this care, certain displacements did occur, as no permanent wind-bracing was fixed as yet, and the superstructure could deviate under a strong wind from the center line of the bridge.

Sun Radiation

It made a material difference whether the sun was shining on the east side or the west side of the bridge. In the morning for example, the tubular members expanded on the side facing the sun, effecting therefore a temporary bend away from the sun, which was then reversed again in the evening. The wind bracings on the other hand could not be measured exactly in advance, as neither top nor bottom members had ever been laid together on the plane in their relative position; so the ties had to be templeted "in situ". At last though, all junctions of the cantilevers' first bay were as fixed as if they were resting on the solid masonry piers down below.

For building out the second bay of the cantilevers, the internal viaduct doubled as a convenient working platform. It took a team of two steam cranes – called "the twins". Just as the "Jubilee Crane" high up took care of all work above the viaduct, "the twins" served everything below that level. In fact, they lifted all material directly out of the barges up to the viaduct, whence the "Jubilee" brought it up to the top members. The erection of bays 3, 4, 5 and 6 was simply routine work; not only had the distances between the girders become less, but also the men had become very skillful in their task. What appeared at the outset insurmountably difficult and hazardous, was now done in all those exposed positions with great ease. In the end, the rivetting gangs were following the erecting gangs so closely, that the men were frequently working side-by-side in the far ends *(Fig. 98).*

The Central Girders – Two Bridges within the Bridge

By July 1889, the three central towers, each with their pairs of cantilevers, stood finally facing each other; only the two central girders were still missing. *(Fig. 97)* The latter resembled rather ordinary railway bridges

Wasser – je zwei feste Punkte ereicht, die ihr volles Endgewicht tragen konnten. Zum Weiterbau der Obergurte wurde der «*Jubilee*-Kran» ständig nachgeführt, was auch den Bau der Druck-Diagonalen erleichterte; mit der Vernietung ihrer oberen Anschlüsse erhielt man ein zweites Paar fester Knotenpunkte. – Trotzdem gab es immer wieder Abweichungen, denn die Windaussteifungen fehlten noch, und der Oberbau konnte sich bei starkem Wind aus der Mittelachse verschieben.

Sonneneinstrahlung

Nicht unerheblich war auch, ob die Sonne von Osten oder Westen schien. Die Rohrträger dehnten sich z. B. morgens an der Sonnenseite aus und bewirkten eine vorübergehende Krümmung in der Gegenrichtung, was sich aber abends wieder umkehrte. Andererseits konnte man die Windaussteifungen nicht vorher ermitteln, da Ober- und Untergurte nie vorher in ihrer relativen Lage flach ausgelegt waren, sondern man mußte sie «in situ» mit Paßstücken einmessen. Schließlich waren aber doch alle Knotenpunkte des ersten Auslegerfeldes so fest verankert, als säßen sie unten auf den Mauerpfeilern.

Für den freien Vorbau des zweiten Auslegerfeldes bot der innere Eisenbahnviadukt eine gute Arbeitsbühne und nahm ein Gespann von zwei Dampfkränen auf – «*the twins*»: Während der «*Jubilee*-Kran» in luftiger Höhe alle Arbeiten oberhalb des Viaduktes besorgte, bedienten die Doppelkräne alles darunter: Sie hievten das Material direkt aus den Lastkähnen zum Viadukt, von wo es der «*Jubilee*» bis zu den Obergurten hinaufholte. – Der Weiterbau der Felder 3, 4, 5 und 6 war dann nur noch Routine-Arbeit: Nicht nur die Abstände zwischen den Trägern waren immer kürzer geworden, auch die Männer waren sehr geschickt in ihrer Arbeit geworden: Was anfangs unüberwindlich schwierig und gefährlich zu sein schien, geschah nun auf all den exponierten Arbeitsstellen mit großer Selbstverständlichkeit: Die Vernietungs-Trupps folgten den Montage-Trupps so dicht, daß die Männer in den Träger-Enden oft Seite an Seite arbeiteten *(Abb. 98).*

Die Mittelträger – Zwei Brücken in der Brücke

Im Juli 1889 standen sich endlich die drei Pfeilertürme mit ihren Auslegerpaaren fertig gegenüber – nur die beiden Mittelträger fehlten noch *(Abb. 97).* Sie ähneln ganz normalen Eisenbahnbrücken mit gebauchtem Obergurt und unterscheiden sich dank mäßiger Spannweite von 350 Fuß (106,68 m) durch ihre Zierlichkeit vom Rest des Bauwerkes *(Abb. 100).* Die freien

with slightly curved top members, yet by their moderate span of 350 ft (106.68 m) and their slenderness, they posed quite a contrast to the rest of the structure *(Fig. 100)*. The free cantilever ends were intended to have a camber of 10 inches (254 mm) when completed; to this end, each section had to be set higher by as much as it was likely to deflect with the addition of the remaining sections. So by blending calculation with judgement, the camber was doubled to a total of 20 inches (508 mm) altogether, and in the end the cantilevers did indeed settle at the designed camber when no load was on the bridge. Assembled from steel plates, the U-shaped top and bottom members of the central girders put no more weight on the cantilever ends than was absolutely necessary – which was precisely Baker's intent. Still, this light-weight "bridge within a bridge" was by 105 ft (32 m) longer than, for example, the high girders of Barlow's new Tay Bridge spanning 245 ft (74.68 m), completed two years before. In view of the dimensions of the Forth Bridge, anything done previously appeared now insignificant. But unlike the high girders of the Tay Bridge, the central girders had not been pre-assembled at the shore, floated to the site and raised into the structure; instead they were built forward piece by piece in extension of the cantilevers, until both ends met in the center. Thus the very advantages of the cantilever principle were fully used, even for those final sections: the closing phase was indeed a precise and carefully planned operation.

Much thought was given to the articulation of the sliding end-supports of the two central girders. At the cantilevers extending from the Queensferry and Fife towers, the central girders were fixed and may thus have been considered rigid extensions of those very cantilevers, whereas at the two Inchgarvie cantilevers, longitudinal movement was provided by an expansion joint of 2 ft (61 cm) each. Here, two 9-ton "rocking posts" descending from the top members of the central girders came to rest via some kind of built-in "knuckles" or half-balls, inside steel castings shaped like cups or sockets, recessed into the hollow end posts of the cantilevers. Thus the weight at these ends was transmitted from the top members of the central girders directly to the bottom members of the cantilevers, without impeding their longitudinal movement in any way. At a rise of 1°F in temperature, this expansion, as was observed, amounted to 1/160 inch for each 100 ft (30.5 m) of the bridge's length, or – when applied to the bridge's total length of 5,330 ft (1624 m), to an expansion of 0.33 inch per 1°F (1,5 cm per 1°C) rise in temperature. Assuming a yearly temperature range of max. 108°F (60°C) (from +104°F [40°C] in sum-

Ausleger-Enden sollten eine Überhöhung von 10 Inch (254 mm) im Endzustand haben. Dafür mußte aber jedes Feld immer um soviel höher angesetzt werden, als es wahrscheinlich nach Anfügung weiterer Felder absinken würde. Indem man also Annahme mit Berechnung verband, verdoppelte man die Überhöhung auf insgesamt 20 Inch (508 mm), und schließlich trafen sich die Ausleger tatsächlich in der vorgesehenen Überhöhung, wenn die Brücke unbelastet war. Aus Flacheisen zusammengebaute U-förmige Ober- und Untergurte der Mittelträger belasten die freien Ausleger-Enden nicht weiter als unbedingt nötig – was ja ganz im Sinne Baker's lag. Dennoch ist diese leichte «Brücke in der Brücke» immer noch um 105 Fuß (32 m) länger als z. B. die Hochträger der neuen Tay-Brücke Barlow's mit ihren 245 Fuß (74,68 m) Spannweite, zwei Jahre vorher vollendet: Angesichts der Dimensionen der Forth-Brücke erschien alles Vorangegangene bedeutungslos. Allerdings hat man die Mittelträger nicht wie die Hochträger der Tay-Brücke am Ufer vormontiert, eingeschwommen und schließlich in die Brücke gehoben; vielmehr hat man sie in Fortsetzung der Ausleger stückweise vorgebaut, bis sich beide Enden in der Mitte trafen: Die Vorteile des Ausleger-Prinzips wurden so auch für diese letzten Teilstücke voll genutzt: Die Schlußphase war eine präzise und gutgeplante Operation.

Sehr überlegt ging man an die Ausbildung der Gleitlager beider Mittelträger: Sie sind an den äußersten Queensferry- und Fife-Enden fest verankert und damit Fortsetzungen der Ausleger; dagegen sind sie an den beiden Inchgarvie-Enden beweglich aufgehängt, mit einer Dehnungsfuge von je 2 Fuß (61 cm): Hier greifen zwei 9 Tonnen schwere *«rocking posts»* (Schaukelpfosten) mit Kugelgelenken in zwei große Stahlguß-Pfannen innerhalb von Hohlkästen am Ausleger-Ende ein. Sie übertragen so das Gewicht von den Obergurten der Mittelträger direkt auf die Untergurte der Ausleger, ohne deren Längsdehnung zu verhindern.
Bei 1°F Temperaturanstieg betrug diese Wärmedehnung, wie man beobachtete, 1/160 Inch je 100 Fuß (30,5 m) Brückenlänge. Metrisch und auf Celsius-Grade umgerechnet, ergibt das eine Ausdehnung von 1/35 cm je 30,48 m Brückenlänge bei 1°C Temperaturanstieg – oder, auf die gesamte Brückenlänge von 1.624 m bezogen, eine Wärmedehnung von 1,5 cm je 1°C Temperaturanstieg. Legt man eine jährliche Temperatur-Differenz von max. 60°C (+40° im Sommer bis –20°C im Winter) zugrunde, ergibt das eine Gesamtausdehnung von 90 cm. Spätere Nachmessungen ergaben eine Gesamtausdehnung von nur 23,44 Inch bzw. 59,5 cm und wurden als typisch für 70% aller Fälle angesehen, weil gemeinhin in Schottland mit ge-

98 >
The Fife Tower at North Queensferry nearly completed. The "Jubilee Crane" (to the left) with attached house has, during its travelling assembly, almost reached the northern end-pier of the viaduct. At the south cantilever (to the right), the assembly of the central girder is just being started [29, plate XVIII].

98 >
Der «Fife-Pfeilerturm» bei North Queensferry nahezu vollendet. Der «Jubilee-Kran» (links) mit aufgesetztem Wohnhaus ist auf seiner Montage-Fahrt fast am nördlichen Viadukt-Endpfeiler angekommen. Am Südausleger (rechts) beginnt bereits der Vorbau des Mittelträgers [29, Tafel XVIII].

mer to −4°F [−20°C] in winter), this would amount to a total expansion of 35.64 inches (90 cm). Measurements taken later showed a total expansion of only 23.44 inches (59.5 cm), which was held to be indicative for 70% of the cases, since generally in Scotland the temperature range is less than 108°F (60°C). But to be on the safe side, this figure of 23.44 inches (59.5 cm) was doubled to 46.88 inches (119 cm), or almost 4 ft (1.22 m) of total expansion, and all slide-bearings were thus designed accordingly.

Additional rotating half-bearings took up limited lateral deflections caused by wind pressure and sun radiation. A kind of coupling-fork, not unlike that between railway carriages, placed in the centerline of the bridge on either end of the central girders, allowed for a controlled horizontal rotation. If, for instance, a heavy gust of wind hit just one cantilever but not the other, the resulting deflection could take with it one end of the central girder, and let the other one rotate a bit in its coupling support. The only roller bearings of

ringeren Jahres-Temperaturdifferenzen als 60°C zu rechnen war. Um sicher zu gehen, hat man dann den Wert von 59,5 cm nocheinmal verdoppelt auf 119 cm Gesamtausdehnung, und alle Gleitlager entsprechend ausgelegt.

Zusätzliche Drehlager sorgen für kontrollierte Querbewegungen als Folge von Winddruck oder Sonneneinstrahlung. So liegt an beiden Enden des Mittelträgers zentrisch auf der Brückenachse eine Art Kupplungsgabel, wie etwa beim Eisenbahn-Waggon, und erlaubt eine begrenzte horizontale Verschwenkung: Wenn z. B. ein Windstoß nur auf einen Ausleger, aber nicht auf den nächsten trifft, kann die entstehende Seitenbiegung ungehindert ein Ende des Mittelträgers mitnehmen und das andere um das Drehlager etwa rotieren lassen. – Die einzigen Rollenlager konventioneller Art liegen unter den festverankerten Auslegern an beiden Brücken-Enden und erlauben eine maximale Längsdehnung von je 2 Fuß (61 cm).

Solange aber die mittlere Lücke nicht geschlossen ist,

99
Inchgarvie south cantilever completed, while the first bay of the "suspended" central girder is being started, with "Jubilee Crane" installed on top; September 3, 1889 [8, p. 19].

99
Inchgarvie-Südausleger vollendet, Baubeginn des ersten «aufgehängten» Mittelträger-Feldes, mit aufgesetztem «Jubilee-Kran», 3. September 1889 [8, S. 19].

conventional make were placed underneath the fixed cantilevers at the bridge's far ends, allowing for a maximum expansion of 2 ft (61 cm) at either end pier.
As long as the gap in the centre could not be closed, the central girders were simply, for the time being, assembled as extensions of either cantilevers, even at the flexible ends. Here, the combined rotating/sliding supports were by-passed temporarily by heavy ties, only to be released again at the last moment. By now the "Jubilee Cranes" on their way downhill the cantilevers' top members, could be moved across to the top booms of the central girders (Fig. 99). That is to say, the cranes now relieved of their suspended platforms in order to save weight, began to even climb up the curved central girders towards the open gap in midspan. What looked hazardous enough before, now appeared perfectly crazy, since the cranes were sitting atop the two free-floating beginnings of central girders which had never been designed as cantilevers

werden die Mittelträger einfach als Fortsetzung der beiden Ausleger weitergebaut – also auch am flexiblen Ende: Dort überbrückt man das kombinierte Dreh-/Gleitlager mit temporären Zugankern, um es erst ganz zuletzt wieder freizugeben. Die «Jubilee-Kräne» sind inzwischen soweit auf den Obergurten heruntergefahren, daß man sie auf die neuen Obergurte der Mittelträger vorschieben kann (Abb. 99). D.h., die Kräne – inzwischen ohne Arbeitsbühne zwecks Gewichtseinsparung – beginnen jetzt auch noch die Kurve der Mittelträger hinaufzuklettern, gegen die offene Lücke in der Mitte zu: Was vorher schon gefährlich genug aussah, erscheint jetzt aberwitzig – sitzen die Kräne doch auf den freischwingenden Anfängen von Mittelträgern, die selbst nie als Ausleger geplant waren! So stehen sich zuletzt je zwei dieser Kräne – sie waren vor Monaten oben von den Pfeilertürmen abgefahren – in der Mitte an den letzten noch offenen Teilstücken gegenüber, links und rechts

themselves! In the end, two of those cranes – after they had left months earlier the towers on top – stood facing each other across the last gap in the centre, to the left and to the right of the Inchgarvie Tower. In September of 1889, a workman had a ladder suspended from the jibs of the cranes and chanced a first crossing over the 200 ft (61 m) deep chasm, from the southern to the Inchgarvie cantilever.

Forces of Nature and "Keystone"

Meanwhile there was a race on among the platers, as to which team would be first to close the connection at the still free-standing Inchgarvie tower. Already it was possible to walk from the bridgeheads far to the north and south, without interruption, almost up to the lonely colossus on its rocky island. Yet this last bridging was totally dependent on the whims of the weather. And indeed in this decisive end phase, the forces of nature were being put to use to a far greater extent than ever before.

First at the Queensferry side, a point was reached that left a small gap of just 4 inches (10.2 cm) between the opposing bottom booms. It was already September, 1889, the temperature just 16°C, and considerable expansion during the summer months had to be anticipated. This meant putting the central girder into a certain tension stress during the cool season. Slot-hole plates attached to the bottom members were set to coincide at a pre-determined point with rivet holes of their counterparts; but the temperature refused to rise for days beyond 12°C, so the 1 $^7/_8$-inch bolt-holes were barely half open. Even hydraulic pressure only gave an elongation of $^3/_8$ inch, about equal to a rise of 1.5°C. After a long wait, on the afternoon of October 10, 1889, a bright southwesterly sun at last produced a temperature of 13°C – and the bolt-holes opened up on the west side just long enough to allow the bolts to be quickly inserted. But on the shaded east side, the opening was still $^1/_4$ inch short and could not be enlarged even by the strongest hydraulic pull; so some rugs soaked in naphtha were stuffed into the U-shaped bottom booms, 60 ft (18.29 m) along either side of the gap, and set on fire! At once, this bottom boom also expands to full length and all the 23 bolts can be closed.

The top booms had wedge-shaped gaps of 10 inches (25.4 cm) between the web plates. In order to sustain the camber of the bottom booms, it would be essential to put in the final wedges at the lowest possible temperature, thus closing the top boom like a rigid arch with a "keystone". Here now the strong camber of the cantilevers was of benefit in that the following day,

vom Inchgarvie-Turm. Im September 1889 hängt ein Arbeiter eine Leiter an die Kranbäume und balanciert als erster über den 200 Fuß (61 m) tiefen Abgrund, vom südlichen auf den Inchgarvie-Ausleger hinüber.

Naturkräfte und «Schlußstein»

Inzwischen gibt es einen Wettlauf unter den Stahlarbeitern, welche Gruppe zuerst den Anschluß an den noch freistehenden Inchgarvie-Pfeilerturm schaffen würde. Schon kann man von den fernen Brückenköpfen im Norden und Süden ohne Unterbrechung bis kurz vor diesen einsamen Koloß auf seiner Felseninsel gelangen. Aber die letzte Überbrückung war völlig von den Launen der Witterung abhängig – wie überhaupt in dieser entscheidenden Schlußphase die Naturkräfte viel stärker als je zuvor genutzt werden.

Zuerst ist auf der Queensferry-Seite ein Punkt erreicht, wo in der Mitte nur noch eine Öffnung von 4 Inch (10,2 cm) zwischen den Untergurten bleibt. Es ist bereits September 1889, die Temperatur nur 16°C und man muß die erheblichen Wärmeausdehnungen der Sommermonate vorwegnehmen. – Das bedeutet, den Mittelträger in der kühlen Jahreszeit in eine gewisse Zugspannung zu versetzen. Die Langloch-Platten der Untergurte sollen sich an vorausbestimmter Stelle mit den Nietlöchern der Gegenseite decken, aber die Temperatur steigt tagelang nicht über 12°C, und die 48 mm Bohrlöcher sind erst halb offen. Auch hydraulischer Druck erreicht nur eine Verlängerung von 1 cm, was etwa 1,5°C entspricht. Nach langem Warten kommt am Nachmittag des 10. Oktober 1889 endlich eine kräftige Südwest-Sonne mit 13°C hervor – und die Bohrlöcher öffnen sich auf der Westseite gerade weit genug, um schnell die Bolzen hindurchzuschieben. Auf der schattigen Ostseite fehlen immer noch 6 mm und können durch keine noch so starke Hydraulik verlängert werden. Da stopft man ölgetränkte Lappen auf 60 Fuß (18,29 m) Länge in den U-förmigen Untergurt und zündet sie an! Im Nu dehnt sich der Ostgurt auf volle Länge, und alle 23 Nieten können geschlossen werden.

Bei den Obergurten hatten die Stegplatten einen keilförmigen Ausschnitt von 10 Inch (25,4 cm). Um die Überhöhung der Untergurte zu halten, hätte das Keilstück unbedingt bei tiefstmöglicher Temperatur eingepaßt werden müssen, damit ein steifer Bogen mit «Schlußstein» entsteht. Hier macht sich jetzt die starke Überhöhung der Ausleger bezahlt, so daß man die Keilstücke trotz ungenügender Schrumpfung der Obergurte hart einschlagen und verschrauben kann. Damit hängt der Mittelträger erstmals und selbständig zwischen den Auslegern. Sorgfältige Vermes-

100
The delicate central girder spanning 350 ft (106.68 m) is suspended between the two cantilevers near the south shore, viewed from the southwest. The cantilever principle becomes very clear in this photograph – a minimum of suspended mass, supported free-floating by the large cantilever piers. In the center of the river behind, Inchgarvie Island appears with the ancient castle and the remnants of foreign workers' housing, from the time of construction.

100
Der zierliche Mittelträger mit seinen 350 Fuß (106,68 m) Spannweite hängt zwischen den beiden Auslegern am Südufer, Blick von Südwesten. Das Prinzip des Kragträgers wird in dieser Aufnahme besonders deutlich – ein Minimum an aufgehängter Masse, freischwebend getragen von den großen Pfeilertürmen. In der Flußmitte darunter die Insel Inchgarvie mit der alten Burg und den Resten der seinerzeitigen Arbeiter-Behausungen.

and despite insufficient shrinkage of the top booms, the key-plates could be driven down hard and bolted up at once. Thus the southern central girder was for the first time freely suspended as a connecting link between the cantilevers. A careful levelling showed an actual camber of 3 $^7/_8$ inches (9.8 cm) *(Fig. 100)*.
The northern central girder still needed some time, but on October 15, 1889, a gangway was pulled across to enable the directors of the FBR to walk across the bridge from end to end. The last section could only be closed in a dramatic finale. Because of the late season, the camber of the bottom booms had been set less high, so there was now hardly any camber left. Only on November 14, 1889, did the temperature drop low enough to allow the wedges to be driven in. After a slight rise in temperature in the late morning, the temporary connecting ties were thought to be relieved of tension and one began to remove their bolts. But whether the thermometer had given a false reading or

sungen ergeben eine tatsächliche Überhöhung von 9,8 cm *(Abb. 100)*.
Der nördliche Mittelträger ist noch nicht soweit, doch zieht man am 15. Oktober 1889 eine Gangway hinüber, damit die Direktoren der *Forth Bridge Railway Company* ihre Brücke zum erstenmal in ganzer Länge durchschreiten können! – Das letzte Teilstück läßt sich nur in einem dramatischen Schlußakt schließen: Wegen der späten Jahreszeit hatte man nämlich die Überhöhung der Untergurte etwas geringer angesetzt, so daß sie im November schon praktisch keine Überhöhung mehr hatten. Erst am 14. November 1889 fällt das Thermometer tief genug, um die oberen Keilstücke einzusetzen. Als es am späten Vormittag wieder wärmer wird, beginnt man, die dann vermeintlich entlasteten Hilfs-Zuganker abzuschrauben – und ob es nun an einer Fehlanzeige des Thermometers lag oder daran, daß die Ausleger noch nicht genug Zeit gehabt hatten, sich voll auszudehnen – die Zuganker

Two Rare Points of View

Zwei seltene Blickpunkte

101
This view, showing the direction of travel across the double-track rail-bed, remains the privilege of the engine driver (200 trains daily crossing the bridge) – or of those invited bridge-walkers who get to experience a guided tour by British Rail [19, p. 231].

101
Der Blick in Fahrtrichtung auf den zweispurigen Schienenweg bleibt dem Lokführer vorbehalten (200 Züge gehen täglich über die Brücke) oder jenen geladenen Brückengängern, die eine Führung durch *British Rail* erleben dürfen [19, S. 231].

102
The rarely shown underbelly of the Forth Bridge – every-day view to the 40 men of the bridge's permanent maintenance staff [14, p. 31].

102
Die selten gezeigte Untersicht der Forth-Brücke – alltäglicher Anblick für die 40 ständigen Männer vom Brückenunterhalt [14, S. 31].

whether the cantilevers did not have enough time to expand fully – the plate-ties suddenly sprang loose with a bang like that of a heavy gun, shearing the remaining 36 bolts! A shaking reverberated throughout the entire structure and was felt even at the far end. However, nobody got hurt, and no girders fell – contrary to what was stated in some papers. Instead, the central girder swang to and fro a bit in its rocking posts – "as freely as if it had been freed in the most natural manner." With this thrilling act of nature, the last gap was finally closed.

springen plötzlich mit dem Donnerknall einer schweren Kanone davon und reißen 36 noch nicht gelöste Schrauben mit sich! Ein Schütteln geht durch die gesamte Konstruktion und ist noch am fernen Ende zu spüren... trotzdem wird niemand verletzt und entgegen den Zeitungsmeldungen stürzt auch kein Träger herab: Statt dessen schwingt der Mittelträger noch ein paarmal in seinen «*rocking posts*» hin und her – «*as freely, as if it had been freed in the most natural manner*». Mit diesem Naturereignis ist auch das letzte Teilstück endgültig geschlossen.

Loading Tests

On January 21, 1890, two trains side by side entered the bridge from the south end. Measuring almost 1,000 ft (300 m) in length, their 100 fully loaded coal wagons and 6 locomotives of 72 tons each amounted to a total weight of 1,800 tons. In the course of several tests, the locomotives moved slowly up to the $3/4$ point of the northern central girder, the rear engine reaching the mid point of the Inchgarvie tower. Thus they had gotten to the least favourable point of loading for the Inchgarvie north cantilever. Careful measuring by Fowler and Baker at various points gave these results: the vertical columns of the central tower were drawn north by $1^1/_4$ inches (3.2 cm), and those of the Fife tower drawn $1/2$ inch (1.3 cm) to the south; the Inchgarvie north cantilever deflected by $6^7/_8$ inches (17.5 cm) and the Fife south cantilever by $2^1/_2$ inches (6.4 cm), while the Inchgarvie south cantilever rose by $3^1/_2$ inches (8.9 cm). Though a deflection of $6^7/_8$ inches (17.5 cm) may appear extensive, this actually is a small amount

103 a
Full view of the Forth Bridge from the northwest. In front North Queensferry, in the centre Inchgarvie Island underneath the first span, to the right in the back South Queensferry with the "Hawes Inn" (formerly called "Hawes Hotel" with its bar, indicated by arrow). This photograph is taken from the same position as the one used for *Fig. 103 b,* and shows in the foreground the changes that have taken place since 1903. The north shore was very barren in those days, almost void of vegetation. Among today's new dwellings, the two stone houses can still be made out to the left, as in *Fig. 103 b.*

103 a
Gesamtansicht der Forth-Brücke von Nordwest. Vorne North Queensferry, in der Mitte unter dem ersten Brückenbogen Inchgarvie Island, rechts hinten South Queensferry mit «Hawes Inn» (früher Hawes Hotel mit seiner Bar, durch Pfeil gekennzeichnet). Die Aufnahme ist vom selben Standort wie *Abb. 103 b* aufgenommen und zeigt die Veränderungen seit 1903 im Vordergrund: Das Nordufer war seinerzeit sehr karg und fast ohne Vegetation. Zwischen den heutigen Neubauten erkennt man links zwei Steinhäuser, die auch schon in *Abb. 103 b* zu sehen sind.

103 b
Post card of the Forth Bridge around 1903 – seen from the same position as in *Fig. 103 a*, but with a sailing vessel before the bridge. The two stone houses in the left foreground are still standing today [15, p. 87].

103 b
Postkarte der Forth-Brücke von 1903 – vom gleichen Standort wie *Abb. 103 a* und mit einem Segler vor der Brücke. Die beiden Steinhäuser links vorne stehen heute noch [15, S. 87].

considering the size of the structure. The other findings too, were fully within the calculated allowances. This happy outcome and the fact that the bridge had already endured heavy gales, before and after the tests, without any harm, moved the strict inspectors of the Board of Trade to the following conclusion:

"This great undertaking, every part of which we have seen at different stages of its construction, is a wonderful example of thoroughly good workmanship with excellent materials, and both in its conception and execution is a credit to all who have been connected with it." [14, p. 23]

Official Opening

The opening on March 4, 1890, was a great day, though a wet and windy one, for South Queensferry. At 11:30 a.m. the Royal Party arrived by special train at Forth Bridge Station (Dalmeny today), to be received by Fowler, Baker, Arrol and the directors of the FBR gathered there. A two hour program commenced with an inaugural train ride across the bridge to Inverkeithing, in front a train-load of dignitaries, behind the Royal Train. At North Queensferry everyone went by boat for a lengthy sight-seeing trip of the miraculous structure from sea-level, to be concluded by circling Inchgarvie Island. For the return trip across the bridge, the Royal Train went ahead, to stop twice; first at the northern central girder, in order for the Prince of Wales

Belastungsprobe

Am 21. Januar 1890 fuhren zwei Züge nebeneinander von Süden her in die Brücke ein. Mit einer Zuglänge von 300 m, 100 vollbeladenen Kohlewaggons und 6 Lokomotiven von 72 Tonnen ergibt das ein Gesamtgewicht von 1800 Tonnen. Im Laufe mehrerer Tests fuhren die Lokomotiven langsam bis zum 3/4-Punkt des nördlichen Mittelträgers, mit dem Zugende in der Mitte des Inchgarvie-Pfeilers: Damit hatten sie den ungünstigsten Belastungspunkt für den Inchgarvie-Nordausleger erreicht. Fowler und Baker machten genaue Messungen an verschiedenen Punkten und fanden folgende Werte: Die Vertikalrohre des Mittelpfeilers bogen sich 1,25 Inch (3,2 cm) nach Norden und die des Fife-Pfeilers 1/2 Inch (1,3 cm) nach Süden; das Nordende des Inchgarvie-Auslegers senkte sich um 6 – 7/8 Inch (17,5 cm), das Südende des Fife-Auslegers um 2,5 Inch (6,4 cm); dagegen hob sich das Ende des Inchgarvie-Südauslegers um 3,5 Inch (8,9 cm) an: Wenngleich eine Durchbiegung des Auslegers von 17,5 cm hoch erscheint, ist das angesichts der Größe des Bauwerkes ein geringer Wert. Auch die übrigen Ereignisse lagen alle innerhalb der errechneten Toleranzen.

Dieses günstige Ergebnis und der Umstand, daß die Brücke bereits schwere Stürme, vor und nach den Tests, schadlos überstanden hatte, veranlaßten die gestrengen Inspekteure der IHK zu folgendem Resumé:

to drive home the last of 7 million rivets, which was gilded for the occasion; (Fig. 104) and again at the south cantilever, for a remarkably short opening speech by the Prince:
> "Ladies and Gentlemen, I now declare the Forth Bridge open."

The blustery weather was already blowing the top hats off the visitors' heads and long speeches were left for the historic banquet to follow. For this, a wise strategy had chosen the large model-loft at South Queensferry, the black floor of which had brought forth for seven years the many scale drawings and templets for the bridge. So this was a most fitting place for the subsequent conferral of a baronetcy to Fowler, and knighthood to Baker and Arrol (Fig. 105).

The Workmen

Since the beginning of the Railway Age it had become a practice to marshall entire troops of "navvies" for the construction of tracks, tunnels and bridges. Once the then largest job site of the world was established, men looking for work came streaming in from all over, to live in lodgings at the Forth or in the surrounding villages of Dalmeny, Kirkliston, Inverkeithing or Dunfermline. Soon they became known as "the Briggers", not unlike the "riggers" of today's oil rigs. The briggers were easily recognised by their swaggering gait, stalking about in their bell-bottomed trousers with red spotted scarves or wearing large iron rings fashioned from scraps of steel found on the bridge [15, p. 64]. In all their prowess at playing and drinking, they earned themselves a reputation for their hard work and readiness to help others in distress, as is seafarers' way. They came in about equal numbers from Scotland, England and Ireland, and though the Irish furnished fewer skilled workers, they were known to be very reliable, conscientious and hard-working men. Quite a few, in their search for work, had brought their families with them – some descendants still living "in the shadow of the bridge" up to this day. But some others had just come for the free passage by paddle steamer, as it became quickly known all over the country that "workers" got a free ride across the Forth; so someone who had just arrived at the south side, would ask for a job on the north side, and vice-versa, until a stop had to be made to this kind of tourism. And of course there were some who just stayed for a week or two, doing little but eating and collecting their pay packets – "as black sheep are found everywhere... and hundreds were mere birds of passage."

Soon the number of workmen could no longer be accomodated in the hamlets along the Forth. Special

«Dieses großartige Werk, das wir in all seinen verschiedenen Bauphasen verfolgt haben, ist ein wunderbares Beispiel vorzüglicher Bauausführung mit ausgezeichneten Materialien und ist sowohl in seinem Entwurf als auch in der Durchführung ein Glanzstück für alle, die daran beteiligt waren.» [14, S. 23]

Feierliche Eröffnung

Die Brücken-Eröffnung am 4. März 1890 war ein großer Tag – wenn auch ein nasser und windiger – für South Queensferry: Um 11 Uhr 30 traf die «Royal Party» mit Sonderzug in Forth Bridge Station (heute Dalmeny) ein, begrüßt von Fowler, Baker, Arrol und den versammelten Direktoren der Forth Bridge Railway Company. Ein 2-Stunden-Programm begann mit einer ersten Fahrt über die Brücke nach Inverkeithing – vorneweg eine Zugladung voller Würdenträger, dahinter der königliche Sonderzug. Anschließend begab man sich in North Queensferry an Bord für eine ausgiebige Besichtigungsfahrt des Wunderwerkes vom Wasserspiegel aus, gefolgt von einer Ehrenrunde um Inchgarvie herum. Auf dem Rückweg über die Brücke fuhr jetzt der königliche Zug voraus und hielt zweimal: Zuerst im nördlichen Mittelträger, damit der Prince of Wales dort per Knopfdruck die letzte von 7 Millionen Nieten einschlagen konnte – sie war extra für den Anlaß vergoldet worden; (Abb. 104) und wieder im Südausleger für eine denkbar kurze Eröffnungsansprache des Prinzen:
> «Ladies and Gentlemen, I now declare the Forth Bridge open.»

Das ungemütliche Wetter blies schon die Zylinder von den Köpfen und man ließ lange Reden für später. Für das historische Festbankett hatte eine kluge Regie jenen großen Modell-Saal in South Queensferry gewählt, dessen schwarzer Boden 7 Jahre lang die maßstabsgetreuen Kreidezeichnungen und Schablonen für die Brückenteile hervorgebracht hatte. Und das war auch der angemessene Ort für die anschließende Verleihung der Adelstitel «Baronet» an Fowler und «Sir» an Baker und Arrol (Abb. 105).

Die Arbeiter

Seit den Zeiten des frühen Eisenbahnbaues war es üblich geworden, ganze Heerscharen von Erdarbeitern – sogenannte «navvies» – für Schienen-, Tunnel- und Brückenbauten anzuheuern. Mit der Einrichtung der damals größten Baustelle der Welt strömten Arbeitswillige von überall her und hausten in Unterkünften am Forth oder in den umliegenden Dörfern Dalmeny,

104
The opening ceremony of the Forth Bridge on March 4, 1890. The Prince of Wales pushing the button to drive home the last rivet. In his company Sir William Arrol, Lord Rosebery, Sir Benjamin Baker, the Marquis of Tweeddale, and others. – The grip at the top hats is not meant as a salute, but to prevent their being blown off by the wind [15, p. 55].

Die Eröffnung der Forth-Brücke am 4. März 1890. Der Prince of Wales drückt den vergoldeten Knopf zur letzten Nietverbindung. Daneben Sir William Arrol, Lord Rosebery, Sir Benjamin Baker, der Marquis of Tweeddale und andere. – Der Griff zu den Zylindern ist kein Salut, sondern soll die Hüte gegen den Wind sichern [15, S. 55].

Kirkliston, Inverkeithing oder Dunfermline. Bald waren sie als «*the Briggers*» bekannt – ähnlich den «Riggers» heutiger Ölbohrplattformen. «*Briggers*» erkannte man sofort am Gang, wie sie mit ausgestellten Hosenbeinen herumstolzierten, dazu rotgetupfte Halstücher oder große Eisenringe trugen, die sie sich aus Stahlschrott gefertigt hatten [15, S. 64]. Bei aller Spiel- und Trinkfreudigkeit war man des Lobes voll über die Männer, ihren Arbeitseinsatz und ihre Hilfsbereitschaft bei Gefahr, wenn sie sich nach Seefahrer-Art gegenseitig beistanden. Sie kamen zu etwa gleichen Teilen aus Schottland, England und Irland, und obwohl die Iren weniger Facharbeiter stellten, galten sie als sehr zuverlässige, gewissenhafte und arbeitsame Leute. Nicht wenige hatten auf der Suche nach Arbeit ihre Familien mitgebracht – einige Nachfahren leben heute noch «*in the shadow of the Bridge*». Aber manche waren auch nur wegen des freien Raddampfer-Dienstes gekommen, denn es hatte sich bald im Land herumgesprochen, daß man am Forth als «Arbeiter» freie Überfahrt hatte: War jemand gerade am Südufer angelangt, fragte er nach Arbeit am Nordufer und umgekehrt, so daß dem Touristenstrom bald Einhalt geboten werden mußte. Und natürlich gab es auch einige, die nur 1 – 2 Wochen blieben und wenig taten außer essen und Lohntüten abholen – «*as black sheep are found everywhere... and hundreds were mere birds of passage*».

Bald konnte die Zahl der Arbeiter nicht mehr in den Ortschaften am Forth untergebracht werden. Man setzte eigene Arbeiterzüge zwischen dem Südufer und Edinburgh bzw. zwischen dem Nordufer und Dunfermline ein – immer zwei Züge bei Schichtwechsel morgens und abends in beiden Richtungen. Mehrere hundert Männer kamen so aus Leith (dem heutigen Hafen Edinburghs), mußten ihr Haus morgens um 4.00 Uhr verlassen, um vor 6.00 Uhr bei der Arbeit zu sein, und kamen erst um 19.00 Uhr wieder nachhause – eine Schicht dauerte 12 Stunden! Aber viele zogen diese Lebensweise den übervölkerten «Krähennestern» des Arbeitsplatzes vor. Wenn man wegen schlechten Wetters die Arbeit abbrechen mußte, wurden die Züge herantelegraphiert – um die Männer nicht für den Rest des Tages der «zärtlichen Barmherzigkeit der Freudenhäuser» auszusetzen – «*of which there were in this place far too many*».

Viele Facharbeiter kamen auch von den Schiffswerften am Clyde und der Industrie-Region von Mittel-Schottland. Sie waren sehr geübt im Formen und Vernieten von Stahlplatten, von meist viel komplizierterer Bauart als den für die Brücke benötigten. Auch Streiks gab es schon – so nach einem Unfall im Juni 1887, der einzig durch die Nachlässigkeit einiger weniger ver-

105 Front cover of the luncheon menu for the Opening Ceremony on March 4, 1890. Behind the notabilities portrayed here, Edinburgh's Waverly Station can be made out at the bottom of the Castle. A symbolic "train of the times" piloted by the engine "Progress" carries the illustrator's fanciful heading, "Through Carriage, Aberdeen-New York, via Tay Bridge, Forth Bridge, Channel Tunnel and Alaska" [15, p. 53]. As far as the Channel Tunnel goes, it would take another century before British and French miners would shake hands over champagne to celebrate the first breakthrough.

Titelblatt der Speisekarte zur Eröffnungsfeier vom 4. März 1890. Hinter den dargestellten Honoratioren ist Edinburghs Waverly Station zu Füßen des Burgberges angedeutet. Ein «Zug der Zeit» mit der symbolischen Lokomotive «Progress» an der Spitze trägt – in der Fantasie des Zeichners – die Aufschrift: «Direktwagen Aberdeen–New York über Tay-Brücke, Forth-Brücke, Kanal-Tunnel und Alaska» [15, S. 53]. – Was den Kanal-Tunnel betrifft, so wird es noch ein Jahrhundert dauern, bevor britische und französische Tunnelarbeiter mit einem Glas Champagner den ersten Durchstich feiern können.

workers' trains were laid on from the south shore to Edinburgh and from the north shore to Dunfermline – always two every morning and two every night when the shifts changed. Several hundred men even came from Leith (today Edinburgh's harbour), leaving their house early in the morning at 4 o'clock, in order to be at work by 6 a.m., only to return home at 7 p.m. as one shift lasted 12 hours! But many preferred this way of life to the over-populated "crows nests" at the work place. When the weather was too bad for the men to work, the trains were called back by telegraph in order not to leave the men for the rest of the day "to the tender mercy of the public houses ... of which there were in this place far too many."

Many of the skilled workers came from the Clyde shipyards and the industrial belt of mid-Scotland. They were quite experienced in fashioning and rivetting steel plates of rather more complicated shapes than those needed for the bridge. There were even some strikes – such as following an accident in June of 1887, caused solely by the negligence of a few. During the collapse of a movable platform caused by forced lifting, two men and a boy were killed; thereupon "agitators" called for a strike demanding an increase of a penny per hour as a danger bonus – which would have amounted to a raise of 15–20% all around. But this demand came from those who did not work on the bridge themselves but in the yards away from all the danger, and after a week, most of the strikers were glad to be allowed back to work.

It was evident to everyone that accidents had to be expected in such a difficult and novel task – but not that

106
A group of navvies, in the age of picks, shovels and wheelbarrows, in May 1889. The clothing demarcation clearly identifies different working grades: Anyone in brass buttons somehow belongs to the railway crew, while overalls and bowler hat immediately proclaim some authority [15, p. 37].

106
Eine Gruppe von Erdarbeitern – «navvies» – aus der Zeit von Pickel, Schaufel und Schubkarren, im Mai 1889. Die Kleiderordnung markiert deutlich die Rangunterschiede: Alles, was Messingknöpfe trägt, gehört irgendwie zur Eisenbahn, während Anzug und «bowler hat» (Melone) sogleich eine gewisse Autorität bedeuten [15, S. 37].

three-quarters of all serious accidents should clearly be due to preventible causes. In the main, the accidents resulted from that kind of carelessness and indifference to danger which fatal familiarity breeds. Young briggers for example, would jump freely from girder to girder while working several hundred feet above the water. During the daily work on stagings cluttered with material and scrap – and without the safety helmets of today – even a bolt dropping from that height could be fatal. Special clearing gangs for the platforms were organised, coupled with stern safety measures – such as hand rails and wire netting below exposed points, as well as life boats constantly stationed underneath the cantilevers – but it was to little avail.

Drunkeness proved to be a perennial problem. At one time Sir William Arrol remarked there had not been a day on his visits to the site "without seeing drunken briggers hanging about", so he had more foreign workers hired because they would not drink as much as the Scots. Also he referred to the bar at South

schuldet worden war: Beim Absturz einer Arbeitsbühne hatte es wegen unsachgemäßer Hebung 3 Tote gegeben, worauf «Agitatoren» zum Streik aufriefen und 1 Penny mehr Stundenlohn als Gefahrenzulage verlangten – was einer Lohnsteigerung von 15 – 20 % entsprochen hätte. Die Forderung kam aber von denen, die selbst nie auf der Brücke gearbeitet hatten, sondern auf festem Lande in den Werkstätten – und nach einer Woche waren die meisten froh, wiederkommen zu dürfen.

Daß bei einer so schwierigen und neuartigen Aufgabe mit Unfällen zu rechnen sein würde, war allen klar – nicht aber, daß 3/4 aller schweren Unfälle eindeutig vermeidbar gewesen wären. Der Hauptgrund lag in jener Art von Sorglosigkeit und Indifferenz gegenüber der Gefahr, wie es die Gewöhnung mit sich bringt: Junge «briggers» pflegten z.B. frei und mehrere hundert Fuß über Wasser von Träger zu Träger zu springen. Bei der täglichen Arbeit mit den Unmengen von Material und Schrott auf den Arbeitsbühnen – und ohne den heute üblichen Sturzhelm – konnte bereits

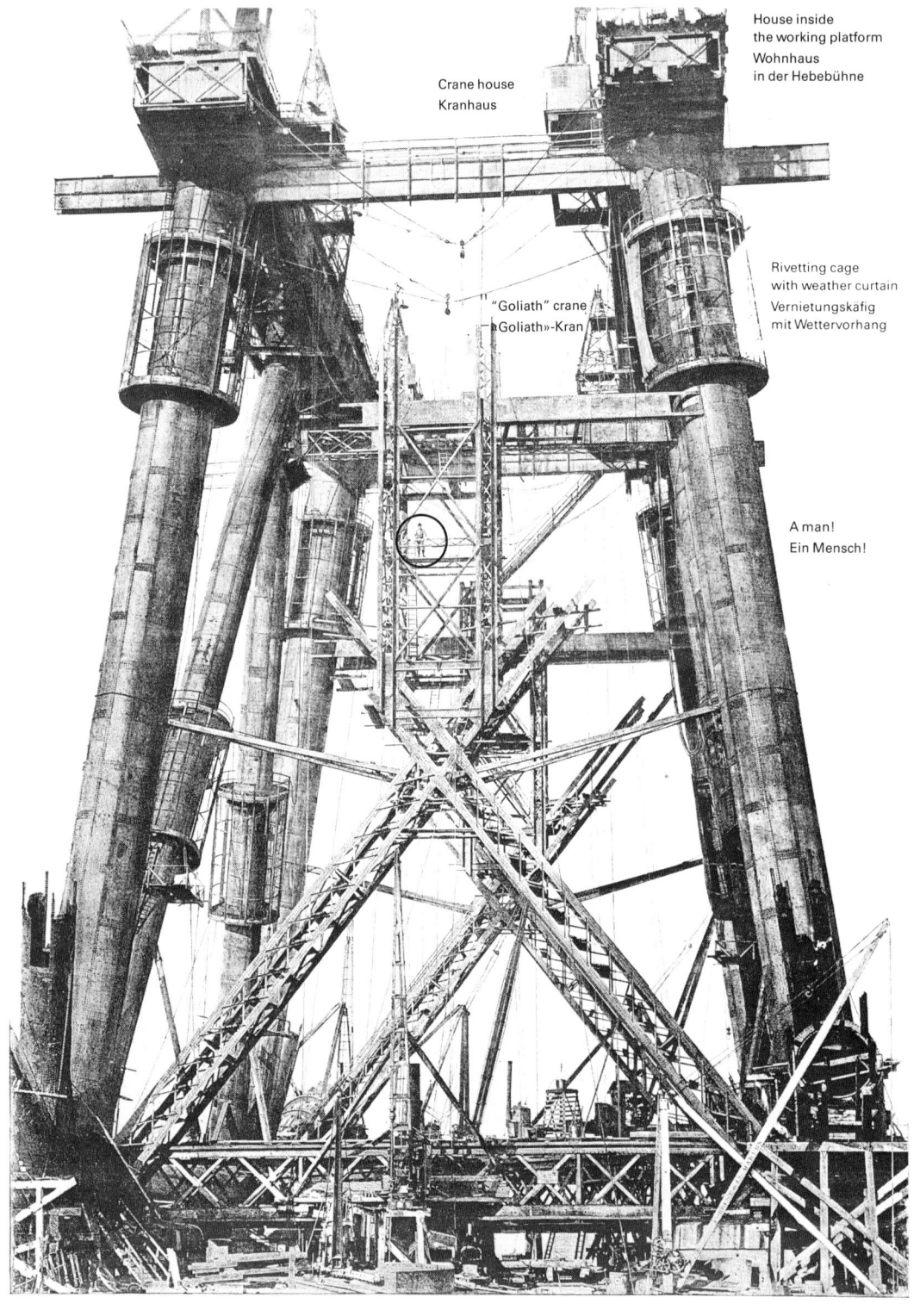

Crane house / Kranhaus
House inside the working platform / Wohnhaus in der Hebebühne
Rivetting cage with weather curtain / Vernietungskäfig mit Wettervorhang
"Goliath" crane / »Goliath«-Kran
A man! / Ein Mensch!

To Illustrate the Working Conditions

107
The dangerous nature of the working conditions is much more apparent in the semi-finished state of the central towers, than in the completed one. The assembly of the superstructure is progressing in every conceivable direction, freely upwards, and without the use of any auxiliary scaffolding. But the number of fatal accidents is also rising continually with increasing height – and will eventually claim a total of 57 lives [59, plate X].

Zur Illustration der Arbeitsbedingungen

107
Die Gefährlichkeit der Arbeitsbedingungen wird im unfertigen Zustand der Pfeilertürme viel deutlicher als im fertigen. Die Montage des Oberbaues geht in alle möglichen Richtungen frei nach oben und fast ohne Hilfsgerüste. Die Zahl der tödlichen Unfälle steigt kontinuierlich mit zunehmender Bauhöhe – auf insgesamt 57 [59, Tafel X].

108 a
"Hawes' Hotel" in South Queensferry in an old post card [15, p. 92].

108 a
«Hawes' Hotel» in South Queensferry auf alter Postkarte [15, S. 92].

108 b
The same "Hawes Inn" today [14, p. 1].

108 b
Dasselbe «Hawes Inn» von heute [14, S. 1].

"The Hawes" dates back to the 16th century and was made famous in Robert Louis Stephenson's *Kidnapped*. It was the site of that bar which Mr. Arrol used to call the "curse of the works"; and Baker thought in 1887, that the Hawes Inn was "flourishing too well for being in the middle of our works, its attractions proving irresistible for a large proportion of our 3,000 workmen..." and many dead and injured would have escaped – had it not been for the whiskey of the Hawes Inn. The bar-keeper was quoted as saying that, on pay day, he had "200 glasses sitting ready filled with liquor for the men to swallow and get outside to make room for their neighbours". Adjoining the pretty garden hemmed by hawthorns, behind the house, a sick and accident ward had been installed.

«The Hawes» stammt aus dem 16. Jahrhundert und wurde berühmt durch Robert Louis Stevensons «Kidnapped». Es war der Ort, dessen Bar Arrol den «Fluch dieses Werkes» nannte; und Baker meinte 1887, das Hawes Inn gedeihe wegen seiner Lage mitten im Arbeitsgelände viel zu prächtig, denn seine Anziehungskraft auf den Großteil der 3 000 Arbeiter sei unwiderstehlich... und viele würden noch leben – *«had it not been for the whisky of the Hawes Inn»*. Vom Schankwirt ist der Ausspruch überliefert, er hätte am Zahltag 200 Glas Whisky fertig gefüllt an der Theke stehen gehabt, damit die Männer das Zeug herunterspülen und gleich wieder Platz für ihre Kameraden machen konnten. – Am Hagedorn-umsäumten Garten hinter dem Haus hatte man die Unfall- und Krankenstation eingerichtet.

Queensferry as the "curse of the works" *(Fig. 108).* Along with the danger of the work, the consumption of whisky was a major cause for the relatively high number of deaths (57), and some 600 injured during the seven years of construction, when compared to the Tay Bridge figures.

One remark by Arrol about the German workmen shall not be missed here. Speaking at the annual social gathering of his Tay Bridge workmen in 1887, he referred to the international competition in all departments of industry, and continued:

> "The Germans, for instance, are a quiet, sober, plotting people, and unless our men stick to their work with the same steadiness and energy, it will be almost impossible for us to overcome competition." [8, p. 95]

Moreover, Arrol's Tay Bridge workmen (he had gotten the contract for the second Tay Bridge designed by Barlow) – proved that they not only could get along without drinking, but that the work there was better disciplined and more safely performed than at the Forth Bridge.

The attacks in the press about frequent accidents and insufficient safety devices – in a single three-month period in 1887, there had been seven deaths – culminated in an editorial in the *Dunfermline Journal:* "The monthly slaughter cannot be tolerated; the warnings of the government inspectors seem to be blown over the cantilevers like morning mist". The demand to have special safety inspectors appointed was strongly opposed by the engineers and contractors; they thought this would unduly delay the progress of construction. In the end the government inspectors had to agree that it was quite difficult to convince many of the men to observe even the most elementary safety precautions.

Already with commencement of construction in 1883, the FBR had set up a Sick and Accident Club. The compulsory contribution consisted of one hour's pay per week, to which the contractors gave a further 200 Pounds annually.

This common fund proved to be a veritable blessing to the men, and even more so to their wives and children, who would otherwise have been left without any help and care. In 1888 at the time the workforce was at its strongest (comprising 4,600 men) on the average 99 men per week were receiving sick allowances or compensation for accidents. Wages on the whole were above average, and since most of the work was done on a piecework basis, an able and steady worker could double or triple his normal time wages.

Finally, our chronicler Westhofen credits a new invention, the steel wire rope, with having prevented even

eine Schraube aus großer Höhe tödlich sein. Eigene Räumtrupps für die Bühnen wurden organisiert, verbunden mit strengen Sicherheitsauflagen – z. B. für Handgeländer und Schutznetze unter exponierten Lagen, dazu ständige Rettungsboote unter den Auslegern – aber ohne viel Erfolg.

Dazu erwies sich Trunkenheit als ein Dauerproblem; Sir William Arrol bemerkte einmal, er hätte auf seinen Baustellenbesuchen keinen Tag erlebt *«without seeing drunken briggers hanging about»* – und er habe dann mehr Fremdarbeiter einstellen lassen, weil sie nicht so arg tränken wie die Schotten. Auch bezeichnete er die Bar in South Queensferry *(Abb. 108)* als einen «Fluch dieses Werkes»: Neben der Arbeitsgefahr war der Whisky-Konsum eine Ursache für die im Vergleich zur Tay-Brücke hohe Zahl von 57 Todesopfern während der 7 Jahre Bauzeit, und den ca. 600 Schwer- und Leichtverletzten. Eine Bemerkung Arrol's über die deutschen Arbeiter soll hier nicht fehlen: Er sprach auf der Jahresfeier seiner Tay-Brücken-Belegschaft 1887 über den internationalen Wettbewerb auf allen Gebieten der Industrie und bemerkte:

> «Die Deutschen z. B. sind ruhige, nüchterne, arbeitsame Leute, und wenn sich unsere Männer nicht mit derselben Stetigkeit und Energie ihrer Arbeit widmen, wird es für uns schier unmöglich sein, im Wettbewerb zu bestehen.» [8, S. 95]

Im Übrigen bewiesen seine Arbeiter am Tay – Arrol hatte den Zuschlag für die zweite Tay-Brücke nach den Plänen Barlow's erhalten – daß sie nicht nur ganz gut ohne Trinken auskamen, sondern daß auch die Werkarbeit dort besser und sicherer vonstatten ging als am Forth.

Die Attacken in der Presse über häufige Unfälle und mangelnde Sicherheitsvorkehrungen – 1887 hatte es in einem Vierteljahr 7 Tote gegeben – gipfelten in einem Leitartikel des *«Dunfermline Journal»*: «Das monatliche Gemetzel kann nicht länger hingenommen werden; die Warnungen der Regierungsinspektoren scheinen über die Kragarme geweht zu werden wie Morgendunst.» Die Forderung, besondere Sicherheitsinspekteure zu ernennen, stieß auf scharfen Widerstand der Ingenieure und Unternehmer; sie meinten, das würde den Baufortschritt ungebührlich verzögern. Am Ende mußten die IHK-Inspekteure einräumen, daß es sehr schwierig sei, viele der Männer zur Beachtung elementarster Vorsichtsregeln zu überreden.

Schon bei Baubeginn 1883 war eine werkseigene Kranken- und Unfallversicherung aufgebaut worden. Der Pflichtbeitrag betrug einen Stundenlohn pro Woche, wozu die Unternehmer noch 200 Pfund jähr-

more lives and time from being lost. Whenever a single wire commenced to give way, it invariably gave "ample warning by a crackling noise" to the attendant of the lift or crane, everytime, whereas the iron chains used previously would just break without any warning. Besides, these had four to five times the weight of a steel wire rope, given equal strength.

The Everlasting Painting Job

This job is not everlasting in the sense that it gives protection from corrosion eternally; it is simply a job which never ends. The painting work begins during the dry season at one end of the bridge and by the time the other end has been reached in a rough cycle of four to six years, one has to go right back – as the saying goes – to where one started; the job never ends. Of the total of 40 men who are constantly engaged in the maintenance of the bridge, no less than 16 painters are steadily at work. The paint is supplied, up to this day, by the firm Craig & Rose of Leith. Craig & Rose had won the original painting contract, and ever since they have been manufacturing their "Forth Bridge Oxide of Iron Brushing Paint" – with justifiable pride and according to their own specifications. The paint has to cover a total area of 145 acres ($1/2$ km^2) four coatings of which weigh 50 tons. Over two coats of red lead already mentioned, a priming coat of "dark chocolate brown" is applied, whereas the finishing coat is of a "bright Indian or Persian red" which, however, darkens considerably within a short time. Considering the bridge's exposed position engulfed by sea-spray and winds, this Sisyphean job is highly necessary – albeit it can be rather relaxing to watch.

The Creators of the Bridge

Sir John Fowler (1817–1898)

Sir John Fowler's unparalleled career as railway engineer coincides with the 50 most important years of this profession. Following the great names of the cast-iron age, such as Telford, Trevithick, Watt, Rennie, there came a new generation now, that of the steel builders. Born in Sheffield in 1817, he went to work as an apprentice for a hydraulic engineer, at the age of 17. It was a time when after the close of the Napoleonic wars, the whole world became the customer of British industry. Subsequently he worked with various railway engineers in laying out England's first railway lines; on the line from London to Brighton there is scarcely a bridge which was not designed by him. His great versatility was typical for a time when

lich beisteuerten. Diese Gemeinschaftskasse erwies sich als ein wahrer Segen für die Männer, und noch mehr für ihre Frauen und Kinder, die sonst ohne jede Hilfe und Fürsorge geblieben wären. 1888, zur Zeit der höchsten Belegschaft mit 4600 Mann, bezogen durchschnittlich 99 Leute pro Woche Kranken-oder Unfallgeld. Die Löhne lagen höher als üblich, und da die meisten Arbeiten nach Stücklohn vergütet wurden, konnte es ein geschickter und stetiger Arbeiter auf das 2 – 3fache seines normalen Zeitlohnes bringen.

Schließlich dankt es der Chronist Westhofen einer neuen Erfindung – dem geflochtenen Stahlseil – daß nicht noch mehr Menschenleben und Zeit verloren gingen: Wenn ein Seil einmal ausgeleiert war, gab es dem Aufzugs- oder Kranführer jedesmal «*ample warning by a crackling noise*», wogegen die früher benutzten eisernen Ketten ohne Vorankündigung zu brechen pflegten. Zudem hatten sie bei gleicher Tragkraft das 4 – 5fache Gewicht eines Drahtseiles.

Der ewige Anstrich

Dieser ist nicht ewig im Sinne eines immerwährenden Rostschutzes – sondern «ewig», weil er nie aufhört: Die Streicharbeiten beginnen in der trockenen Jahreszeit an einem Ende, und ist man schließlich – so geht die Rede – im Turnus von 4 bis 6 Jahren am anderen Ende angelangt, muß man vorne wieder anfangen – so daß die Arbeit niemals aufhört. Von den insgesamt 40 Mann des Brückenunterhaltes sind allein 16 Maler ständig am Werk. Die Farbe dazu liefert bis heute die Firma Craig & Rose aus Leith: Sie hatte seinerzeit den Zuschlag erhalten und stellt noch immer ihre rote «*Forth Bridge Oxide of Iron Brushing Paint*» her – nicht ohne Stolz und nach Haus-Rezept. Die Farbe muß eine Gesamtfläche von 145 *acres* – über 1/2 km^2 – bedecken und wiegt in 4 Schichten allein 50 Tonnen. Über zwei Lagen von Mennige kommt eine Grundierung in «dunklem Schokolade-Braun» und darauf ein Deckanstrich von «hellem Indisch- oder Persisch-Rot», das allerdings rasch nachdunkelt. Bei der exponierten Lage mit salzhaltigem Wasser und Wind ist diese Sysiphus-Arbeit dringend notwendig – und dennoch recht geruhsam anzuschauen.

Die Schöpfer der Brücke

Sir John Fowler (1817 – 1898)

Sir John Fowler's beispiellose Karriere als «*railway engineer*» fällt in die wichtigsten 50 Jahre dieses neuen Berufes: Nach den großen Namen der Gußeisen-Zeit, Telford, Trevithick, Watt, Rennie, kam eine

knowledge was gained less from theory than out of practical experience, every step marking a separate experiment on a working scale. There simply were no text-book examples!

He started an independent career at the age of 26, planning and building a number of new railways, when "Railwaymania" was just emerging and whole fortunes were made and lost in a single night – a time which only men of iron constitution came through unscathed. Wild promoters and speculators were dealing in railway shares, and ever new proposals were being brought before Parliament, or defeated by competing engineers. Fowler himself fought through quite a few such proposals – an example is his "Great Grimsby Railway" which came to stand for work of solid and successful engineering. So adept had Fowler become at this, that on one occasion in presenting his ever latest railway proposal to the House of Commons, he was given the summary approval by Lord Devon who exclaimed: "What, Great Grimsby again! Go it, Great Grimsby!"

Fowler's reputation now required his permanent residence in London, where commissions – of a magnitude hard to imagine today – were flowing in from all directions: At least 25 railway lines, along with railway stations, bridges and water works. But the great event in London, in 1860, was seen in his building the Metropolitan Railways (today's Metropolitan Line of London's underground system). The difficulties of such a novel undertaking were enormous, and engineers of the highest eminence were uttering gloomy forebodings that this underground railway could never be built; that if it could, it would not work; and that if it did, still nobody would travel by it. Despite all this nay saying, the confidence placed in Mr. Fowler was unshakable, and the volume of traffic after completion surpassed even the keenest expectations – notwithstanding Lord Palmerston in his opening speech remarking to Fowler: "I intend to keep above ground as long as I can." Subsequently, the construction of the "Inner Circle Line" was initiated; but Fowler's far-sighted plan for a comprehensive "Outer Circle Line" was rejected – whereby London had lost for all times the chance for an outer ring of tube railways.

When Fowler became President of the Institution of Civil Engineers, he called attention to the fact that "the exclusive position hitherto held by the English engineers" was not likely to be held forever in the world, as indeed French and German engineers were pulling up fast. Being a real all-round-man he stood for that type of the early railway engineer who was equally versed in surveying, geology, the design of bridges, tunnels and water works, and who ideally also had "some

109
Sir John Fowler (1817–1898)
[29, plate I].

109
Sir John Fowler (1817–1898)
[29, Tafel I].

neue Generation, die der Stahlbauer. 1817 in Sheffield geboren, ging er mit 17 in die Lehre bei einem Wasserbau-Ingenieur – zu einer Zeit, als nach dem Ende der napoleonischen Kriege die ganze Welt zum Kunden britischer Industrie wurde. Es folgten Mitarbeit bei verschiedenen Eisenbahn-Ingenieuren und am Bau der ersten Bahnlinien Englands; auf der London-Brighton-Strecke gibt es kaum eine Brücke, die nicht von ihm ist. Seine große Vielseitigkeit war typisch für die Zeit, in der neue Erkenntnisse weniger aus der Theorie als aus der praktischen Erfahrung kamen und jeder Schritt ein eigenes Experiment im Maßstab 1:1 bedeutete: Es gab noch keine «textbook examples»!

Mit 26 bereits selbständig, plante und baute Fowler eine ganze Reihe neuer Eisenbahnlinien, als die «Railwaymania» gerade anfing und ganze Vermögen in einer Nacht verspielt wurden – eine Zeit, die nur Männer mit eiserner Konstitution unbeschadet überstanden: Wilde *promoters* und Spekulanten handelten mit Eisenbahn-Aktien, und immer neue Pläne wurden im Parlament von konkurrierenden Ingenieuren propagiert bzw. befehdet. Fowler selbst focht mehrere solcher Vorlagen durch – wie z.B. seine «Great Grimsby Railway» zu einem Inbegriff solider und erfolgreicher Ingenieursarbeit geworden war. Das ging soweit, daß Fowler bei der Vorlage der x-ten neuen Bahnlinie im House of Commons von Lord Devon mit dem Ausruf «What, Great Grimsby again! Go it, Great Grimsby!» die pauschale Billigung bekam.

Fowler's Ruf erforderte nun seinen Wohnsitz in London, und Aufträge – heute unvorstellbaren Ausmaßes – kamen von allen Seiten: Mindestens 25 Eisenbahnlinien, dazu Bahnhöfe, Brücken, Wasser-

knowledge of architecture, and a taste for appropriate decoration". Finally, he saw the engineer not just as a man of technical skill to the service of capitalists, but as a member of an honourable and noble profession, which could not lend itself to enterprises which did not give fair promise of being beneficial to the world, and to the advancement of civilisation. Fowler's great successes were based on his combination of practical thinking with an extraordinary talent in dealing with people.

He went to India as a consulting engineer, and from 1868 onwards he made several trips to Egypt where he worked on various major projects, such as the famous Sudan Railway. Probably his most daring scheme was that for a ship incline to pass from the lower to the upper level of the Assuan cataracts. The boats were to be floated upon a cradle running along a 3-km railway, and to be hauled over land by hydraulic engines of 400 hp. Here, the force of the cataract itself was to propel the engines – an ingenious piece of *"perpetuum mobile"*, which was thwarted however "by jealous foreign rivals". Although most of Fowler's projects devised for Egypt were not carried out in the end, they were not in vain, as in 1885 he was knighted – "for important services and guidance to Her Majesty's Government in connection with Egypt".

In retrospect, the Forth Bridge – though Fowler's best known work – is just one of many, when seen in the context of his overwhelming career. The struggle in Parliament for this bridge had been extremely stubborn, and the final approval would hardly have come about without Fowler's perseverance; he had managed to remove the last doubts regarding this extraordinary undertaking by his proposal that the inspectors should report to Parliament every three months on the progress of the work and the quality of material and workmanship – all these reports later on were of the highest praise, of course. Fowler and Baker kept personal control over the entire construction process, Fowler being mainly responsible for the masonry viaducts and Baker for the actual superstructure.

Sir Benjamin Baker (1840–1907)

Whereas Fowler was the widely-experienced and well-versed practitioner of railway engineering, Baker may be described as the young theoretician of bridge design in Britain. When he began his career, Fowler had already been actively engaged in many major works for more then 20 years. Following the era of the great pioneers, engineering now began to change into a science, which replaced the more or less experimental pursuit by precise calculations. The rich experience of

werke. Das große Ereignis in London aber war sein Bau der *Metropolitan Railways* – die heutige *Metropolitan Line* der Londoner U-Bahn, 1860. Die Schwierigkeiten einer so neuartigen Sache waren enorm, und Ingenieure höchsten Ansehens versicherten immer wieder, daß diese U-Bahn nie gebaut werden könne, und wenn ja, daß sie nie funktionieren würde, und wenn doch, daß niemand damit fahren würde. Trotzdem vertraute man Mr. Fowler und der Verkehr nach Fertigstellung übertraf alle Erwartungen – selbst wenn Lord Palmerston noch bei seiner Eröffnungsrede zu Fowler sagte: «*I intend to keep above ground as long as I can.*» Es folgte der Bau der «*Inner Circle Line*» in London, jedoch wurde sein weitsichtiger Plan für eine umfassende «*Outer Circle Line*» abgelehnt – womit London für alle Zeiten die Chance eines äußeren U-Bahn-Ringes verlor.

Als Präsident des britischen Ingenieur-Verbandes erinnerte Fowler daran, daß die «exklusive Stellung des englischen Ingenieurs» keinen Ewigkeitsanspruch auf der Welt habe, ja, französische und deutsche Ingenieure rasch aufholten. Als wirklicher *all-round-man* trat er für jenen Typus des frühen Eisenbahn-Ingenieurs ein, der gleichermaßen in Vermessung, Geologie, Brücken-, Tunnel- und Wasserbau versiert sein mußte, aber auch «*some knowledge of architecture... and a taste for appropriate decoration*» haben solle. Auch sah er ihn nicht als einen bloßen Technik-Handlanger von Kapitalisten, sondern als Mitglied eines ehrenwerten und noblen Berufes, der sich nicht für Geschäfte hergeben könne, die nicht Gemeinwohl und Zivilisation zu fördern versprächen. Fowler's große Erfolge beruhten auf seiner Kombination praktischen Denkens mit einem großen Talent im Umgang mit Menschen.

Als Konsultant ging Fowler nach Indien und ab 1868 mehrmals nach Ägypten, wo er an verschiedenen Großprojekten tätig war, so am Bau der berühmten Sudan-Bahn. Sein wohl kühnstes Projekt war ein schräger *Schiffsaufzug über die Assuan-Fälle:* Die Schiffe sollten auf einem schlittenartigen Gestell aufschwimmen, das auf einem 3 km langen Schienenweg von stationären 400 PS Hydraulik-Motoren hochgezogen würde. Dabei sollte der Wasserfall selbst für den Antrieb der Motoren sorgen – ein geniales Stück «*perpetuum mobile*», das von «eifersüchtigen ausländischen Rivalen» zu Fall gebracht wurde! Auch wenn die meisten von Fowler's Ägypten-Projekten nicht ausgeführt wurden, waren sie nicht umsonst, denn 1885 machte ihn die Königin zum «*Sir – for important services and guidance to Her Majesty's Government in connection with Egypt*».

Die Forth-Brücke endlich – wenn auch Fowler's be-

110
Sir Benjamin Baker
(1840–1907)
[29, plate II].

110
Sir Benjamin Baker
(1840–1907)
[29, Tafel II].

the predecessors was now combined with the technical possibilities of rapid industrialisation in all fields. Born in Glamorganshire in 1840, Baker began to work at the age of 16 (which was Fowler's age, when he started out as a surveyor) as an apprentice in the well-known South-Wales Iron Works at Neath Abbey. There, already around 1790 the first cast-iron pumping engines had been built, as well as strange locomotives with spur gearing and racks. After spending several years in various engineering offices, Baker gained practical experience in iron smelting, surveying and brickwork, and finally entered the office of W. H. Wilson in London, to work as an assistant in the construction of Victoria Station.

Eventually, Baker joined Fowler's London office in 1862, to become his partner in 1875. It was a most fruitful collaboration which was to last until Folwer's death in 1898. Among numerous other projects, Baker took an active part in the construction of the two underground railways. On the side, he constantly published articles on the design of bridges, especially regarding the cantilever principle. His articles on "Long-Span Bridges" of 1867 were republished also in America, Germany, Austria and Holland. As time went on he became engaged in many important bridge projects in all parts of the world, and also in an expertise on the early Channel Bridge-project (as promoted and designed by Schneider & Hersent). Baker even took upon himself the task of preserving and strengthening three great historical bridges by the legendary Telford; namely, the Menai Suspension Bridge *(Fig. 4)*, the Buildwas cast-iron arch bridge, and the masonry arch bridge across the Severn near Gloucester. He succeeded in thwarting

kanntestes Werk – ist im Lichte seiner überwältigenden Laufbahn nur eines von vielen. Der Kampf im Parlament um die Brücke war außerordentlich hartnäckig, und ohne das Durchhaltevermögen Fowler's wäre die Genehmigung kaum zustande gekommen; letzte Zweifel an einem so unerhörten Vorhaben räumte er mit dem Vorschlag aus, alle 3 Monate Inspektionsberichte über Baufortschritt und Arbeitsqualität vorlegen zu lassen – und all diese «reports» waren natürlich später voll höchsten Lobes. Während der ganzen Bauzeit überwachten Fowler und Baker persönlich alle Arbeiten, Fowler verantwortlich für die Mauerwerks-Viadukte und Baker für den eigentlichen Brücken-Oberbau.

Sir Benjamin Baker (1840 – 1907)

War Fowler der allseits versierte und gestandene Praktiker des Eisenbahnbaus, so kann man Baker als den jungen Theoretiker des Brückenbaus in England bezeichnen. Als er seine Laufbahn begann, war Fowler schon über 20 Jahre an vielen Großbauten tätig gewesen. Nach dem Zeitalter der großen Pioniere wurde das Ingenieurwesen nun zu einer Wissenschaft, welche die mehr oder weniger experimentellen Vorgehensweisen durch exakte Berechnung untermauerte: Man verband die reiche Erfahrung der Alten mit den technischen Möglichkeiten rapider Industrialisierung auf allen Gebieten. In Glamorganshire geboren, fing Baker – so wie Fowler als Vermesser begonnen hatte – mit 16 Jahren als Lehrling in den bekannten *South Wales Ironworks* von *Neath Abbey* an. Dort waren bereits um 1790 die ersten gußeisernen Pumpmaschinen gebaut worden – sowie später absonderliche Zahnrad-Lokomotiven. Nach mehreren Jahren in verschiedenen Ingenieurbüros kam Baker über praktische Erfahrung in Eisengießerei, Vermessung und Maurerhandwerk nach London, um unter W. H. Wilson am Bau der *Victoria Station* mitzuwirken.

Schließlich kam Baker 1862 zu Fowler's Büro in London und wurde 1875 sein Partner – eine fruchtbare Zusammenarbeit, die bis zu Fowler's Tod 1898 dauern sollte. Neben vielen anderen Projekten leitete Baker dort den Bau der beiden U-Bahnen. Daneben veröffentlichte er regelmäßig Artikel zum Brückenbau, insbesondere über das Kragträger-Prinzip: Seine Arbeit über «Long-Span Bridges» von 1867 erschien auch in Amerika, Deutschland, Österreich und Holland. Schließlich war er an zahlreichen Brückenprojekten in der ganzen Welt tätig, sogar an einem Gutachten über eine frühe Großbrücke über den Ärmelkanal (propagiert von Schneider und Hersent). Baker fiel bereits die denkmalpflegerische Aufgabe zu,

the planned demolition of the Buildwas Bridge and the attempts to change to the appearance of the Menai and Severn Bridge of his great predecessor.

Apart from these tasks, Baker was also working as a consulting engineer, and in 1878 he helped the contractor John Dixon with the perilous task of transporting the large obelisk, "Cleopatra's Needle", from Egypt to England, and once in London, to bring it safely back into the vertical again. This was followed later by Baker's consulting work at the Assuan Dam in Egypt, 1894–1902. In all his enthusiasm for the Forth Bridge, Baker always drew attention to the collective background of a century of manifold and anonymous ingenuity – and did so in a manner marked by wholesome understatement, barely concealed pride and missionary zeal, so characteristic of the era:

> "The merit of a design, if any, will be found, not in the novelty of the principle underlying it, but in the resolute application of well-tested mechanical laws and experimental results to the somewhat difficult problem offered by the construction of so large a bridge across so exposed an estuary as the Firth of Forth.... Where no precedent exists, the successful engineer is he who makes the fewest mistakes."

Baker was no friend of the popular delusion that a great work must necessarily be the work of a great man, but – in fact there were plenty of other people around "in England, Europe and America", who were just as capable of building a Forth Bridge (quod est demonstrandum). From 1883 to 1890, Baker virtually lived at the site, together with the men on the Forth. Like Fowler, he later became President of the Institute of Civil Engineers, as well as laying the groundwork for the British Standards Institution – equivalent to the German DIN-Standards.

In contributing articles to the *Encyclopedia Britannica*, Baker dealt professionally with everything that was fascinating by virtue of its novelty and difficulty, such as ship-railways, tunnelling, and even guns for the War Department; the time of narrow specialisation had not arrived as yet. At the opening ceremony of the Forth Bridge on March 4, 1890, he was knighted by the Prince of Wales; Sir Baker riposted dryly, in memory of poor Sir Thomas Bouch:

> "I do not believe in astronomy being a safe guide for practical engineering."

Sir William Arrol (1839–1913)

Sir William Arrol – the third of the lot – was, like so many entrepreneurs of his time, a self-made man. He rose from the humblest ranks to become one of Brit-

drei historische Brücken des legendären Telford zu sichern; die Menai-Hängebrücke *(Abb. 4)*, die Buildwas-Gußeisen-Bogenbrücke und die Mauerwerk-Bogenbrücke über den Severn bei Gloucester; es gelang ihm auch, den geplanten Abriß bzw. Änderungen am Erscheinungsbild der Menai- und der Severn-Brücke seines großen Vorgängers zu verhindern.

Daneben arbeitete Baker auch als beratender Ingenieur und half 1878 John Dixon bei dem gefährlichen Geschäft, den großen Obilisken – «*Cleopatra's Needle*» – von Ägypten nach England zu schaffen und dann auch noch in London wieder heil in die Senkrechte zu bringen! Dazu kam später Baker's beratende Tätigkeit am Assuan-Damm in Ägypten, 1894–1902. – Bei allem Enthusiasmus für die Forth-Brücke wies Baker immer wieder auf den kollektiven Hintergrund eines Jahrhunderts vielfältiger und anonymer Könnerschaft hin – mit jener Mischung von bekömmlichem *understatement,* kaum verhohlenem Stolz und Sendungsbewußtsein, welche die Epoche kennzeichnet:

> «Das Verdienst eines Entwurfes findet sich, wenn überhaupt, nicht in der Neuartigkeit der ihm zugrundeliegenden Prinzipien, sondern in der beherzten Anwendung wohlerprobter mechanischer Gesetze und experimenteller Ergebnisse auf das einigermaßen schwierige Problem, welches die Konstruktion einer so großen Brücke über einen so weiträumigen Meeresarm wie den Firth of Forth darstellt... Wo es keinen Präzedenzfall gibt, ist derjenige Ingenieur erfolgreich, welcher die wenigsten Fehler macht.»

Baker war kein Freund jener populären Illusion, daß ein großes Werk automatisch die Tat eines großen Mannes sein muß, sondern daß es «in England, Europa und Amerika» noch viele andere Leute gäbe, die ebensogut eine Firth-of-Forth-Brücke hätten bauen können (quod est demonstrandum). Von 1883–1890 wohnte Baker praktisch auf der Baustelle mit den Leuten am Forth. Später wurde er so wie Fowler Präsident des britischen Ingenieur-Verbandes und begründete das Regelwerk der *British Standards Institution* – die britischen DIN-Normen.

Als Mitarbeiter der «Encyclopedia Brittannica» befaßte sich Baker mit allem, was in seiner Neuartigkeit und Schwierigkeit faszinierte, wie z.B. Schiffs-Schienenbahnen, Tunnelbauten, ja Kanonen für's Kriegsministerium, denn es war noch nicht die Zeit engen Spezialistentums. Bei der Eröffnung der Forth-Brücke am 4. März 1890 wurde Benjamin Baker vom Prince of Wales geadelt: Trocken bemerkte Sir Baker, in Erinnerung an den armen Sir Thomas Bouch:

111
Sir William Arrol
(1839–1913)
[29, p. 70].

111
Sir William Arrol
(1839–1913)
[29, S. 70].

ain's most respected contractors. Born in Houston, Renfrewshire, as the fourth of nine children, he was the son of a cotton spinner, who was already unable to work when Arrol was still in his infancy. So at the age of nine, he too entered the cotton mill, and when the family moved to Paisley, he apprenticed himself to a blacksmith at the age of fourteen. At the same time he was attending nightschool in the shop of a weaver who taught mathematics. Four years later he went to England as a journeyman, as there was no work to be had in Scotland – mending porridge pots and kitchen utensils on his way to setting up his own business. After returning to Scotland, whether working as a blacksmith, fitter or boilermaker, his work was always characterised "by the touch of originality which distinguishes the man born to command from the one made only to obey". With energy, industry and versatile talent, Arrol was searching for ever new and bolder ways out of the grooves prescribed by established handicraft. Before long he became foreman of a large bridge and boiler works in Glasgow; at last, in 1868, he did set up his own business – with the pitiful capital of 85 Pounds, his life's savings. Boldly launching himself as a "Contractor and Repairing-Engineer" he purchased – a mere 20 years before the Forth Bridge was completed – his first engine and the few tools he could afford. In 1872 he set up his Dalmarnock Works, made a name for himself within three years, and got the contract for three bridges. His Clyde Bridge at Bothwell for the NBR proved that continuous cantilever girders could very well be assembled on land, whereafter they would be rolled out from pier to pier across the water, thus saving a lot of time, money

«Ich halte die Astronomie für keine sichere Anleitung zum praktischen Ingenieur-Bau.»

Sir William Arrol (1839 – 1913)

Sir William Arrol – der Dritte im Bunde – war wie so viele berühmte Unternehmer seiner Zeit ein *«self-made man»*. Er brachte es von einfachsten Verhältnissen zu einem der angesehensten Bauunternehmer Großbritanniens. Er wurde in Houston, Renfrewshire, als viertes von neun Kindern und Sohn eines Baumwollspinners geboren. Als Arrol noch im Kindesalter war, wurde sein Vater bereits arbeitsunfähig. So ging er selbst mit 9 Jahren in die Spinnerei, und als die Familie nach Paisley zog, arbeitete er mit 14 als Lehrling in einer Schmiede. Gleichzeitig bildete er sich weiter und besuchte in der Werkstatt eines Webers, der Mathematik lehrte, die Abendschule. Vier Jahre später ging er als Wandergeselle nach England, da Arbeit in Schottland nicht zu bekommen war – und reparierte Porridge-Töpfe und Küchengeräte für seinen Traum eines selbständigen Unternehmers. Nach Schottland zurückgekehrt, zeichneten sich seine Arbeiten – gleich ob als Schmied, Monteur oder Boilermacher, immer «durch jenen Hauch zum Originellen aus, der eine Führerpersönlichkeit vom bloßen Handwerker unterscheidet». Mit Energie, Fleiß und Vielseitigkeit suchte Arrol immer neue und kühnere Wege aus den Gleisen des alt-etablierten Handwerks, wurde Werkmeister eines großen Brücken- und Boiler-Werkes in Glasgow und machte sich schließlich 1868 selbständig – mit dem lächerlichen Kapital von 85 Pfund, seinen ganzen Ersparnissen. Als «Bau-Unternehmer und Reparatur-Ingenieur» kaufte er sich – nur 20 Jahre vor Vollendung der Forth-Brücke – seine ersten Maschinen und die paar Werkzeuge, welche er sich leisten konnte. 1872 gründete er seine Dalmarnock-Werke, machte sich in drei Jahren einen Namen und erhielt drei Brückenaufträge. Seine Clyde-Brücke in Bothwell für die *NBR* bewies, daß man die Durchlaufträger auch an Land montieren und sie dann von Pfeiler zu Pfeiler über das Wasser vorschieben konnte, was viel Zeit, Geld und Arbeitskraft sparte. Und mit der Clyde-Brücke in Broomielaw für die *CR* erfand er jene speziellen Bohr- und Nietmaschinen, die dann beim Bau der Forth-Brücke eine so entscheidende Rolle spielten. Da man noch nicht schweißen konnte, suchte man ständig nach besseren Vernietungsmethoden. Mit Arrol's Erfindung schaffte man 7 Vernietungen je Minute, zum halben Preis der früheren Hand-Nietung.
1873 erhielt Arrol den Auftrag der *Forth-Bridge-Company* für den Bau von Sir Thomas Bouch's Hängebrücke. Als die Sache mit dem Einsturz der Tay-Brücke

and workforce. And with the Clyde Bridge for the CR at Broomielaw, he invented those special drilling and rivetting machines which later were to play such a decisive role during the construction of the Forth Bridge. As the technique of steel welding had not been invented as yet, one was constantly searching for improved rivetting methods. Arrol's invention made it possible to close 7 rivets per minute, at half the cost of the old hand rivetting.

In 1873, Arrol won the contract to carry out Sir Thomas Bouch's design for the suspension bridge across the Forth. When construction was stopped following the failure of the Tay Bridge, Arrol in 1882 was given also the contract for reconstructing the Tay Bridge, in addition to his contract for the new Forth Bridge by Fowler and Baker. Thus, the two greatest bridge-building projects of the world were placed into Arrol's hands – an inconceivable amount of responsibility for the former cotton spinner! An impression of his strenuous weekly working schedule may be fitting here; he kept to this routine for the full time it took to complete both bridges:

He would rise every Monday morning at 5, to be at his Glasgow works by 6, where he would carefully go over all the work on the two great structures with his staff. He would then walk to the train station, stopping on the way for a quick breakfast at Miss Cranston's restaurant (having watched him for years she thought he was a sailor, until a friend told her of his true identity.) At 8:45 a.m. he boarded the train to the Forth Bridge and worked there for the remainder of Monday and Tuesday, before catching the Dundee train to arrive at the Tay Bridge about 11 p.m. At 6 on Wednesday morning he was at the site, returning to Glasgow late that night, in order to be back at his works by 6 on Thursday morning. But the same day he would again travel to the Forth and on to the Tay at night. Finally on Friday night, he would take the sleeper for the long trip to London, in order to spend Saturday discussing progress with the consulting engineers. He would then return to Scotland on the Saturday night train, unless business in London demanded him to stay through Sunday, when he made sure not to miss hearing the famous preacher at Spurgeon's Tabernacle. Today we would call someone like William Arrol a classic "workaholic" since his creative power and stamina appear to have taken on super-human dimensions.

Subsequently, the world-wide acclaim for the Forth and Tay Bridges brought in further contracts to build the Tower Bridge in London (opened 1894), the Nile Bridge in Cairo (1904–1908) and the Wear Bridge in Sunderland (1905–1909). By the end of the century, his firm was the largest structural engineering com-

eingestellt wurde, erhielt Arrol 1882 auch den Auftrag für Fowler's und Baker's neue Forth-Brücke, sowie für den Wiederaufbau der Tay-Brücke: Damit waren die beiden größten Brückenbauprojekte der Welt in Arrol's Hand – eine kaum vorstellbare Fülle von Verantwortung für den ehemaligen Baumwollspinner! Hierzu ein Bild seines anstrengenden Wochenablaufes, den er für die gesamte Bauzeit beider Brücken aufrechterhielt:

Nach dem Aufstehen um 5.00 Uhr war er jeden Montagmorgen um 6.00 Uhr im Glasgower Werk, um alle Arbeiten und Pläne mit seinen Leuten durchzugehen. Dann pflegte er auf dem Weg zum Bahnhof für ein rasches Frühstück in Miss Cranston's Restaurant einzukehren (sie beobachtete ihn über Jahre und dachte, er sei wohl ein Seemann, bis ihr jemand verriet, wer er wirklich war). Um 8.45 Uhr bestieg er den Zug zur Forth-Brücke und arbeitete dort den restlichen Montag und Dienstag, um dann noch mit dem letzten Abendzug nach Dundee zur Tay-Brücke zu fahren, mit Ankunft dort um 23.00 Uhr. Mittwochmorgen um 6.00 Uhr war er an der Baustelle und fuhr spät abends zurück nach Glasgow, um am nächsten Morgen wieder im Werk zu sein. Aber gleichen Tages fuhr er nochmals zum Forth und abends weiter zum Tay, um Freitag abends schließlich den Schlafwagen für die Fahrt nach London zu besteigen und dort am Samstag zur Beratung mit den beteiligten Ingenieuren zusammenzutreffen. Danach fuhr er den langen Weg zurück nach Norden, wenn nicht die Geschäfte sein Bleiben auch sonntags in London erforderten – dabei versäumte er aber nicht, den berühmten Prediger von Spurgeon's Tabernacle zu hören: Heute würde man William Arrol als einen klassischen «workaholic» bezeichnen und seine Schaffenskraft erscheint fast übermenschlich.

Später verschaffte ihm die weltweite Anerkennung für die Forth- und Tay-Brücke weitere Aufträge zum Bau der Tower-Brücke in London (1894 eröffnet), zu einer Nil-Brücke in Kairo (1904 – 1908) und zur Wear-Brücke in Sunderland (1905 – 1909). William Arrol war in der Tat so etwas wie der Traum eines Viktorianers – tatkräftig, erfinderisch, fromm, und hatte sich mit Ideenreichtum und Fleiß zu Ruhm und Adelstitel emporgearbeitet. Das Vertrauen von Fowler und Baker in Arrol's Fähigkeiten war so groß, daß sie bedenkenlos alle Bauaufgaben in seine Hände legen konnten – «and needless zu say, Mr. Arrol has justified the confidence placed in him».

Es ist nur angemessen, diese Darstellung der Forth-Brücke und ihrer Erbauer mit den Worten des Chronisten Wilhelm Westhofen von 1890 auch zu beenden:

«Es ist kein Wunder, daß sich die Ingenieure alter Schule mit solcher Vielseitigkeit von einem

pany in the country, employing some 5,000 men. William Arrol indeed seemed to embody the ideal Victorian; he was energetic, inventive, God-fearing, and his breadth of imagination and diligence had brought him fame and a knighthood. The trust placed into Arrol's abilities by Fowler and Baker was such that they could leave all tasks to him without worry – "and needless to say, Mr. Arrol has justified the confidence placed in him."

It seems only fitting here to conclude this account of the Forth Bridge and its creators with another observation by our tireless chronicler Westhofen, of 1890:

> "It is no wonder that those engineers of the old school can turn from one subject to another with so much versatility when we consider what an education they had. Instead of having professors to fill them with ready digested knowledge like the young men of the present day, they were moved from one position of responsibility to another, and their intellects were hardened and invigorated by constant work. Every step they took was an experiment on a working scale, and every fact they learned was imprinted on their memories by the toil and trouble it had cost."

A Few Figures

Total length of the bridge including viaducts:	8,296 ft (2528.62 m)
Length without viaducts:	5,330 ft (1624.58 m)
Two main spans:	1,710 ft each (521.21 m)
Length of cantilever:	680 ft (207.26 m)
Maximum height:	343 ft (104.55 m)
Clear height for shipping:	150 ft (45.72 m)
Weight:	50,958 tons
Weight per meter of bridge length:	31,5 tons/m
Total number of rivets:	7 million
Thickest pack of steel plates rivetted up:	9 inches (22.9 cm)
Area of paint in 4 coatings:	145 acres (580 000 m^2)
Duration of construction:	7 years
Maximum number of people employed in 1888:	4,600 men
Construction costs including approach railways:	2,549,200 pounds
Construction costs for beginnings of suspension bridge:	250,000 pounds
	2,799,200 pounds
Parliamentary, engineers' and interest fees:	378,006 pounds
Total costs as of January 1, 1890:	3,177,206 pounds
(of which 1/3 is accounted for by wages)	
Wage per ton of weight for the superstructure:	6 £ 17 s. 9 p.
Average hourly wage:	5 pennies
Deaths:	57 lives

Conclusions

In order to arrive at some kind of resumé regarding the significance of *both* of the bridges, we have to separate their practical usefulness from their wider impact. The full service to the general public was in fact felt only from 1890 onwards, when the Forth Bridge had been completed and the Tay Bridge reconstructed in its present form. Only then did there exist a direct and continuous railway line over land (instead of by ferries) from Edinburgh to Aberdeen. As to the political impact of transportation, the accessibility of northern Scotland was of course eminently important to the industrialisation of the country, and for providing access to modern jobs for a poor population. Considering the situation, however, from a nationwide perspective, some geographic facts must not be overlooked: In terms of population distribution, the lucrative lines were situated for the most part to the south of the line Edinburgh-Glasgow, namely on the long north-south connections to middle and southern England, especially along the two direct coastal lines of the NBR and the CR to London. It almost appeared as if those new bridges were ahead of their time; the railway connections were by no means improved so dramatically as might have been expected.

The reason for this primarily had to do with the condition of the British railways. Even 50 years after their introduction, the trains between Edinburgh and London were still quite slow. There were no reliable emergency braking systems, and to the safety consciousness of the traffic companies, 40 m.p.h. (64 km/h) was considered a quite acceptable average speed. Only the private competition between the great rivals of the east and west coasts stirred up the scene; therefore, it might be useful to take a look here at this curious period of the "races" of 1895. Back in 1887, the east coast companies on their *Flying Scotsman* still needed 9 hours to travel from London to Edinburgh. By 1888, "east coast" and "west coast" began to literally race each other along this run. By the end of the year, Edinburgh could be reached already in less than 8 hours, Perth in 11 and Aberdeen in just under 14 hours. But then the "east coast" decided that $8^1/_2$ hours was fast enough for this run – as any further reduction would be unsafe and too costly. The rolling stock, the tracks and especially the lengthy interim stops, caused by over-

Ein Schlußwort

Will man abschließend ein Fazit über die Bedeutung beider Brücken ziehen, so ist zwischen ihrem praktischen Nutzen und der weiteren Wirkung zu unterscheiden. Der unmittelbare Nutzen für die allgemeine Öffentlichkeit begann eigentlich erst ab 1890, als die Forth-Brücke fertig und die Tay-Brücke in ihrer neuen Form wiederaufgebaut war. Erst von da an bestand eine direkte und durchgehende Eisenbahnverbindung über Land (anstatt über Fähren) von Edinburgh nach Aberdeen. Verkehrspolitisch war natürlich die Erschließung Nordschottlands von großer Bedeutung für die Industrialisierung des Landes, und damit für den Zugang einer armen Bevölkerung zu modernen Arbeitsplätzen. Betrachtet man aber die Situation im landesweiten Rahmen, kommt man um einige geografische Tatsachen nicht herum: Bevölkerungsmäßig lagen die lukrativen Eisenbahnstrecken größtenteils südlich der Linie Edinburgh – Glasgow, d.h. auf den langen Nord-Süd-Verbindungen nach Mittel- und Süd-England, insbesondere auf den beiden direkten Küstenlinien der *NBR* und *CR* nach London. Fast schien es so, als ob die neuen Brückenbauten ihrer Zeit voraus waren. Denn die Zugverbindungen auf den Langstrecken wurden damit keineswegs so dramatisch verbessert, wie man das hätte erwarten können.

Das lag zuallererst am Zustand der britischen Eisenbahnen. Denn 50 Jahre nach ihrer Einführung waren die Geschwindigkeiten zwischen Edinburgh und London immer noch sehr langsam. Es gab noch keine zuverlässigen Notbremsanlagen, und für das Sicherheitsbewußtsein der Verkehrsgesellschaften galten 40 Meilen/Stunde (64 km/h) als gute Durchschnittsgeschwindigkeit. Erst der private Wettbewerb zwischen den großen Rivalen der Ost- und Westküste brachte Bewegung in die Szene – weshalb ein Rückblick auf jene kuriose Periode der «races» von 1895 mit zur Brückengeschichte gehört. Noch 1887 fuhren die Ostküsten-Gesellschaften mit dem «*Flying Scotsman*» in 9 Stunden von London nach Edinburgh. Ab 1888 begannen sich «Ostküste» und «Westküste» regelrechte Wettläufe auf der Strecke zu liefern. Mit Jahresende konnte man Edinburgh bereits in unter 8 Stunden erreichen, Perth in 11 und Aberdeen in knapp 14 Stunden. Aber dann entschied die «Ostküste», daß 8 1/2

loaded or badly coordinated timetables, had become a source of constant aggravation.

The result of this voluntary restriction was that with the opening of the Forth Bridge in 1890, the shorter route northward did not offer any advantage. The west coast companies – the "London & Northern" and the CR – had an express train already in 1889, which reached Aberdeen in 12 hours and 50 minutes, so they beat their arch-rivals by half an hour. Also the Post Office was bent on sending mail by the fastest way possible to the north and gave the lucrative contract to the "west coast" until 1895. But on July 1, 1895, the "east coast" lifted their speed embargo and reached Edinburgh with the night train from London in 8 hours, and Aberdeen in 11 hours and 20 minutes. To get even, the "west coast" reduced the load on their train and already on July 15, 1895, offered a travel time of exactly 11 hours for the London-Aberdeen run – a new record! Thus the race was on in full force and culminated in a spectacular "exhibition run" by the "west coast" with a travel time from London to Aberdeen of just 8 hours and 32 minutes – an incredibly short time! But this could only be achieved by using a light-weight train of 68 tons carrying just a handful of passengers – but also *two* firemen who probably also had to race each other in shoveling coal. Excesses such as this one were of no practical value to regular service; still, they demonstrated what was technically possible. By comparison, today's Inter-City *The Aberdonian* makes the London–Aberdeen run in 7 hours and 26 minutes, according to the 1980/81 timetable. In other words, it took another 85 years to cut the regular travel time by one hour!

It is understandable that the railway companies soon returned to their old timetables, following the extensive wear and tear on the material, and the missed connections by passengers, on whose goodwill the private companies depended. It is less understandable why the companies did so little to improve the general railway traffic. Some major stations of the country were in a pitiful condition, as was even Edinburgh's Waverly Station. Despite a three-fold increase in rail traffic, this important station still found itself in the condition of 1860. There are drastic descriptions of the chaotic conditions prevailing on the overcrowded wooden platforms just 4 feet (1.20 m) wide, where bewildered crowds of passengers and porters were staggering about between trains that had come in late. It was only in 1900, that Waverly Station acquired the grand air of today.

Many other major stations were little more than run-down wooden shanties. The passengers were at the mercy of this service, often being stuck for a long time

Stunden schnell genug für diese Strecke waren – denn jede weitere Fahrzeitverkürzung wäre unsicher und zu teuer: Das Zugmaterial, die Schienenwege und besonders die langen Zwischenaufenthalte wegen überlasteter und schlecht abgestimmter Fahrpläne hatten sich als ständiges Ärgernis erwiesen.

Das Ergebnis der freiwilligen Selbstbeschränkung war, daß mit Eröffnung der Forth-Brücke 1890 die kürzere Route nach dem Norden keinen Zeitvorteil mehr brachte. Denn die Westküsten-Gesellschaften – die «London and Northern» sowie die «Caledonian» – hatten schon 1889 einen Schnellzug nach Aberdeen in 12 Stunden 50 Minuten eingeführt und so die Rivalen um 1/2 Stunde geschlagen. Auch die Post wollte ihre Sendungen mit dem schnellstmöglichen Zug nach Norden bringen und vergab den lukrativen Auftrag bis 1895 an die «Westküste». Zum 1. Juli 1895 aber hob die «Ostküste» ihr Geschwindigkeits-Embargo auf und erreichte Edinburgh mit dem Nachtzug ab London in 8 Stunden, und Aberdeen in 11 Stunden 20 Minuten. Im Gegenzug reduzierte die «Westküste» das Gewicht ihrer Züge und bot bereits am 15. Juli 1895 eine Fahrzeit von genau 11 Stunden für die Strecke London–Aberdeen – einen neuen Rekord! Der Wettstreit war jetzt in vollem Gange und gipfelte in einem aufsehenerregenden Vorführungslauf – «*exhibition run*» – der «Westküste» mit einer Fahrzeit London–Aberdeen von nur 8 Stunden 32 Minuten – eine kaum vorstellbar kurze Zeit! Das war aber nur möglich mit einem Leichtzug von 68 Tonnen Gesamtgewicht und nur einer Handvoll Passagieren, dafür aber mit zwei Heizern – die wohl um die Wette schaufeln mußten. Solche Exzesse hatten keinerlei praktischen Wert für den normalen Betrieb, zeigten aber, was technisch möglich war. Zum Vergleich gelangt der Inter-City «*The Aberdonian*» im Fahrplan von 1980/81 in 7 Stunden 26 Minuten von London nach Aberdeen: Es dauerte also noch 85 Jahre, um die Fahrzeit im Normalbetrieb um eine gute Stunde zu verkürzen!

Es ist verständlich, daß die Bahngesellschaften nach den teuren Wettfahrten, dem hohen Verschleiß an Material, den verpaßten Anschlüssen der Passagiere – von deren Wohlwollen die Privatgesellschaften abhängig waren – bald wieder zu den alten Fahrplänen zurückkehrten. Weniger verständlich ist, warum die Gesellschaften so wenig zur allgemeinen Verbesserung des Eisenbahnwesens taten; denn wichtige Stationen des Landes waren in erbärmlichem Zustand, so auch Edinburgh's Waverly Station. Trotz Verdreifachung des Zugverkehrs befand sich dieser wichtige Bahnhof immer noch im Ausbauzustand von 1860; es gibt drastische Beschreibungen der chaotischen Zustände auf den nur 1,20 m breiten und über-

in motionless, unheated trains – and these without any lavatories, except in a few First Class long-distance coaches or private saloons. British rail travel was quite an ordeal, and for a long time it was behind the standard achieved on the continent; these conditions were never mentioned by the Victorian writers in their multitudinous handbooks, purporting to tell "all that the reader needed to know about railway travel". So quite a few preferred to travel by sea along the coast – a natural advantage which Central Europe, on the other hand, was not in a position to offer. The railway companies, and especially the NBR, enjoyed a singularly bad press, and people began to question the true merit of the "Great Bridge". Since the races of 1895 had merely been stunts, they could not be taken as test-runs for realistic timetables in Anglo-Scottish rail traffic. Therefore, in June 1896 the rivals reached an agreement on a minimum $8^1/_2$ hour schedule for the daytime trains from London to Edinburgh, and London to Glasgow respectively. Meanwhile the night trains were still going for some time at speeds which could only be matched again another 40 years hence. In 1897 the night train of the "east coast" took 7 hours and 45 minutes from London to Edinburgh, while the daytime trains, despite their extended schedule, could barely keep to within the $8^1/_2$ hours.

It seems that the "west coast" in the struggle for passengers was acting with more flexibility and inventiveness. In 1898 for the first time they put a dining car on their Glasgow to London morning train, saving on the 20-minute refreshment stop at Preston. But instead of taking those 20 minutes off the travel time, the CR used it to ease the overall schedule – whereby their trains indeed began to stay within the prescribed $8^1/_2$ hours. On the "east coast" there was a similar stop at York, but the NBR agreed to put a dining car on the *Flying Scotsman* only if the 20 minutes were really taken off the travel time. Thus by December 1900, a further quarter of an hour was finally shaved off both long-distance runs – resulting in times of $8^1/_4$ hours which were kept until World War I, when rail traffic generally deteriorated again. When in 1923 the $8^1/_4$-hour-schedule could be achieved again, the railway companies were already being reorganised: the east coast companies found themselves joined up with others in the "London and North Eastern Railway" (L.N.E.R.), whereas the west coast was merged into the "London Midland and Scottish" (L.M.S.). Despite several experiments in using non-stop trains, the $8^1/_4$-hour-timetables remained in force – that is to say, until 1932 one had not gotten beyond an average speed of 47.6 m.p.h. (76.6 km/h)! Only in 1936 did the *Flying Scotsman* first make a non-stop run from Kings Cross

füllten Holzbahnsteigen, wo Horden von Fahrgästen und Trägern ratlos zwischen den verspäteten Zügen herumirrten. Erst ab 1900 bot Waverly Station das großzügige Bild von heute.

Aber viele wichtige Stationen waren nichts weiter als verlotterte Holzbuden. Die Passagiere waren diesem Service ausgeliefert, saßen oft lange in ungeheizten und bewegungslosen Zügen fest – und das ohne jegliche Toiletten, mit Ausnahme der wenigen 1. Klasse-Langstrecken-Waggons oder privater Salonwagen. Britische Eisenbahnfahrten waren eine wahre Strapaze und blieben noch lange hinter dem auf dem Kontinent erreichten Stand zurück – Umstände, welche die viktorianischen Berichterstatter in ihren mannigfaltigen Handbüchern «mit allem Wissenswerten zum Eisenbahnverkehr» unerwähnt ließen. So zog mancher Reisende den Seeweg entlang der Küste vor – ein natürlicher Vorteil, den Mitteleuropa wiederum nicht bieten konnte. Die Bahngesellschaften, besonders die *NBR*, genossen eine ausnehmend schlechte Presse, und die Leute begannen sich zu fragen, ob die «große Brücke» sich denn wirklich gelohnt habe. Dabei waren die Wettläufe von 1895 reine «stunts» (Kraftakte) gewesen, also keine Testläufe für realistische Fahrpläne im anglo-schottischen Zugverkehr. So einigten sich im Juni 1896 die Rivalen auf eine Minimalfahrzeit von 8 1/2 Stunden für die Tageszüge von London nach Edinburgh bzw. London nach Glasgow. Wogegen die Nachtzüge noch einige Zeit weiter mit Geschwindigkeiten fuhren, die man erst 40 Jahre später wieder erreichen sollte. 1897 fuhr der Nachtzug der «Ostküste» in 7 3/4 Stunden von London nach Edinburgh, während die Tageszüge trotz verlängerter Fahrzeit selbst ihre vereinbarten 8 1/2 Stunden kaum einhalten konnten.

Es scheint, daß die «Westküste» im Kampf um Fahrgäste flexibler und erfinderischer vorging: 1898 hängte sie erstmalig einen Speisewagen an ihren Morgenzug von Glasgow nach London und sparte damit den 20-minütigen Zwischenaufenthalt für Erfrischungen in Preston. Anstatt aber die 20 Minuten von der Gesamtfahrzeit abzuziehen, benutzte die *CR* diese Zeitspanne, um den Fahrplan zu entkrampfen – womit ihre Züge erstmals pünktlich innerhalb der 8 1/2 Stunden liefen. Auf der Oststrecke gab es einen ähnlichen Zwischenstop in York, aber die *NBR* erklärte sich nur dann zum Einsatz eines Speisewagens auf ihrem «*Flying Scotsman*» bereit, wenn die 20 Minuten wirklich von der Fahrzeit abgezogen würden. Mit Dezember 1900 hatte man so schließlich eine weitere Viertelstunde von den beiden Langstrecken gekappt, so daß fortan eine Fahrzeit von 8 1/4 Stunden galt – und zwar bis zum 1. Weltkrieg, als sich der gesamte Zugverkehr

to Waverly Station in 7¼ hours, and in 10 hours 28 minutes from London to Aberdeen. In 1937 the L.N.E.R. offered a 6 hour "Coronation Summer Service" from Kings Cross to Waverly, which the L.M.S. answered with a 6½ hour "Coronation Scot Service" to Glasgow – thus these epigones were still marvelling in their exhibition runs, some 50 years after the first race. But then the end of the steam age was fast approaching, and after World War II, nationalisation meant the formation of "British Rail", and thus the end of any competition. So much for this digression on time gains on long-distance lines, which are determined not by bridges but by the entire transport system.[15]

Apart from the obvious practical usefulness of the Forth Bridge, its further, less tangible, impact on the era cannot be overestimated. This structure had long since become a monument in its own right; today it is regarded quite simply as a masterpiece of Victorian engineering. Already upon its completion, it was hailed as a national symbol of progress, and by some even as the eighth wonder of the world. It fascinated this epoch so strongly because it manifested the Victorians' deepest convictions – namely that mankind could master anything by hard work, resoluteness and application of the most modern technology, indeed it could master even nature itself. In this way, the bridge hit very precisely upon the spirit of the age. Still, in all admiration for its magnitude, one was far less certain in judging its aesthetic value: too unfamiliar were those forms, though elegant when seen from afar, but over-sized and massive in detail – and the entire project was executed without the slightest attempt at artistic embellishment and ornamentation of which the Victorians were usually so fond in their technical structures and even in machines. Yet the opinions held about the justification of such after-the-fact embellishments were already changing after 1890. A letter written to John Fowler by Royal Academician Alfred Waterhouse, is very much to the point:

> "One feature especially delights me – the absence of all ornament. Any architectural detail borrowed from any style would have been out of place in such a work. As it is, the bridge is a style unto itself: the simple directness of purpose with which it does its work is splendid and invests your vast monument with a kind of beauty of its own, differing though it certainly does from all the beautiful things I have ever seen." [14, p. 28]

15 Anyone who has attempted, like this writer, to travel quickly from, for example, Holyhead to Newcastle will appreciate that connections running *laterally* across this country are even worse than the north-south ones.

Words like these may remind us of Adolf Loos, living about that time, in his radical rejection of any ornamentation. And for our time, Kenneth Clark has firmly placed the Forth Bridge into the tradition of the great artistic and philosophical achievements of mankind; his classic work *Civilisation* shows the Forth Bridge on the back cover. What had been to the art critic John Ruskin the crude practicality of an industrial product, was to Clark a graceful hymn to a new pragmatism; its soaring girders demonstrated the essence of this era much more vividly than any other structure of this unusually inventive decade. To the general public, the Forth Bridge became an institution equal to the Eiffel Tower, and thus, transcending any practical purpose, a mythic symbol of the time. This is beautifully summarised by Hamilton Ellis:

> "Since then this superb bridge has stridden the Firth of Forth, changeless in a changing world, tremendous, most grandly austere. Its beauty is unconscious, as in so many Victorian structures. Not one of its designers, builders or sponsors remarked on any ornamental quality, yet there it stands in all its giant grace, today a part of the land-and-seascape on which it was imposed, and scenically as full of moods as a mountain. It should be seen at sunrise; it should be seen in the evening; it should be seen in a storm; it should be seen when a white sea mist drifts up the firth, hiding all but the tops of the towers..."[16]

In its telling lines of stresses, its three-fold rhythm of a rising and falling profile, all the while self-reliant and well settled in the Firth like a matter of course, this structure remains a jewel of the era. –

The former Tay Bridge by comparison was to appear far less imposing, even timid, fraily reduced to the mere limits of stability, and heterogeneous by appearance since its many parts simply did not mesh into the kind of united whole found in the Forth Bridge. Still, it remains the merit of Thomas Bouch to have actually realised the first bridging across the Tay. Even if others before him had been advancing this idea with a conviction similar to those who later on would argue in favor of tunelling the Channel, there still remains a big difference between intent and action. Bouch was the first to come forth with a practicable plan for this project – though it was not blessed with a long lifespan. Today we can gain only an approximate idea of the actual impact of this structure, by the illustrations and drawings shown here. Yet in its day, it was no doubt an unheard of and extremely bold undertaking – combining a very slender profile with a minimal use of con-

[16] C. Hamilton Ellis: The North British Railway. Ian Allan, 1955.

solch nachträglicher Verschönerung begann sich nach 1890 bereits zu wandeln: Ein Brief an John Fowler von Alfred Waterhouse, einem Mitglied der königlichen Akademie, macht dies deutlich:

> «Ein Merkmal freut mich ganz besonders – das Fehlen jeglichen Ornamentes. Jedes architektonische Detail, das man von irgendeinem Stil entlehnt hätte, wäre an einem solchen Werke fehl am Platze gewesen. So wie sie ist, ist die Brücke ein Stil ganz für sich: Die einfache Direktheit, mit der sie ihre Arbeit tut, ist herrlich und verleiht Ihrem gewaltigen Monument eine ganz eigene Art von Schönheit, wenngleich sie sich von all den schönen Dingen, die ich je gesehen habe, gewiß unterscheidet.» [14, S. 28]

Worte, die uns an den etwa zeitgleichen Adolf Loos erinnern mögen, mit seiner radikalen Ablehnung jeglichen Ornamentes. Und für unsere Zeit hat Sir Kenneth Clark die Forth-Brücke fest in die Tradition der großen künstlerischen und ideellen Errungenschaften der Menschheit gestellt: Sein Hauptwerk *«Civilisation»* trägt die Forth-Brücke im Umschlagbild. Was für den Kunstkritiker John Ruskin noch die derbe Gewöhnlichkeit eines Industrieproduktes war, sah Clark als einen eleganten Hymnus auf den neuen Pragmatismus, welcher mit himmelstürmenden Stahlträgern das Wesen der Epoche viel besser zum Ausdruck brachte als irgendein anderes Bauwerk dieser außerordentlich produktiven Jahre. Im allgemeinen Empfinden der Bevölkerung wurde die Forth-Brücke zu einer ähnlichen Institution wie etwa der Eiffel-Turm, und so, über jeden praktischen Zweck hinaus, zu einem Symbol der Zeit. Dazu schreibt treffend Hamilton Ellis:

> «Seit jenem Tage überschreitet diese herrliche Brücke den Firth of Forth, unverändert in einer sich ändernden Welt, riesenhaft und in großartiger Strenge. Ihre Schönheit ist, wie bei so vielen viktorianischen Bauwerken, unbewußt. Keiner ihrer Schöpfer, Erbauer oder Geldgeber scherte sich um ornamentales Beiwerk – aber hier steht sie in all ihrer graziösen Riesenhaftigkeit, heute selbst ein Teil der Land- und Meeresszene, in die sie hineingestellt war, und dabei landschaftlich so voller Stimmungen wie ein Berg. Man sollte sie bei Sonnenaufgang sehen, sollte sie abends sehen und auch in einem Sturm erleben; man sollte sie sehen, wenn ein weißer Meeresdunst den Firth heraufzieht und sie ganz verhüllt, bis auf die Spitzen der Türme...»[16]

Mit seinem gut ablesbaren Kräfteverlauf, dem dreifachen Rhythmus eines steigenden und fallenden

[16] C. Hamilton Ellis: The North British Railway. Ian Allan, 1955.

struction material. Here we are confronted by that type of Victorian engineering structure which strides across any natural obstacle with daring impertinence; its Spartan reduction of masses to the barest essentials makes no attempt to accomodate features found in nature. Further to this general appearance of the first Tay Bridge, there is an impression of crude matter-of-factness in detail, as by well proven pioneer's habit; a structural principle tried elsewhere – here that of the railway viaduct – is summarily applied to the new task. Thus the Tay Bridge is a telling example of how this era has achieved pinnacles of technology by basically using rather ordinary means: nobody would speak here in earnest of sophistication in the sense of a mature performance – but perhaps more of labouring persistence and resoluteness, to embark on this great first step with modest means.

The tightening-up of safety requirements by the Board of Trade, after the disaster, took some of the carefree boldness out of the engineering structures during the last two decades of the 19th century. The new creations grew heavier, more substantial, and thus of course safer, but just that much less stirring to the imagination than those earlier, daring lightweights had been! The change in the exterior appearance was to change in turn also the perception and appreciation of such structures.

Today, in contemplating long-span bridges, one may argue over which form is more appropriate in the context of the natural environment. Should the bridge be that kind of accomodating structure which manipulates and models the necessary volume in such a way as to resemble the landscape's contours and, in the end, become in itself a natural extension of nature? Should it be that superimposed "apparatus" which – quite oblivious to Morris' negative reaction – derives its form solely from the rational flow of stresses, and which, by virtue of its startling contradiction to natural forms, might even lend wings to the beholder's sense of reason? Bridges – more so than many other structures – elicit conflicting opinions. To what extent can they be regarded as rational solutions to specific engineering problems? Or, on the other hand, should they express an innate respect for the natural environment – to the point in some cases where the structure even strives to enhance that environment? A synthesis between nature and technology is rarely found in major structures today. In the case of bridges, it is usually less the actual structure itself than the long cuttings into the soil, and the approach viaducts, which can spoil a grand overall impression. *Fig. 122* illustrates how different yet equally favourable such a

Linienwerkes, dabei eigenständig und doch wie selbstverständlich in die Meeresbucht eingebettet, bleibt dieses Bauwerk ein Glanzstück der Epoche.

Im Vergleich dazu muß die seinerzeitige Tay-Brücke weit weniger imposant, ja zaghaft wirken, spindelig abgemagert bis an die Grenze der Standsicherheit und im Erscheinungsbild heterogen – die vielen Teile ergeben nicht jene schlüssige Ganzheit wie im Falle der Forth-Brücke. Aber es bleibt das Verdienst von Thomas Bouch, die erste Brücke über den Tay wirklich gebaut zu haben. Selbst wenn andere vor ihm die Idee dazu mit ähnlicher Überzeugung vertreten hatten, wie man später einmal von der Untertunnelung des Ärmelkanals sprechen sollte, bleibt doch ein großer Unterschied zwischen Idee und Verwirklichung. Bouch war der erste, der einen baureifen Plan für diesen Brückenschlag realisiert hat – auch wenn seinem Werk noch keine lange Lebens-Dauer beschieden war. Wir können uns heute nur noch anhand der hier gezeigten Abbildungen und Zeichnungen ein ungefähres Bild von der tatsächlichen Wirkung dieser Brücke machen. Für ihre Zeit war sie jedenfalls ein unerhörtes und höchst wagemutiges Unterfangen – das einen hohen Schlankheitsgrad mit geringstmöglichem Materialeinsatz verband: Hier begegnet uns jener Typus viktorianischer Ingenieurbauten, die mit kecker Unverfrorenheit über alle natürlichen Hindernisse hinwegschreiten; in ihrer spartanischen Reduktion jeglicher Masse auf das Notwendigste machen sie erst gar nicht den Versuch, sich den Naturformen anzugleichen. Dazu kommt im Erscheinungsbild der ersten Tay-Brücke noch der Eindruck von ruppiger «Gewöhnlichkeit» im Detail, hat man doch in bewährter Pionier-Manier ein anderswo erprobtes Konstruktionsprinzip – hier das des Eisenbahn-Viaduktes – kurzerhand auf die neue Aufgabe übertragen. D. h. die Tay-Brücke ist ein sprechendes Beispiel dafür, wie jene Epoche mit im Grunde sehr gewöhnlichen Mitteln zu technischen Höchstleistungen gefunden hat: Von «Raffinesse» im Sinne einer reifen Leistung wird hier niemand ernstlich sprechen wollen – eher schon von bemühter Beharrlichkeit und Entschlossenheit, mit bescheidenen Mitteln diesen ersten großen Schritt zu wagen.

Die nach dem Einsturz verschärften Sicherheitsbestimmungen des *Board of Trade* nahmen den Ingenieurbauten der beiden letzten Jahrzehnte des 19. Jahrhunderts ihre kecke Unbekümmertheit. Die neuen Erzeugnisse wurden schwerer, solider, damit natürlich sicherer, aber eben auch weniger phantasiebeflügelnd als der wagemutige Leichtbau der frühen Empiriker! Mit dem Wandel der äußeren Erscheinung haben sich später auch die Wahrnehmung und Bewertung solcher Bauwerke gewandelt.

112
Today's two bridges across the Firth of Forth. In front, the Rail Bridge by Fowler and Baker; behind, the modern Road Suspension Bridge by Mott, Hay & Anderson (the firm originally cofounded by Sir Benjamin Baker and Sir Basil Mott), in association with Freeman, Fox & Partners, completed in 1964. The contractor of the Road Bridge was, among others, the firm of Sir William Arrol & Co Ltd. of Glasgow, once again. – Whereas the railway bridge had been built to stand solidly and immobile against the wind, the road bridge was designed to be more flexible through the use of high-tension steel cables and prefabricated components – and all this was accomplished by employing just one tenth of the workforce needed for the former bridge [13, p. 152].

112
Die zwei Brücken über den Firth of Forth von heute. Vorne die Eisenbahnbrücke von Fowler und Baker, dahinter die moderne Straßen-Hängebrücke von Mott, Hay & Anderson, assoziiert mit Freeman, Fox & Partners, 1964 fertiggestellt. Sir Benjamin Baker hatte noch zusammen mit Sir Basil Mott das erstere Ingenieurbüro begründet. Und Bauunternehmer war, neben anderen, wieder einmal die Firma Sir William Arrol & Co. Ltd. aus Glasgow. – Während die Eisenbahnbrücke noch massiv und unverrückbar fest gegen den Wind gebaut wurde, ist die Straßenbrücke flexibel ausgelegt, dank hochbelastbarer Stahlseile und vorgefertigter Teile – und das alles mit nur einem Zehntel der Arbeitskräfte ausgeführt [13, S. 152].

synthesis may appear. Between the Forth Rail Bridge of 1889 (to the left) and the modern Road Bridge of 1964 right next to it lie 75 years, yet each solution to the same problem has been developed convincingly out of the possibilities of its time.

As to the question of the visual impact of long-span bridges upon the viewer, this issue is less related to bridges' inner rules of stability, than to their easy readability: if the human eye gets neither hints regarding the scale of a structure, nor some insight into the flow of stresses, which most definitely are at work inside, then a relationship, or indeed something like an inner rapport, is very difficult to achieve with such pragmatic works. The long straight lines of many modern prestressed concrete bridges – such as the Kochertal Autobahn Bridge – may serve as an example here. By their expedient reduction to a universally exchangeable type, the relationship to a specific place and thus to the viewer gets lost. Actually, the minimalising of volumes is characteristic of modern structures – whether they be bridges or buildings. But if this minimalisation is carried to the extreme of actually suppressing the obvious interplay of forces, the structures deteriorate to more abstract products, to static patterns; in all their "efficiency", they no longer find any equivalent or rapport to forms which might be recognisable in human terms. "Recognisable" in this context refers to forms (such as growth patterns of any organic material, or even the crystalline structures of anorganic matter) which, because of their presence in nature, can evoke a certain congeniality and affinity to man. In short, the logically thought-out form of design can not remain an end in itself if this form is to touch man not just as a rational being, but in his sensory whole.

Granted, gradual accustomation makes many things easier to tolerate, even in the realm of utilitarian structures that are less successful – and 100 years indeed make for a long life-span of a major technical structure. But this Rail Bridge across the Firth of Forth has never ceased to fill users and admirers, since its completion in 1890, with a sense of wonder and affection. In many ways, the image of the bridge had become more important than its practical reality – it had itself become an icon of strength, permanence and reliability. Murray [15, p. 99] describes the structure as a symbol of Scotland – comparable to the Pyramids or the Statue of Liberty symbolising the countries in which they stand. This bridge sums up the spirit of Scotland, in all her sternness, her rugged beauty and her ingenuity. And it may be regarded as one of the few structures which appeals to man so strongly on both the rational and emotional levels. Even in the new cen-

Heute mag man bei der Betrachtung von Großbrücken darüber streiten, welches Erscheinungsbild der natürlichen Umgebung besser gerecht wird: Das «angepaßte» Bauwerk, das die notwendige Masse dergestalt gliedert und modelliert, bis sie selbst wie eine mehr oder weniger natürliche Fortsetzung der Natur erscheint, sich also in Form und Masse dem Körper der Landschaft angleicht. Oder jene hineingestellte «Apparatur» – die unbeschadet der negativen Wertung von Morris – ihre Form einzig dem rationalen Kräfteverlauf dankt und gerade durch ihren augenfälligen Gegensatz zur Naturform den Vernunftsinn des Betrachters ausgesprochen beflügelt. Denn besonders beim Anblick von Brückenbauten scheiden sich die Geister – darüber nämlich, was gerade noch verstandesmäßig als eine sinnvoll-rationale Lösung gelten darf, und was andererseits ein starkes Gefühl für den Eigenwert der natürlichen Umgebung widerspiegelt, ja möglicherweise dieses Naturerlebnis noch zu steigern vermag: Eine solche Synthese zwischen Landschaft und Technik ist bei Großbauwerken selten – zumal da bei Brücken weniger ihre eigentliche Konstruktion als die langen Geländeschnitte und Anfahrtsviadukte die große Gesamtwirkung zunichte machen können. – Wie verschieden und doch gleichermaßen gelungen eine solche Synthese aussehen kann, vermittelt *Abb. 112:* Zwischen dem Bau der Eisenbahnbrücke über den Forth von 1889 (links) und der modernen Straßenhängebrücke daneben von 1964 liegen 75 Jahre: Beide Lösungen für dieselbe Aufgabe sind jeweils aus den Möglichkeiten ihrer Zeit schlüssig entwickelt.

Die Frage, wie weitgespannte Brücken heutiger Bauart auf uns wirken, hat weniger mit ihren inneren Gesetzen der Tragfähigkeit zu tun, als mit deren Anschaulichkeit: Wenn dem menschlichen Auge weder Hilfen zum Maßstab noch Einsicht in die Ableitung der Kräfte geboten werden, die doch ohne Zweifel im Bauwerk wirksam sind, läßt sich nur schwer ein Bezug oder gar so etwas wie Anteilnahme für solche Zweckbauten herstellen. Die langen parallelen Geraden manch moderner Spannbetonbrücken – etwa der Kochertal-Autobahnbrücke – sind dafür ein Beispiel: In ihrer Reduzierung auf einen austauschbaren Universaltyp geht der spezielle Bezug auf einen Ort und den Betrachter verloren. An sich gilt ja die Minimierung der Querschnitte als ein Kennzeichen moderner Konstruktionen – gleich ob Brücken oder Gebäude. Wenn aber diese Minimierung soweit geht, das sinnfällige Spiel der Kräfte zu verleugnen, werden die Konstruktionen zu abstrakten Gebilden, zu statischen Schemata, die bei aller «Effizienz» keine Entsprechung mehr in menschlich nachvollziehbaren Formen finden. «Nach-

tury to come – and we hope the bridge will see out these next 100 years in full measure – the happy union of function and beauty will prevail.

vollziehbar» in diesem Zusammenhang wären Formen, die durch ihr Auftreten in der Natur, etwa in der Struktur organischen Wachstums, im Knochenbau, oder auch in den kristallinen Strukturen anorganischer Substanzen, eine Vertrautheit und Affinität zur menschlichen Gestalt (die selbst ein Teil der Natur ist) hervorrufen. Die logisch erdachte statische Form kann nicht Selbstzweck bleiben, wenn sie den Menschen in seiner sinnhaften Ganzheit und nicht nur als rationales Wesen berühren soll.

Nun macht zwar die Gewöhnung vieles erträglicher, gerade auch bei weniger gelungenen Zweckbauten – und 100 Jahre sind gewiß eine lange Lebensdauer für ein technisches Großbauwerk. Aber die Eisenbahnbrücke über den Firth of Forth hat seit ihrer Vollendung 1890 nicht aufgehört, die Zeitgenossen in staunende Ergriffenheit zu versetzen. Auf vielfältige Weise war ihr Bild wichtiger als ihre praktische Wirklichkeit geworden – nämlich zum Sinnbild von Stärke, Dauer und Zuverlässigkeit. Murray[15, S. 99] nennt sie ein Symbol für Schottland – vergleichbar den Pyramiden oder der Freiheitsstatue als Sinnbilder der Länder, in denen sie stehen: Die Brücke verkörpert den Geist von Schottland, in all seiner Strenge, seiner rauhen Schönheit und seinem Erfindungsreichtum. Und sie darf als eine der ganz wenigen Konstruktionen gelten, in denen beide Bereiche menschlicher Wahrnehmung – der rationale und der emotionale – zu einer glücklichen Einheit finden. Auch im kommenden Jahrhundert, das wir dem Fortbestand der Brücke wünschen wollen, wird diese Konstante unserer Wahrnehmung ihre Gültigkeit behalten.

Bibliography
Bibliographie

1. Beckett, Derrick: Great Buildings of the World: Bridges. The Hamlyn Publishing Group Ltd., London, New York, Sydney, Toronto, 1969.
2. Cronshaw, Andrew: Old Dundee, Picture Postcards. Mainstream Publishing Co., Edinburgh, 1988.
3. Durie, Alastair, and Mellor, Roy: George Washington Wilson and the Scottish Railways. Dalesman Publishing Co., Clapham; Aberdeen University Library, 1988.
4. Eyth, Max: Hinter Pflug und Schraubstock. Skizzen aus dem Taschenbuch eines Ingenieurs. Deutsche Verlags-Anstalt, Stuttgart und Berlin, 1918. Inzwischen neu herausgegeben mit dem Untertitel: Die Abenteuer eines Ingenieurs im vorigen Jahrhundert. DVA Stuttgart, 1986.
5. Feldhusen, Gernot: Tand, Tand ist das Gebilde von Menschenhand… In: Fachmagazin Bausubstanz, Heft 3, Mai 1987.
6. Grothe, Albert, Publ. Nr. 1: The Tay Bridge – Its History and Construction. John Leng & Co., Dundee, 1878.
7. Grothe, Albert, Publ. Nr. 2: The Tay Bridge. In: Good Words, 1878.
8. Hammond, Rolt: The Forth Bridge and its Builders. Eyre & Spottiswoode, London, 1964.
9. Hansen, Hans Jürgen: Sensationen von gestern – Geschichte heute. Bildreportagen aus den Jahren 1848 – 1900. Verlag Gerhard Stalling, Oldenburg und Hamburg, 1976.
10. Hartung, Giselher und Behnisch, Günter: Eisenkonstruktionen des 19. Jahrhunderts. Schirmer/Mosel, München, 1983.
11. Hayden, Martin: The Book of Bridges. Marshall Cavendish, London & New York, 1976.
12. Inglis, John: A Victorian Edinburgh Diary. Ed. by Ena Vaughan. The Ramsay Head Press, Edinburgh, 1984.
13. Jurecka, Charlotte: Brücken. Historische Entwicklung – Faszination der Technik. Anton Schroll & Co., Wien und München, 1986.
14. Mackay, Sheila: Bridge Across The Century: The Story of the Forth Bridge. Moubray House Press, Edinburgh, 1986.
15. Murray, Anthony: The Forth Railway Bridge. Mainstream Publishing Co., Edinburgh, 1983.
16. Ordnance Survey, Maps: Landranger 54, Dundee to Montrose, 1987; and: Routemaster 4, Central Scotland & Northumberland, 1987.
17. Overman, Michael: Man the Bridgebuilder. Priory Press Ltd., Hove, 1975.
18. Peters, Tom F.: Transitions in Engineering: Guillaume Henri Dufour and the Early 19th Century Cable Suspension Bridges. Birkhäuser Verlag, Basel/Boston, 1987.
19. Pottgießer, Hans: Eisenbahnbrücken aus zwei Jahrhunderten. Birkhäuser Verlag, Basel/Boston/Stuttgart, 1985.
20. Prebble, John, Publ. Nr. 1: The High Girders. Secker & Warburg, London, 1956.
21. Prebble, John, Publ. Nr. 2: The Tay Bridge Disaster. In: History is my Witness. Ed. by Gordon Menzies, BBC, London, 1976.
22. Report of the Court of Inquiry on the Tay Bridge Disaster, June 1880. Circulated to members of both Houses of Parliament. Available in the Library of the Institution of Civil Engineers, London.
23. Rolt, L. T. C.: Victorian Engineering. Alan Lane The Penguin Press, London, 1970.
24. Rudolf, Jochen: «Paper Pushers» und «Metal Bashers», Eifersüchteleien zwischen Industrie und Finanz. Frankfurter Allgemeine Zeitung vom 12. 1. 1987.
25. Sager, Peter: Schottland. DuMont Kunst-Reiseführer. Köln, 1988.
26. Thomas, John: The Tay Bridge Disaster – New Light on the 1879 Tragedy. David & Charles, Newton Abbot, 1972.
27. Voss, Hans: Neunzehntes Jahrhundert. Reihe: Epochen der Architektur. Umschau Verlag, Frankfurt.
28. de Vries, Leonard: Victorian Inventions. John Murray, London, 1971.
29. Westhofen, Wilhelm: The Forth Bridge. Reprinted from Engineering Magazine, London, 1890.
30. Zug der Zeit – Zeit der Züge. Deutsche Eisenbahn 1835 – 1985. Band 2. Siedler Verlag, Berlin, 1985.

Table of Illustrations	Verzeichnis der Abbildungen

1	"Tay Bridge Disaster"	1	«Tay Bridge Disaster»
2	"Dee Bridge Disaster"	2	«Dee Bridge Disaster»
3	Railway Bridge across the Firth of Tay	3	Eisenbahnbrücke über den Firth of Tay
4	Menai Suspension Bridge by Telford	4	Menai-Hängebrücke von Telford
5	Sir Thomas Bouch	5	Sir Thomas Bouch
6	Location map of both railway bridges	6	Übersichtskarte zur Lage der Eisenbahnbrücken
7	The triumvirate of British railway engineers	7	Das Triumvirat der britischen Railway Engineers
8	Max Eyth	8	Max Eyth
9	"Behind plow and bench-vice"	9	«Hinter Pflug und Schraubstock»
10	View of Dundee	10	Blick auf Dundee
11	Elevation of Tay Bridge	11	Seitenansicht der Tay-Brücke
12	Point of transition of girders	12	Der Punkt des Übergangs der Fachwerkträger
13	Robert Stephenson's Britannia Bridge	13	Robert Stephenson's Britannia-Brücke
14	Floating double-caissons	14	Schwimmende Doppelsenkkästen
15	Period illustration of Tay Bridge	15	Zeitgenössische Darstellung der Tay-Brücke
16	Hexagonal masonry pier	16	Sechseckiger Mauerwerkskörper der Pfeiler
17	Longitudinal and transverse elevation of tall piers	17	Längs- und Queransicht der Hochpfeiler
18	One of the "High Girders" on staging	18	Einer der großen Mittelträger auf der Arbeitsbühne
19	Hydraulic lifting of girders	19	Hydraulische Hebung der Brückenträger
20	Trial runs at south shore	20	Probeläufe am Südende
21	Total view of Tay Bridge from north shore	21	Gesamtaufnahme der Tay-Brücke vom Nordufer
22	Esplanade of Dundee	22	Die Esplanade von Dundee
23	Official opening of Tay Bridge	23	Offizielle Freigabe der Tay-Brücke
24	Harbour of Dundee	24	Der Hafen von Dundee
25	Railway Network of North British Railway	25	Das Streckennetz der North British Railway
26	Locomotive Nr. 224	26	Die Lokomotive Nr. 224
27	Map showing last approach to Dundee	27	Kartenausschnitt des Anfahrtsweges nach Dundee
28	Total view of Tay Bridge from south shore	28	Gesamtaufnahme der Tay-Brücke vom Südufer
29	"Magdalen Green" in Dundee	29	«Magdalen Green» in Dundee
30	Today's Tay Bridge	30	Heutige Tay-Brücke
31	Cross-section of Tay Bridge	31	Querschnitt der Tay-Brücke
32	Illustration of disaster of December 28, 1879	32	Einsturz-Szene vom 28. 12. 1879
33	Paddle steamer ferry "Dundee"	33	Raddampfer-Fähre «Dundee»
34	Tay Bridge after disaster	34	Tay-Brücke nach dem Einsturz
35	Scene of accident at Firth of Tay	35	Der Unglücksort am Firth of Tay
36	Tay Bridge after disaster from south shore	36	Tay-Brücke nach dem Einsturz vom Südufer
37	Tay Bridge after disaster from Wormit	37	Tay-Brücke nach dem Einsturz von Wormit
38	Central girder after recovery	38	Der geborgene Mittelträger
39	Assumed position of train	39	Rekonstruktion der Position des Zuges
40	Ill-fated engine Nr. 224	40	Die Unglückslokomotive Nr. 224
41	Position of girders and train at riverbed	41	Lage der Träger und des Zuges am Flußgrund
42a	Demolition of Bouch's Tay Bridge	42a	Abbau der Bouch'schen Brücke
42b	The same scene today	42b	Dieselbe Szene heute
43a	View from Tay Bridge towards Wormit	43a	Blick von der Tay-Brücke Richtung Wormit
43b	House in Wormit today	43b	Haus in Wormit heute
44	Comparing old and new cross-section of bridge	44	Vergleich des alten und neuen Brückenquerschnittes
45	Shifting of old girders	45	Versetzen der alten Träger
46	South curve of Tay Bridge today	46	Südkurve der Tay-Brücke heute
47	Sir Thomas Bouch in later life	47	Sir Thomas Bouch in späteren Jahren
48	Bouch's former residence	48	Der ehemalige Wohnsitz von Bouch

49	Sketch by John Waddell	49	Die Skizze von John Waddell
50	End of bridge's remaining northern half	50	Ende der stehengebliebenen Nordhälfte
51	Long line of piers' stumps	51	Die lange Reihe der Pfeilerstümpfe
52	Panorama photograph of Tay Bridge today	52	Panoramafoto der Tay-Brücke heute
53	Sir Samuel Brown's Brighton Chain Pier	53	Sir Samuel Brown's Brighton Chain Pier
54	Failure of Basse-Chaine Suspension Bridge	54	Einsturz der Basse-Chaine-Hängebrücke
55	Railway line Edinburgh-Dundee	55	Die Strecke Edinburgh–Dundee
56	Early design by James Anderson	56	Der frühe Entwurf von James Anderson
57	Anna and James Anderson	57	Anna und James Anderson
58	Thomas Bouch's railway ferry	58	Thomas Bouch's Eisenbahnfähre
59	Design of suspension bridge by Sir Thomas Bouch	59	Hängebrücken-Entwurf von Sir Thomas Bouch
60	Bouch's masonry pier with lighthouse	60	Bouch's Ziegelpfeiler mit Leuchtturm
61	Ink drawing of Bouch's suspension bridge	61	Federzeichnung der Bouch'schen Hängebrücke
62	Alternative designs for suspension bridge across the Forth	62	Alternative Hängebrücken-Entwürfe zur Forth-Brücke
63	Cantilever bridge by Fowler and Baker	63	Auslegerbrücke von Fowler und Baker
64	Wandipore Bridge in Tibet	64	Wandipore-Brücke in Tibet
65	Egyptian stone corbel and lintel combination	65	Ägyptische Kragstein- und Balkenkombination
66	Five examples of Chinese cantilever bridges	66	Fünf Beispiele chinesischer Kragträger-Brücken
67	Road bridge at Srinagar	67	Straßenbrücke in Srinagar
68a	Ritter's continuous-girder bridge	68a	Ritters Durchlaufträger-Brücke
68b	Competition entry for Long-Island Bridge	68b	Wettbewerbsentwurf Long-Island-Brücke
68c	Warthe Bridge near Posen	68c	Warthe-Brücke bei Posen
69	Living model of Forth Bridge	69	Lebendes Modell der Forth-Brücke
70	Site plan of Forth Bridge	70	Lageplan der Forth-Brücke
71	Landscape at Firth of Forth	71	Landschaft am Firth of Forth
72	Southern cantilever tower in profile	72	Profil-Ansicht des südlichen Auslegerturmes
73	Forth Bridge from landing pier at South Queensferry	73	Forth-Brücke vom Pier bei Southqueensferry
74	Baker's wind gauges at Inchgarvie	74	Die Windmeßgeräte Baker's auf Inchgarvie
75	View of Forth Bridge from northwest	75	Nordwest-Ansicht der Forth-Brücke
76	Position of railway viaduct inside	76	Die Lage des inneren Eisenbahnviaduktes
77	"Skewback" at Fife pier	77	«Skewback» am Fife-Pfeiler
78a	Schematic drawing of anchoring the piers	78a	Schema der Pfeilerverankerungen
78b	Schematic drawing of distribution of loads	78b	Schema der Lastverteilung
79	Rocky islet of Inchgarvie	79	Die Felseninsel Inchgarvie
80	Pneumatic caisson	80	Pneumatischer Senkkasten
81a	Inchgarvie caisson	81a	Inchgarvie-Caisson
81b	Air-lock with compression chamber	81b	Luftschleuse mit Druckkammer
81c	Supports for cutting edge	81c	Abstützung der Schneidekante
81d	Excavation at riverbed	81d	Aushub am Meeresgrund
82	Arrol's "hydraulic spade"	82	Der «hydraulische Spaten» Arrol's
83	Lifting of viaduct girders	83	Die Hebung der Viaduktträger
84	Bottom main juncture	84	Unterer Hauptknotenpunkt
85a	Hydraulic rivetting machine	85a	Hydraulische Nietmaschine
85b	Rivetting cage	85b	Vernietungs-Käfig
86	Large drilling machine	86	Die große Bohrmaschine
87	Yard at South Queensferry	87	Das Werkgelände bei South Queensferry
88	Assembly of central tower	88	Montage eines Pfeilerturmes
89	View of central tower at Fife from northwest	89	Nordwest-Ansicht des Pfeilerturmes auf Fife
90	Configuration of top main junction	90	Ausbildung der oberen Hauptknotenpunkte
91	Free assembly of bottom members	91	Freier Vorbau der Untergurte
92	Central tower with plate-tie	92	Mittelturm mit Zugband
93	Couple of "Jubilee Crane" and rivetting crane	93	Doppelgespann aus «Jubilee-Kran» und Niet-Kran
94	Inchgarvie tower freely extending	94	Inchgarvie-Turm im freien Vorbau
95	View of Forth Bridge from northwest	95	Nordwest-Ansicht der Forth-Brücke
96	Construction sequence of erection	96	Fotoserie zum Baufortschritt
97	Considerable extension of cantilever piers	97	Beträchtliche Ausladung der Pfeilertürme
98	Fife Tower at North Queensferry	98	Der Fife-Pfeilerturm bei North Queensferry

99	Inchgarvie south cantilever completed	99	Inchgarvie-Südausleger vollendet
100	Central girder between cantilevers	100	Der Mittelträger zwischen den Auslegern
101	View along the tracks	101	Blick auf die Fahrspur
102	Underbelly of Forth Bridge	102	Untersicht der Forth-Brücke
103a	Total view of Forth Bridge from northwest	103a	Gesamtansicht der Forth-Brücke von Nordwest
103b	Postcard of Forth Bridge around 1903	103b	Postkarte der Forth-Brücke von 1903
104	Opening ceremony of Forth Bridge	104	Die Eröffnung der Forth-Brücke
105	Front cover of luncheon menu	105	Titelblatt der Speisekarte
106	Group of navvies	106	Eine Gruppe von Erdarbeitern
107	Illustration of working conditions	107	Zur Illustration der Arbeitsbedingungen
108a	"Hawes Hotel" in South Queensferry	108a	«Hawes Hotel» in South Queensferry
108b	Today's "Hawes Inn"	108b	«Hawes Inn» von heute
109	Sir John Fowler	109	Sir John Fowler
110	Sir Benjamin Baker	110	Sir Benjamin Baker
111	Sir William Arrol	111	Sir William Arrol
112	Forth Bridge next to new suspension bridge	112	Forth-Brücke neben neuer Hängebrücke

Library of Congress Cataloging-in-Publication Data

Koerte, Arnold.
Two railway bridges of an era : Firth of Forth and Firth of Tay : technical progress, disaster, and new beginning in Victorian engineering = Zwei Eisenbahnbrücken einer Epoche : technischer Fortschritt, Desaster und Neubeginn in der Viktorianischen Ingenieurbaukunst / Arnold Koerte.
p. cm.
English and German.
Includes bibliographical references.
ISBN 3-7643-2444-9 – ISBN 0-8176-2444-9
1. Forth Bridge (South Queensferry, Scotland : Railroad bridge)
2. Tay Bridge (Dundee, Scotland : Railroad bridge)
3. Tay Bridge Disaster, 1879.
4. Railroad bridges–Scotland–Design and construction.
I. Title.
II. Title: Zwei Eisenbahnbrücken einer Epoche.
TG64.S68K64   1992
624'.2'09411–dc20

Deutsche Bibliothek Cataloging-in-Publication Data

Koerte, Arnold:
Two railway bridges of an era : Firth of Forth and Firth of Tay ; technical progress, disaster and new beginning = Zwei Eisenbahnbrücken einer Epoche / Arnold Koerte. – Basel ; Boston ; Berlin : Birkhäuser, 1992
ISBN 3-7643-2444-9 (Basel ...)
ISBN 0-8176-2444-9 (Boston)

This work is subject to copyright. All rights are reserved, whether the whole or part of the material is concerned, specifically those of translation, reprinting, re-use of illustrations, broadcasting, reproduction by photocopying machine or similar means, and storage in data banks. Under § 54 of the German Copyright Law where copies are made for other than private use a fee is payable to «Verwertungsgesellschaft Wort», Munich.

© 1992 Birkhäuser Verlag Basel
Layout: Albert Gomm swb/asg, Basel
Printed in Germany on acid-free paper
ISBN 3-7643-2444-9
ISBN 0-8176-2444-9